Abstract Duality Pairs
in Analysis

Abstract Duality Pairs
in Analysis

Charles Swartz

New Mexico State University, USA

World Scientific

NEW JERSEY · LONDON · SINGAPORE · BEIJING · SHANGHAI · HONG KONG · TAIPEI · CHENNAI

Published by

World Scientific Publishing Co. Pte. Ltd.

5 Toh Tuck Link, Singapore 596224

USA office: 27 Warren Street, Suite 401-402, Hackensack, NJ 07601

UK office: 57 Shelton Street, Covent Garden, London WC2H 9HE

Library of Congress Cataloging-in-Publication Data

Names: Swartz, Charles, 1938–　author.

Title: Abstract duality pairs in analysis / by Charles Swartz (New Mexico State University, USA).

Description: New Jersey : World Scientific, 2017. | Includes bibliographical references and index.

Identifiers: LCCN 2017039625 | ISBN 9789813232761 (hardcover : alk. paper)

Subjects: LCSH: Scalar field theory. | Abelian groups. | Functional analysis.

Classification: LCC QA433 .S875 2017 | DDC 515/.63--dc23

LC record available at https://lccn.loc.gov/2017039625

British Library Cataloguing-in-Publication Data

A catalogue record for this book is available from the British Library.

For any available supplementary material, please visit
http://www.worldscientific.com/worldscibooks/10.1142/10783#t=suppl

Printed in Singapore

Dedicated to the memory of Professor Ronglu Li

Contents

Preface

This text is the result of a collaboration with Professor Ronglu Li when he was a visiting scholar in the mathematics department of New Mexico State University from 1988 to 1990. The idea of abstract duality pairs or abstract triples is to replace the scalar field in the theory of duality pairs of vector spaces with an Abelian topological group. This results in multiple applications to various topics in analysis. Upon his return to his home university, Harbin Institute of Technology, Harbin, China, Professor Li asked Professor Min-Hyung Cho of Kum-Oh National Institute of Technology, Kumi, South Korea, to join our efforts. Our collaboration resulted in 3 preprints which covered such topics as generalizations of the Orlicz–Pettis Theorem on subseries convergent series, the Hahn–Schur Theorem on the equivalence of weak and norm convergence of sequences in l^1, the Uniform Boundedness Principle, the Banach–Steinhaus Theorem and the Mazur–Orlicz Theorem on the continuity of separately continuous bilinear operators. These preprints were never published as written although portions of the results have appeared in various papers in the literature. This exposition contains the results which appeared in the 3 preprints as well as many further developments.

The first chapter starts with the requisite definition of an abstract duality pair or an abstract triple. Examples of abstract triples are then given along with the developments of some of the theories which are required. In particular, we give descriptions of the Dunford, Pettis and Bochner integral and the integral of a scalar function with respect to a vector valued integral and use these to give examples of abstract triples. Spaces of vector valued measures are also considered.

The classical Orlicz–Pettis Theorem asserts that a series in a normed space which is subseries convergent in the weak topology is subseries convergent in the norm topology. Chapter 2 considers generalizations of the Orlicz–Pettis Theorem to abstract triples and gives multiple applications of the generalizations to series in locally convex spaces, topological vector spaces, series of linear operators, the Hahn–Schur Theorem, the Nikodym Convergence and Boundedness Theorems, the Uniform Boundedness Principle and bilinear operators.

A series $\sum_j x_j$ in a topological vector space is bounded multiplier convergent if the series $\sum_{j=1}^{\infty} t_j x_j$ converges for every $\{t_j\} \in l^{\infty}$. Chapter 3 studies bounded multiplier series in abstract triples. Versions of the Orlicz–Pettis Theorem in abstract triples are established as well as versions of the Hahn–Schur Theorem.

If λ is a vector space of scalar sequences, a series $\sum_j x_j$ in a topological vector space is λ multiplier convergent if the series $\sum_{j=1}^{\infty} t_j x_j$ converges for every $\{t_j\} \in \lambda$. Bounded multiplier convergent and subseries convergent series are examples of multiplier convergent series where $\lambda = l^{\infty}$ and $\lambda = m_0$, the space of sequences with finite range, respectively. In Chapter 4 we consider multiplier convergent series in abstract triples. Versions of the Orlicz–Pettis Theorem and the Hahn–Schur Theorem for multiplier convergent series require assumptions on the space of multipliers λ some of which are called gliding hump assumptions. These gliding hump conditions are defined and then versions of the Orlicz–Pettis and Hahn–Schur Theorems are established. Several applications of multiplier convergent series to topics in geometric functional analysis are given.

The classical version of the Uniform Boundedness Principle asserts that a family of continuous linear operators from a Banach space to a normed space which is pointwise bounded is uniformly bounded on bounded subsets of the domain space. In Chapter 5 we give generalizations of the Uniform Boundedness Principle for abstract triples including versions of the Uniform Boundedness Principle which require no completeness or barrelledness assumptions on the domain space. Similarly, Chapter 6 contains versions of the Banach–Steinhaus Theorem for abstract triples. Applications to the Nikodym Convergence Theorem and a summability result of Hahn and Schur are given.

In Chapter 7 we consider biadditive operators from the product of topological groups into another topological group. This is a natural setting for abstract triples. We consider generalizations of the Mazur–Orlicz Theorem on the continuity of separately continuous bilinear operators to this setting. Families of separately continuous bilinear operators are also considered.

In Chapter 8 we consider abstract triples which have a sequence of projections defined on the triple. Many of the examples of abstract triples which consist of sequence spaces, spaces of measures or spaces of integrable functions have natural projections defined on them. We establish a Uniform Boundedness Principle and several uniform convergence results for abstract triples with projections. Applications to sequence spaces, spaces of countably additive vector measures, and the spaces of Bochner and Pettis integrable functions are given.

Finally, we consider weak compactness for abstract triples which involve sequence spaces, spaces of vector valued set functions and spaces of integrable functions.

There are appendices which contain notation, terms and results used in the text which may not be found in standard texts on functional analysis.

Special thanks are due to Pat Morandi who rescued me numerous times with valuable technical assistance.

Chapter 1

Abstract Duality Pairs or Abstract Triples

The notion of abstract duality pairs or abstract triples is a generalization of the notion of dual pairs of vector spaces which is utilized in the development of locally convex spaces. A pair of (real) vector spaces E, F is a dual pair if there exists a bilinear map

$$\langle \cdot, \cdot \rangle : E \times F \to \mathbb{R}$$

which separates points. The idea of abstract duality pairs or abstract triples is to replace the scalar field $\mathbb{R}$ by a Hausdorff Abelian topological group G and to replace the vector spaces E, F by arbitrary sets; in most applications the sets E, F have additional structures. A theory of abstract triples was initially developed by Professor Ronglu Li and Charles Swartz when Professor Li was a visiting scholar in the mathematics department of New Mexico State University from 1988–1990. The development continued after Professor Li returned to his home university, Harbin Institute of Technology, and Professor Li later engaged Professor Min-Hyung Cho of Kum-Oh National Institute of Technology in Kumi, Korea, to join in the development. This resulted in a series of notes on the subject but the notes were never published although the idea of abstract triples has appeared in various places in the literature. In this text we will present the results of the original notes as well as a number of further developments and applications of the theory. The idea has found applications in a number of different areas of analysis. In particular, we give applications to general versions of the Orlicz–Pettis Theorem on subseries convergence of series, various topics in the theory of vector valued measures and vector valued integrals, sequence spaces, multiplier convergent series, the Uniform Boundedness Principle, the Banach–Steinhaus Theorem, the Mazur–Orlicz Theorem on bilinear operators and weak compactness.

In this chapter we will give the definition and a number of examples of abstract triples which will be used in applications in later chapters.

Let E, F be sets and G a Hausdorff, Abelian topological group with

$$b : E \times F \to G;$$

if $x \in E$ and $y \in F$, we often write

$$b(x, y) = x \cdot y$$

for convenience. We refer to E, F, G as an abstract duality pair E, F with respect to G or an abstract triple and denote this by

$$(E, F : G);$$

this notation does not indicate the role of the map b but hopefully this causes no difficulties. In what follows $(E, F : G)$ will denote an abstract triple. Dually, note that $(F, E : G)$ is an abstract triple under the map $\bar{b}(y, x) = b(x, y)$.

There have been several similar abstractions of the duality between vector spaces and Abelian groups which have been used to treat versions of the Orlicz–Pettis Theorem. For example, Blasco, Calabuig and Signes ([BCS]) have considered a bilinear map

$$b : E \times F \to G,$$

where E, F, G are Banach spaces and b is a bilinear operator satisfying continuity conditions. They establish a general version of the theorem for subseries convergent series and apply it to vector integration. There is a more general version given in [Sw7] where E is a vector space, F, G are locally convex spaces and b is a bilinear map and multiplier convergent series are considered. Applications to multiplier convergent series in spaces of operators are given. Swartz ([Sw6]) considered biadditive maps from the product of two Abelian groups into another Abelian group and established versions of the Orlicz–Pettis Theorem. Another generalization is given by Chen and Li ([CL]) where E, F are vector spaces, G is a locally convex space and b is what they call a bi-quasi-homogeneous operator. They consider multiplier convergent series of quasi-homogeneous operators. Li and Wang ([LW]) have considered the case when E is a set and F is a set of G valued functions. They consider operator valued multiplier convergent series where the space of multipliers is vector valued. The case where E, F are vector spaces, G is a locally convex space and b is a bilinear map is considered in [LS3]; general versions of the Orlicz–Pettis Theorem are established and numerous applications are given. A similar treatment is given in Chapter

4 of [Sw4], pages 73–82. Li and Cho ([LC]) have used the general abstract setting above to obtain a generalization of an Orlicz–Pettis result of Kalton; we will consider this result later. Cho, Li and Swartz used the abstract setting to establish subseries versions of the Orlicz–Pettis Theorem and versions of the Banach–Steinhaus Theorem ([CLS1], [CLS2]); these results will be discussed in Chapters 2 and 6. Zheng, Cui and Li ([ZCL]) have also considered abstract duality pairs in spaces with sectional operators and indicated applications to sequence spaces. By considering subseries convergent series only, we are able to treat the case of group valued series in our setup.

Let $(E, F : G)$ be an abstract triple and let

$$w(E, F)$$

be the weakest topology on E such that the family of maps $\{b(\cdot, y) : y \in F\}$ are continuous from E into G; $w(F, E)$ is defined similarly.

If $\sum g_j$ is a formal series in G, the series is *subseries convergent* if the series $\sum_{j=1}^{\infty} g_{n_j}$ converges in G for every subsequence $\{n_j\}$. If σ is an infinite subset of $\mathbb{N}$, we write

$$\sum_{j \in \sigma} g_j = \sum_{j=1}^{\infty} g_{n_j},$$

where the elements of σ are arranged in a subsequence $\{n_j\}$; if σ is finite, the meaning of $\sum_{j \in \sigma} g_j$ is clear.

Definition 1.1. A sequence $\{x_j\} \subset E$ or a (formal) series $\sum x_j$ is $w(E, F)$ subseries convergent if for every $\sigma \subset \mathbb{N}$, there exists $x_\sigma \in E$ such that

$$\sum_{j \in \sigma} x_j \cdot y = x_\sigma \cdot y$$

for every $y \in F$. We symbolically write $\sum_{j \in \sigma} x_j = x_\sigma$ and say that the series $\sum x_j$ is $w(E, F)$ subseries convergent.

Note that we do not assume any algebraic structure on E, the algebraic operations are transferred to G via the map $b : E \times F \to G$; of course, if the set E has sums defined on it we use the usual definition of $\sum_{j \in \sigma} x_j$.

We give some examples which will be employed later.

Example 1.2. Of course, the simplest example of an abstract triple is a pair of vector spaces E, F in duality where G is just the scalar field and the topology $w(E, F)$ is just the weak topology $\sigma(E, F)$ from the duality. In this case, if $\sum x_j$ is $w(E, F)$ subseries convergent, $\sum_{j \in \sigma} x_j$ is the usual weak sum.

Example 1.3. Let Σ be a σ-algebra of subsets of a set S and let $ca(\Sigma, G)$ be the space of all G valued, countably additive set functions defined on Σ. Let $\mathcal{M}$ be a subset of $ca(\Sigma, G)$. Define $b : \Sigma \times \mathcal{M} \to G$ by $b(A, m) = m(A)$ so

$$(\Sigma, \mathcal{M} : G)$$

is an abstract triple. If $\{A_j\}$ is a pairwise disjoint sequence from Σ, then the (formal) series $\sum A_j$ is $w(\Sigma, \mathcal{M})$ subseries convergent with $\sum_{j \in \sigma} A_j = \cup_{j \in \sigma} A_j$. The space $ca(\Sigma, G)$ is used in treating the theorems of Nikodym from measure theory. If one identifies a set A with its characteristic function χ_A, one could also treat the abstract triple

$$(\{\chi_A : A \in \Sigma\}, \mathcal{M} : G)$$

under the map $(\chi_A, m) \to m(A)$; this would avoid using the formal addition of sets, $\sum A_j$. Let λ be a positive measure on Σ and let $ca(\Sigma, G : \lambda)$ be the subspace of $ca(\Sigma, G)$ which consists of the measures m which are λ continuous in the sense that

$$\lim_{\lambda(A) \to 0} m(A) = 0.$$

This subspace and the triple

$$(\Sigma, ca(\Sigma, G : \lambda) : G)$$

is useful in treating the Vitali–Hahn–Saks Theorem.

Example 1.4. Let G be a topological vector space (TVS) and $ba(\Sigma, G)$ the space of all bounded, finitely additive set functions from Σ into G. It will be shown later that if G is a semi-convex topological vector space every member of $ca(\Sigma, G)$ is bounded so that in this case $ca(\Sigma, G)$ is a subspace of $ba(\Sigma, G)$. If G is a normed space, $ba(\Sigma, G)$ has a natural norm defined by

$$\|m\| = \sup\{\|m(A)\| : A \in \Sigma\}$$

which is complete if G is complete (there is another equivalent norm on $ba(\Sigma, G)$ given by the semi-variation (see [DS]IV.10.4 and material later in the chapter)). Then

$$(\Sigma, ba(\Sigma, G) : G)$$

is an abstract triple under the map $(A, m) \to m(A)$.

Example 1.5. Let E, F be Abelian groups such that there exists a biadditive map $b : E \times F \to G$. Then $(E, F : G)$ is an abstract triple; this abstract setting was utilized in [Sw6].

Example 1.6. Let E be a topological space and $C(E, G)$ be the space of all continuous maps from E into G. If $b(x, f) = f(x)$ for $x \in E$ and $f \in C(E, G)$, then

$$(E, C(E, G) : G)$$

is an abstract triple. Also, $(C(E, G), E : G)$ is an abstract triple under the map $(f, x) \to f(x)$. One can treat the space of sequentially continuous functions from E into G in a similar manner.

Example 1.7. Let E, G be topological vector spaces and $L(E, G)$ the space of all continuous linear operators from E into G. Then

$$(L(E, G), E : G)$$

forms an abstract triple under the map $b(T, x) = T(x)$; in this case the topology $w(L(E, G), E)$ is just the strong operator topology. Also, $(E, L(E, G) : G)$ forms an abstract triple under the map $b(x, T) = T(x)$. If W is any subset of $L(E, G)$, then $(W, E : G)$ is an abstract triple under the same mapping. One can treat the space of sequentially continuous linear operators from E into G, $LS(E, G)$, similarly.

Example 1.8. Let λ be a vector space of scalar valued sequences, $\Lambda \subset \lambda$ and G be a TVS.. The β-dual of Λ with respect to G is defined to be

$$\Lambda^{\beta G} = \left\{ \{x_j\} \subset G : \sum_{j=1}^{\infty} t_j x_j \text{ converges for every } \{t_j\} \in \Lambda \right\}.$$

Then

$$(\Lambda, \Lambda^{\beta G} : G)$$

is an abstract triple under the map $(t, x) \to \sum_{j=1}^{\infty} t_j x_j = t \cdot x$.

Example 1.9. Let X be a Hausdorff topological vector space and E be a vector space of X valued sequences which contains the subspace $c_{00}(X)$ of all X valued sequences which are eventually 0. Assume that E has a vector topology under which it is an AK space, i.e., the coordinate projection P_k which sends each sequence $x = (x_1, x_2, ...)$ in E into the sequence with x_k in the kth coordinate and 0 in the other coordinates is continuous and if

$$Q_k = \sum_{j=1}^{k} P_j,$$

then $Q_k x \to x$ in the topology of E for every $x \in E$ (see Appendix B). Let $F = \{Q_k : k \in \mathbb{N}\}$. Then

$$(E, F : E)$$

is an abstract triple under the map $(x, Q_k) \to Q_k x$. This situation covers the case of the sequence spaces $l^p(X)$ and $c_0(X)$ when X is a locally convex space (see Appendix B or Appendix C of [Sw4] for these spaces).

Example 1.10. Let X be a Hausdorff topological vector space and E be a vector space of X valued sequences which contains the subspace $c_{00}(X)$ of all X valued sequences which are eventually 0. Let Y be a topological vector space. The β-dual of E with respect to Y is defined to be

$$E^{\beta Y} = \left\{ \{T_j\} \subset L(X, Y) : \sum_{j=1}^{\infty} T_j x_j \text{ converges for every } x = \{x_j\} \in E \right\}.$$

Then

$$(E, E^{\beta Y} : Y)$$

is an abstract triple under the map $(\{x_j\}, \{T_j\}) \to \sum_{j=1}^{\infty} T_j x_j$.

We will now give example of triples which involve spaces of vector valued, integrable functions. Let X be a Banach space and λ a positive σ-finite measure on the σ-algebra Σ. A function $f : S \to X$ is *scalarly Σ measurable* if $x' \circ f = x'f$ is Σ measurable for every $x' \in X'$ and is *scalarly λ integrable* if $x'f$ is λ integrable for every $x' \in X'$. Suppose f is scalarly λ integrable. We then have a linear mapping $F : X' \to L^1(\lambda)$ defined by

$$F(x') = x'f.$$

Proposition 1.11. *The linear operator F is continuous.*

Proof. First suppose that λ is finite and set

$$A_k = \{t \in S : \|f(t)\| \le k\}$$

so $A_k \uparrow S$. Set $f_k = \chi_{A_k} f$ and define $F_k : X' \to L^1(\lambda)$ by

$$F_k(x') = x'f_k.$$

F_k is obviously linear and is also continuous since

$$\|F_k(x')\|_1 = \int_{A_k} |x'f| \, d\lambda \le \|x'\| \, k\lambda(A_k).$$

Since $x'f_k \to x'f$ pointwise and $|x'f_k| \le |x'f|$, the Dominated Convergence Theorem implies

$$\|F_k(x') - F(x')\|_1 \to 0.$$

Hence, F is continuous by the Banach–Steinhaus Theorem.

If λ is σ-finite, let $B_k \in \Sigma$ with $\lambda(B_k) < \infty$ and $B_k \uparrow S$. Set $g_k = \chi_{B_k} f$ and define $G_k : X' \to L^1(\lambda)$ by

$$G_k(x') = x'g_k.$$

By the paragraph above G_k is continuous and

$$\|G_k(x') - F(x')\|_1 \to 0.$$

Again, the Banach–Steinhaus Theorem gives that F is continuous. □

We use the operator F to define the Dunford integral. The transpose operator of F,

$$F' : L^\infty(\lambda) \to X'',$$

is given by

$$F'(g)(x') = g(F(x')) = \int_S gx'f\,d\lambda.$$

The *Dunford integral* of f over $A \in \Sigma$ with respect to λ is defined to be the element

$$\int_A f\,d\lambda = F'(\chi_A) \in X''$$

so

$$\left(\int_A f\,d\lambda \right)(x') = \int_A x'f\,d\lambda$$

for $x' \in X'$.

Let

$$D(\lambda, X)$$

be the space of all Dunford integrable functions from S into X. The space $D(\lambda, X)$ has a natural norm defined by

$$\|f\| = \sup \left\{ \int_S |x'f|\,d\lambda : \|x'\| \le 1 \right\}.$$

Note that the norm is finite by the continuity of F; in general, the norm is not complete ([BS] 5.13).

There is another equivalent norm on $D(\lambda, X)$ defined by

$$\|f\|' = \sup\left\{\left\|\int_A f d\lambda\right\| : A \in \Sigma\right\}.$$

Indeed,

$$\|f\|' = \sup\left\{\left|x'\int_A f d\lambda\right| : \|x'\| \leq 1, A \in \Sigma\right\}$$

$$\leq \sup\left\{\int_S |x'f|\, d\lambda : \|x'\| \leq 1\right\} = \|f\|,$$

and if $\|x'\| \leq 1$, set $P = \{t \in S : x'f(t) \geq 0\}$ and $N = \{t \in S : x'f(t) < 0\}$ and note

$$\int_S |x'f|\, d\lambda = x'\int_P f d\lambda - x'\int_N f d\lambda \leq 2\|f\|'$$

so

$$\|f\| \leq 2\|f\|'.$$

Hence, the two norms are equivalent.

If $g : S \to \mathbb{R}$ is bounded and measurable, the product gf is scalarly integrable with

$$\|gf\| = \sup\left\{\int_S |gx'f|\, d\lambda : \|x'\| \leq 1\right\} \leq \|g\|_\infty \|f\|.$$

Example 1.12.

$$(D(\lambda, X), L^\infty(\lambda) : X)$$

is an abstract triple under the bilinear map $(f, g) \to \int_S gf d\lambda$. Note that the bilinear operator is continuous by the inequality above.

The Dunford integral has the unpleasant feature that the integral of an X valued function has its values in the bidual X''. Pettis singled out the functions whose integrals have values in X. The scalarly integrable function $f : S \to X$ is *Pettis integrable* if its Dunford integral $\int_A f d\lambda \in X$ for all $A \in \Sigma$. We give an example of a Dunford integrable function which is not Pettis integrable.

Example 1.13. Let μ be counting measure on $\mathbb{N}$. Define $f : \mathbb{N} \to c_0$ by $f(k) = e^k$. Let $x' = \{t_k\} \in c_0' = l^1$. Then $x'f(k) = t_k$ so f is scalarly integrable with

$$\int_A x'f d\mu = \sum_{k \in A} t_k = x'(\chi_A)$$

for $x' \in X$. Hence,

$$\int_A f d\mu = \chi_A$$

and when A is infinite, $\int_A f d\mu \notin c_0$. Thus, f is Dunford integrable but not Pettis integrable. Note the indefinite integral, $\int f d\lambda$, is not norm countably additive but is weak* countably additive.

Denote the space of all Pettis integrable functions by

$$P(\lambda, X).$$

If $f \in P(\lambda, X)$ and g is a Σ simple, scalar valued function, the product gf is obviously Pettis integrable and

$$\|gf\| \le \|g\|_\infty \|f\|.$$

If $g \in L^\infty(\lambda)$ pick a sequence of simple functions $\{g_k\}$ which converges uniformly to g. Then

$$\|g_k f - gf\| \le \|g_k - g\|_\infty \|f\| \to 0.$$

In particular,

$$\left\| \int_A g_k f d\lambda - \int_A gf d\lambda \right\| \to 0.$$

Since $\int_A g_k f d\lambda \in X$ for every $A \in \Sigma$, $\int_A gf d\lambda \in X$ and gf is Pettis integrable.

Example 1.14.

$$(P(\lambda, X), L^\infty(\lambda) : X)$$

is an abstract triple under the bilinear operator $(f, g) \to \int_S gf d\lambda$. Note the bilinear operator is continuous as is the case for the Dunford integral.

As was the case for the Dunford integral the norm on $P(\lambda, X)$ is, in general, not complete ([BS] 5.13). We will show later as an application of the Orlicz–Pettis Theorem that the indefinite Pettis integral is norm countably additive. Indeed, it is this property of the indefinite integral which separates the Dunford and Pettis integrals ([DU] II.3.6).

For more information on the Dunford and Pettis integrals, see [DU], [BS].

We next consider the Bochner integral which is the vector analogue of the Lebesgue integral for vector valued functions. A function $f : S \to X$ is *strongly λ measurable* if there exists a sequence of X valued, Σ simple functions $\{f_k\}$ which converge pointwise λ almost everywhere to f in norm.

Proposition 1.15. *If $f : S \to X$ is strongly λ measurable, then the scalar function $\|f(\cdot)\|$ is measurable.*

Proof. If $g = \sum_{j=1}^{n} \chi_{A_j} x_j$ is Σ simple with the $\{A_j\}$ pairwise disjoint, then

$$\|g(\cdot)\| = \sum_{j=1}^{n} \|x_j\| \chi_{A_j}$$

so $\|g(\cdot)\|$ is Σ simple. If f is strongly measurable and $\{g_k\}$ is a sequence of X valued, simple functions converging to f λ almost everywhere, then

$$\|g_k(\cdot)\| \to \|f(\cdot)\|$$

λ almost everywhere so $\|f(\cdot)\|$ is measurable. $\square$

An X valued, Σ simple function $g = \sum_{j=1}^{n} \chi_{A_j} x_j$ is Bochner λ integrable over $A \in \Sigma$ if $\lambda(A \cap A_j) < \infty$ for every j and if this is the case the Bochner integral is defined to be

$$\int_A g d\lambda = \sum_{j=1}^{n} \lambda(A \cap A_j) x_j.$$

From the additivity of λ, the definition of the integral does not depend on the representation of g as a simple function. Note that a simple function g is integrable iff the function $\|g(\cdot)\|$ is integrable and in this case

$$\left\| \int_A g d\lambda \right\| \le \int_A \|g(\cdot)\| \, d\lambda.$$

Definition 1.16. A strongly measurable function $f : S \to X$ is λ Bochner integrable if

(i) there exist a sequence of λ integrable simple functions $\{g_j\}$ such that $g_j \to f$ λ almost everywhere and
(ii) $\lim_j \int_S \|f(\cdot) - g_j(\cdot)\| \, d\lambda = 0$.

The *Bochner integral* of f with respect to λ is defined to be

$$\int_S f d\lambda = \lim_j \int_S g_j d\lambda.$$

The function f is Bochner λ integrable over $A \in \Sigma$ if $\chi_A f$ is Bochner integrable and we define

$$\int_A f d\lambda = \int_S \chi_A f d\lambda.$$

A few remarks are necessary here. First, the integrals in (ii) make sense from Proposition 1.15 since the functions $\|f(\cdot) - g_j(\cdot)\|$ are measurable. Since

$$\left\| \int_S g_j d\lambda - \int_S g_k d\lambda \right\| \leq \int_S \|g_j(\cdot) - g_k(\cdot)\| \, d\lambda$$
$$\leq \int_S \|f(\cdot) - g_j(\cdot)\| \, d\lambda + \int_S \|f(\cdot) - g_k(\cdot)\| \, d\lambda,$$

(ii) implies that $\{\int_S g_j d\lambda\}$ is a Cauchy sequence in X and, therefore, converges so

$$\lim_j \int_S g_j d\lambda$$

exists. Also, the limit is independent of the sequence $\{g_j\}$ so the definition makes sense (if $\{h_j\}$ is another sequence satisfying (i) and (ii) consider the interlaced sequence $\{g_1, h_1, g_2, h_2, ...\}$ which also satisfies (i) and (ii)).

We have a useful criterion for Bochner integrability similar to that for the Lebesgue integral.

Theorem 1.17. *Let* $f : S \to X$ *be strongly measurable. Then* f *is* λ *Bochner integrable iff* $\|f(\cdot)\|$ *is* λ *integrable and then*

$$\left\| \int_A f d\lambda \right\| \leq \int_A \|f(\cdot)\| \, d\lambda.$$

Proof. $\implies$: Let $\{g_j\}$ satisfy (i) and (ii). Then

$$\|f(\cdot)\| \leq \|g_j(\cdot)\| + \|f(\cdot) - g_j(\cdot)\|$$

implies $\|f(\cdot)\|$ is integrable.

$\impliedby$: Let $\{g_j\}$ be a sequence of simple functions which converges λ almost everywhere to f. Put $h_j(t) = g_j(t)$ if

$$\|g_j(t)\| \leq 2 \|f(t)\|$$

and put $h_j(t) = 0$ otherwise. Then each h_j is a simple function with $h_j \to f$ λ almost everywhere and

$$\|h_j(t)\| \leq 2 \|f(t)\|.$$

Since

$$\|h_j(t) - f(t)\| \leq 3 \|f(t)\|$$

and $\|f(\cdot)\|$ is integrable, the Dominated Convergence Theorem implies

$$\int_S \|f(\cdot) - h_j(\cdot)\| \, d\lambda \to 0,$$

so f is Bochner integrable.

Also, the Dominated Convergence Theorem implies

$$\lim \int_S \|h_j(\cdot)\| \, d\lambda = \int_S \|f(\cdot)\| \, d\lambda,$$

so

$$\left\|\int_S f d\lambda\right\| = \lim \left\|\int_S h_j d\lambda\right\| \le \lim \int_S \|h_j(\cdot)\| \, d\lambda = \int_S \|f(\cdot)\| \, d\lambda.$$

$\square$

We note that a version of the Dominated Convergence Theorem holds for the Bochner integral.

Theorem 1.18. (Dominated Convergence Theorem) *Let $f_j : S \to X$ be λ Bochner integrable for $j \in \mathbb{N}$ and suppose $\{f_j\}$ converges to f $\lambda - ae$. If there exists $g \in L^1(\lambda)$ such that*

$$\|f_j(\cdot)\| \le g \; \lambda - ae,$$

then f is λ Bochner integrable and

$$\lim_j \int_S \|f_j(\cdot) - f(\cdot)\| \, d\lambda = 0,$$

in particular,

$$\lim_j \int_S f_j d\lambda = \int_S f d\lambda.$$

Proof. Since $\|f_j(\cdot)\| \to \|f(\cdot)\| \; \lambda - ae$, the Dominated Convergence Theorem implies $\|f(\cdot)\|$ is λ integrable. Theorem 1.17 implies that f is Bochner integrable. Since $\|f_j(\cdot) - f(\cdot)\| \to 0 \; \lambda - ae$ and

$$\|f_j(\cdot) - f(\cdot)\| \le 2g$$

$\lambda - ae$, the Dominated Convergence Theorem implies

$$\lim_j \int_S \|f_j(\cdot) - f(\cdot)\| \, d\lambda = 0.$$

Since

$$\left\|\int_S f_j d\lambda - \int_S f d\lambda\right\| \le \int_S \|f_j(\cdot) - f(\cdot)\| \, d\lambda,$$

it follows that $\lim_j \int_S f_j d\lambda = \int_S f d\lambda$.

$\square$

We note in passing that a Bochner integrable function is Pettis integrable so the Pettis integral is more general than the Bochner integral. Of course, a Pettis integrable function needn't be strongly measurable.

Proposition 1.19. *Let f be Bochner integrable. Then f is Pettis integrable and the two integrals agree.*

Proof. If $x' \in X'$, then $|x'f| \le \|x'\| \, \|f(\cdot)\|$ so $x'f$ is integrable and f is Dunford integrable. We denote the Pettis integral by $P \int$ and the Bochner integral by $B \int$ in what follows. Let $\{g_j\}$ satisfy (i) and (ii) in the definition of the Bochner integral with $\|g_j(\cdot)\| \le 2 \|f(\cdot)\|$ (see Theorem 1.17). The Dominated Convergence Theorem implies

$$\lim x'\left(P \int_S g_j d\lambda\right) = \lim x'\left(B \int_S g_j d\lambda\right)$$
$$= x'\left(B \int_S f d\lambda\right) = \lim \int_S x' g_j d\lambda = \int_S x'f d\lambda$$

so f is Pettis integrable with $B \int_S f d\lambda = P \int_S f d\lambda$. $\qquad\square$

The Bochner integral enjoys many of the properties of the Lebesgue integral including the Dominated Convergence Theorem as noted in Theorem 1.18. One notable exception is that there is no straightforward version of the Radon–Nikodym Theorem for the Bochner (or Pettis) integral. See [DU] for details of the Bochner integral including the Radon–Nikodym Theorem.
Let

$$L^1(\lambda, X)$$

denote the space of all X valued Bochner λ integrable functions; $L^1(\lambda, X)$ has the complete norm

$$\|f\|_1 = \int_S \|f(\cdot)\| d\lambda.$$

Example 1.20. Note that if $g \in L^\infty(\lambda)$ and $f \in L^1(\lambda, X)$, the product gf is Bochner integrable since

$$\|g(\cdot)f(\cdot)\| \le \|g(\cdot)\| \, \|f(\cdot)\| \le \|g\|_\infty \|f(\cdot)\|$$

λ almost everywhere. Then

$$(L^1(\lambda, X), L^\infty(\lambda) : X)$$

is an abstract triple under the bilinear map $(f, g) \to \int_S gf d\lambda$. The bilinear map is continuous since

$$\left\|\int_S gf d\lambda\right\| \le \int_S \|g(\cdot)f(\cdot)\| d\lambda \le \|g\|_\infty \int_S \|f(\cdot)\| d\lambda = \|g\|_\infty \|f\|_1.$$

Dually,

$$(L^\infty(\lambda), L^1(\lambda, X) : X)$$

is an abstract triple under the same bilinear map.

Example 1.21. Similarly, if $L^\infty(\lambda, X')$ is the space of λ essentially bounded X' valued strongly measurable functions with its natural norm, then

$$(L^1(\lambda, X), L^\infty(\lambda, X') : \mathbb{R})$$

is an abstract triple under the continuous bilinear map

$$(f, g) \to \int_S g(t)(f(t)) d\lambda(t).$$

We can also consider results like those above for vector and operator valued functions. Let X, Y be Banach spaces and consider the pair

$$L^\infty(\lambda, X), L^1(\lambda, L(X, Y)).$$

If $f \in L^1(\lambda, L(X, Y))$ and $g \in L^\infty(\lambda, X)$, we first observe that the function $t \to f(t)(g(t))$ is strongly measurable. Suppose first that g is a simple function, $g = \sum_{j=1}^n \chi_{B_j} x_j$, with $\{B_j\}, B_j \in \Sigma$, a partition of S. Then

$$f(\cdot)(g(\cdot)) = \sum_{j=1}^n \chi_{B_j}(\cdot) f(\cdot)(x_j)$$

so $f(\cdot)(g(\cdot))$ is a strongly measurable function. If $g \in L^\infty(\lambda, X)$, there exists a sequence $\{g_k\}$ of simple functions which converges pointwise almost everywhere to g. Then $f(\cdot)(g_k(\cdot)) \to f(\cdot)(g(\cdot))$ almost everywhere so $f(\cdot)(g(\cdot))$ is strongly measurable. Moreover,

$$\|f(t)(g(t))\| \le \|f(t)\| \|g(t)\| \le \|g\|_\infty \|f(t)\|$$

λ almost everywhere implies $f(\cdot)(g(\cdot))$ is Bochner integrable with

$$\left\| \int_S f(\cdot)(g(\cdot)) d\lambda \right\| \le \|g\|_\infty \|f\|_1.$$

Example 1.22. Then

$$(L^\infty(\lambda, X), L^1(\lambda, L(X, Y)) : Y)$$

is an abstract triple under the continuous bilinear mapping $(g, f) \to \int_S f(\cdot)(g(\cdot)) d\lambda$.

It should also be noted that dually, we have the triple

$$(L^1(\lambda, X), L^\infty(\lambda, L(X, Y))) : Y)$$

under the same type of continuous bilinear mapping. Similarly, if $1 < p < \infty$ and $\frac{1}{p} + \frac{1}{q} = 1$, then one may define a triple

$$(L^p(\lambda, X), L^q(\lambda, L(X, Y))) : Y)$$

as above.

It follows from Theorem 1.17 that if f is a bounded, strongly measurable function, then f is Bochner integrable with respect to any finite, countably additive set function. Let $B(\Sigma, X)$ be the space of all bounded, strongly Σ measurable functions with the sup norm and let $ca(\Sigma)$ be the space of all real valued set functions λ on Σ with the variation norm, $|\lambda|$ ([Sw3], 2.2.7). Then we have

Example 1.23.

$$(B(\Sigma, X), ca(\Sigma) : X)$$

is an abstract triple under the map $(f, \lambda) \to \int_S f d\lambda$. Moreover, since

$$\left\| \int_S f d\lambda \right\| \le \|f\|_\infty |\lambda| (S)$$

the bilinear map is continuous. Dually, $(ca(\Sigma), B(\Sigma, X) : X)$ is an abstract triple.

We now consider the space of scalar valued functions which are integrable with respect to a vector valued set function.

First, assume $\nu : \Sigma \to \mathbb{R}$ is bounded and finitely additive. The variation of ν is denoted by $|\nu|$ (see [Sw3] 2.2.7). If $f : S \to \mathbb{R}$ is a Σ simple function, $f = \sum_{k=1}^n a_k \chi_{A_k}$, $\{A_k\}$ pairwise disjoint, the integral of f with respect to ν over A is defined to be

$$\int_A f d\nu = \sum_{k=1}^n a_k \nu(A_k \cap A);$$

the integral is independent of the representation of f as a simple function by finite additivity. Note

$$(*) \quad \left| \int_A f d\nu \right| \le \sum_{k=1}^n |a_k| |\nu(A_k \cap A)| \le \|f\|_\infty |\nu| (A)$$

for $A \in \Sigma$, where $\|\cdot\|_\infty$ denotes the sup-norm. If $f : S \to \mathbb{R}$ is bounded and Σ measurable and $\{f_k\}$ is a sequence of simple functions converging uniformly to f, the integral of f with respect to ν is defined to be

$$\int_A f d\nu = \lim_k \int_A f_k d\nu;$$

note $\{\int_A f_k d\nu\}$ is Cauchy by $(*)$ so the limit exists. The integral is independent of the sequence $\{f_k\}$ (if $\{f_k\}, \{g_k\}$ are two sequences converging uniformly to f consider the "interlaced sequence" $\{f_1, g_1, f_2, g_2, ...\}$). The inequality $(*)$ still holds for f;

$$\left|\int_A f d\nu\right| = \left|\lim_k \int_A f_k d\nu\right| \le \limsup_k \int_A |f_k| \, d \, |\nu|$$
$$\le \limsup_k \|f_k\|_\infty \, |\nu| \, (A) = \|f\|_\infty \, |\nu| \, (A).$$

Next, we define the integral of a scalar valued function with respect to a finitely additive, bounded vector valued set function. Let $m : \Sigma \to X$ be finitely additive and bounded and let $f : S \to \mathbb{R}$ be Σ measurable. We say that f is *scalarly integrable* if f is $x'm$ integrable for every $x' \in X'$.

Definition 1.24. f is m integrable if f is scalarly integrable and for each $A \in \Sigma$ there exists $x_A \in X$ such that

$$\int_A f dx'm = x'(x_A).$$

We write

$$x_A = \int_A f dm$$

so

$$x'\left(\int_A f dm\right) = \int_A f dx'm.$$

We define the semi-variation of m in order to obtain the analogue of $(*)$ for the integral.

Definition 1.25. The semi-variation of m is defined by

$$semi-var(m)(A) = \sup\left\{\left\|\sum_{k=1}^n t_k m(A_k)\right\| : |t_k| \le 1, \{A_k\} \subset \Sigma \text{ a partition of } A\right\}.$$

We compute another useful expression for the semi-variation,

$$\|m(A)\| \le semi - var(m)(A)$$

$$= \sup \left\{ \left| \sum_{k=1}^{n} t_k x' m(A_k) \right| : \|x'\| \le 1, |t_k| \le 1, \right.$$

$$\left. \{A_k\} \subset \Sigma \text{ a partition of } A \right\}$$

$$= \sup \left\{ \sum_{k=1}^{n} |x' m(A_k)| : \|x'\| \le 1, \{A_k\} \subset \Sigma \text{ a partition of } A \right\}$$

$$= \sup\{|x'm|(A) : \|x'\| \le 1\}.$$

Hence,

$$\sup\{\|m(A)\| : A \in \Sigma\} \le semi - var(m)(S) = \sup\{|x'm|(S) : \|x'\| \le 1\}.$$

We have the inequality

$$\sup\{|\nu(B)| : B \subset A\} \le |\nu|(A) \le 2\sup\{|\nu(B)| : B \subset A\}$$

for additive scalar set functions ([Sw3] 2.2.1.7, [DS] III.1.5). Applying this inequality, we have

$$(\#) \ \sup\{|x'm|(S) : \|x'\| \le 1\} \le 2\sup\{|x'm(A)| : \|x'\| \le 1, A \in \Sigma\}$$

$$= 2\sup\{\|m(A)\| : A \in \Sigma\}.$$

Thus,

Theorem 1.26.

$$\|m\| = \sup\{\|m(A)\| : A \in \Sigma\}$$

and

$$semi - var(m)(S) = \sup\{|x'm|(S) : \|x'\| \le 1\}$$

are equivalent norms on

$$ba(\Sigma, X),$$

the space of all bounded, finitely additive set functions from Σ into X.

We have the analogue of the inequality $(*)$ for the integral.

Proposition 1.27. *Let $f : S \to \mathbb{R}$ be bounded and Σ measurable. Then f is m integrable with*

$$(\&) \ \left\| \int_A f dm \right\| \le \|f\|_\infty \, semi - var(m)(A).$$

Proof. First assume that f is m integrable. Then

$$\left\| \int_A f \, dm \right\| = \sup \left\{ \left| \int_A f \, dx'm \right| : \|x'\| \leq 1 \right\} \leq \sup \left\{ \int_A |f| \, d \, |x'm| : \|x'\| \leq 1 \right\}$$
$$\leq \|f\|_\infty \sup\{|x'm|\,(A) : \|x'\| \leq 1\} = \|f\|_\infty \, semi - var(m)(A)$$

so (&) holds.

Next pick a sequence of simple functions $\{g_j\}$ which converge uniformly to f. Clearly every simple function is m integrable so (&) holds for simple functions. From (&),

$$\left\| \int_A g_j \, dm - \int_A g_k \, dm \right\| \leq \|g_j - g_k\|_\infty \, semi - var(m)(A)$$

so

$$\left\{ \int_A g_k \, dm \right\}$$

is Cauchy. Let $x_A = \lim \int_A g_k \, dm$. We claim $x_A = \int_A f \, dm$. Let $x' \in X'$. Then

$$x'(x_A) = \lim x' \int_A g_k \, dm = \lim \int_A g_k \, dx'm = \int_A f \, dx'm$$

justifying the claim. Thus, f is m integrable with (&) holding.   □

Let $B(\Sigma)$ be the space of bounded, Σ measurable functions on S with the sup-norm.

Example 1.28. We have that

$$(B(\Sigma), ba(\Sigma, X) : X)$$

is an abstract triple under the map $(f, m) \to \int_S f \, dm$. Moreover, if $ba(\Sigma, X)$ has one of the norms defined above, the bilinear map is continuous by Proposition1.27. Dually, $(ba(\Sigma, X), B(\Sigma) : X)$ is an abstract triple.

We now consider the case when $m : \Sigma \to X$ is countably additive. We will show later that any such vector measure is bounded (Appendix C) so

$$\|m\| = \sup\{\|m(A)\| : A \in \Sigma\} < \infty.$$

This implies that $ca(\Sigma, X)$ is a subspace of $ba(\Sigma, X)$.

Now suppose $f : S \to \mathbb{R}$ is Σ measurable. Then f is scalarly integrable with respect to m if f is $x'm$ integrable for every $x' \in X'$, where the integral $\int_A f \, dx'm$ is a Lebesgue integral with respect to the countably additive, signed measure $x'm$, and f is integrable with respect to m if f is scalarly integrable with respect to m and for every $A \in \Sigma$ there exists $x_A \in X$ such that

$$x'(x_A) = \int_A f dx' m;$$

we write $x_A = \int_A f dm$ so

$$x'\left(\int_A f dm\right) = \int_A f dx' m.$$

It should be noted that a function can be scalarly integrable and not integrable

Example 1.29. Let Σ be the power set of $\mathbb{N}$ and define a countably additive measure m from Σ into c_0 by

$$m(A) = \sum_{k \in A} (1/k) e^k.$$

Define $f : \mathbb{N} \to \mathbb{R}$ by $f(k) = k$. If $t = \{t_k\} \in l^1$,

$$\int_A f dt m = \sum_{k \in A} t_k$$

so if f were m integrable, we would have

$$\int_A f dm = \chi_A.$$

Thus, f is not m integrable.

Let

$$L^1(m)$$

be the space of all scalar functions which are m integrable (we have used the notation $L^1(\lambda)$ when λ was a positive measure but if we keep in mind that m is a vector measure this should cause no problems). We define a norm on $L^1(m)$ by

$$\|f\|_1 = \sup\left\{\int_S |f| \, d\,|x'm| : \|x'\| \leq 1\right\}$$

$$= \sup\left\{\left|\int f dx' m\right| : \|x'\| \leq 1\right\} = semi - var\left(\int f dm\right).$$

We need to observe that this norm is finite. First, if f is integrable with respect to m, then the indefinite integral of f, $\int f dm$, is countably additive with respect to the weak topology of X since the scalar indefinite integrals $\int f dx' m$ are countably additive. It follows from the Orlicz–Pettis Theorem

which we will establish later (2.11, Theorem 2.9) that the indefinite integral
is norm countably additive (Theorem 2.12) and it will also be shown later
that vector valued countably additive set functions defined on a σ algebra
with values in a Banach space are bounded (Appendix C). Thus,

$$\sup\left\{\left\|\int_A fdm\right\| : A \in \Sigma\right\} = \left\|\int_. fdm\right\|(S) < \infty.$$

By Theorem 1.26,

$$\|f\|_1 = \sup\left\{\int_S |f|\, d\,|x'm| : \|x'\| \le 1\right\} < \infty.$$

From Theorem 1.26, we have

Theorem 1.30. $\|\cdot\|_1$ *and*

$$\|f\|_1' = \sup\left\{\left\|\int_A fdm\right\| : A \in \Sigma\right\} = \left\|\int_. fdm\right\|(S)$$

define equivalent norms on $L^1(m)$.

We will now show that the product of a bounded measurable function
and an integrable function is integrable. Let $f \in L^1(m)$ and $g : S \to \mathbb{R}$ be
bounded and measurable. First, if $h = \sum_{j=1}^n t_j \chi_{A_j}$ is a Σ simple function
with the $\{A_j\}$ pairwise disjoint, then

$$\int_S hfdm = \sum_{j=1}^n t_j \int_{A_j} fdm$$

and

$$\left\|\int_S hfdm\right\| = \left\|\sum_{j=1}^n t_j \int_{A_j} fdm\right\| \le \sum_{j=1}^n \left\|t_j \int_{A_j} fdm\right\|$$

$$\le \|h\|_\infty\, semi-var\left(\int fdm\right) \le \|h\|_\infty \|f\|_1.$$

Pick a sequence of simple functions $\{h_k\}$ which converge uniformly to g.
Then by the inequality above,

$$\left\|\int_A h_j fdm - \int_A h_k fdm\right\| \le \|h_j - h_k\|_\infty \|f\|_1$$

so

$$\left\{\int_A h_j fdm\right\}$$

is Cauchy. Let

$$x_A = \lim \int_A h_j f \, dm.$$

We claim $x_A = \int_A g f \, dm$. Since

$$|h_j f| \le \|h_j\|_\infty |f|,$$

if $x' \in X'$, the Dominated Convergence Theorem implies

$$x'(x_A) = \lim x' \int_A h_j f \, dm = \lim \int_A h_j f \, dx'm = \int_A g f \, dx'm$$

and the claim is established. Thus, the product of bounded measurable functions and integrable functions is integrable. Moreover, we have

$$\left\| \int_S g f \, dm \right\| = \sup_{\|x'\| \le 1} \left| \int_S g f \, dx'm \right|$$

$$\le \sup_{\|x'\| \le 1} \int_S |g f| \, d|x'm| \le \|g\|_\infty \sup_{\|x'\| \le 1} \int_S |f| \, d|x'm| = \|g\|_\infty \|f\|_1 .$$

Example 1.31.

$$(B(\Sigma), L^1(m) : X)$$

is an abstract triple under the bilinear map $(g, f) \to \int_S g f \, dm$. Also, the bilinear map is continuous by the inequality above. Dually,

$$(L^1(m), B(\Sigma) : X)$$

is an abstract triple.

For later use we need a uniform convergence theorem for the integral.

Theorem 1.32. *Let $\{f_j\} \subset B(\Sigma)$ be such that $\{f_j\}$ converges uniformly to a function f. Then*

$$\lim_j \int_S f_j \, dm = \int_S f \, dm.$$

Proof. By Proposition 1.27,

$$\left\| \int_S f_j \, dm - \int_S f_k \, dm \right\| \le \|f_j - f_k\|_\infty \, semi - var(m)(S),$$

so

$$\left\{ \int_S f_j \, dm \right\}$$

is Cauchy. Let $z = \lim_j \int_S f_j \, dm$. We claim that $z = \int_S f \, dm$. If $x' \in X'$,

$$\lim_j x' \left(\int_S f_j \, dm \right) = \lim_j \int_S f_j \, dx'm = \int_S f x' \, dm = x'(z)$$

by the Bounded Convergence Theorem for $x'm$. Thus, $z = \int_S f \, dm$. □

We will also need for later use a Dominated Convergence Theorem for the integral. For this we require some additional properties of the semi-variation.

Proposition 1.33. *The semi-variation is increasing, countably subadditive and continuous.*

Proof. The semi-variation is obviously increasing. Let $\{A_j\}$ be a pairwise disjoint sequence from Σ. Note that if $\{B_i : i = 1, ..., n\}$ is a partition of $A = \cup_{j=1}^{\infty} A_j$, then $\{A_j \cap B_i : i = 1, ..., n\}$ is a partition of A_j. Thus, if $|t_i| \leq 1$,

$$\left\| \sum_{i=1}^{n} t_i m(B_i) \right\| = \left\| \sum_{i=1}^{n} t_i \sum_{j=1}^{\infty} m(A_j \cap B_i) \right\|$$

$$\leq \sum_{j=1}^{\infty} \left\| \sum_{i=1}^{n} t_i m(A_j \cap B_i) \right\| \leq \sum_{j=1}^{\infty} semi - var(m)(A_j)$$

so that

$$semi - var(m)(A) \leq \sum_{j=1}^{\infty} semi - var(m)(A_j).$$

Hence, $semi - var$ is countably subadditive.

Next, suppose there exists $A_j \downarrow \emptyset$ with $semi - var(m)(A_j) > \delta$. Put $n_1 = 1$ and then there exist $x_1' \in X'$, $\|x_1'\| \leq 1$, and $n_2 > n_1$ with

$$|x_1' m(A_{n_1})| > \delta \text{ and } |x_1' m(A_{n_2})| \leq \delta/2.$$

Then we have

$$\sup\{\|m(B)\| \ : \ B \subset A_{n_1} \setminus A_{n_2}\} \geq \left|x_{n_1}' m(A_{n_1} \setminus A_{n_2})\right|$$

$$\geq \left|x_{n_1}'(A_{n_1})\right| - \left|x_{n_1}'(A_{n_2})\right| > \delta/2$$

so there exists $B_1 \subset A_{n_1} \setminus A_{n_2}$ such that $\|m(B_1)\| > \delta/2$. Continuing this construction produces an increasing sequence $\{n_k\}$ and $\{B_k\}$ with

$$B_k \subset A_{n_k} \setminus A_{n_{k+1}} \text{ and } \|m(B_k)\| > \delta/2.$$

This implies m is not countably additive since the $\{B_k\}$ are disjoint. If $\{A_j\} \subset \Sigma$ has limit A, then the inequality

$$semi - var(m)(A) - semi - var(m)(A_j)$$

$$\leq semi - var(m)(\cup_{i \geq j}(A \setminus A_i)) + semi - var(m)(\cup_{i \geq j}(A_i \setminus A))$$

implies

$$semi - var(m)(A) = \lim_{j} semi - var(m)(A_j).$$

□

We now have the necessary machinery to establish the Dominated Convergence Theorem for the integral.

Theorem 1.34. (Dominated Convergence Theorem) *Suppose $f_k, g \in L^1(m)$ are such that*

$$|f_k(t)| \leq |g(t)|$$

for all $k \in \mathbb{N}, t \in S$. If the sequence $\{f_k\}$ converges pointwise to a function f, then $f \in L^1(m)$ and

$$\lim_k \int_S f_k dm = \int_S f dm.$$

Proof. We first claim $\{\int_A f_k dm\}_k$ satisfies a Cauchy condition uniformly for $A \in \Sigma$. Let $\epsilon > 0$. Define a countably additive measure G by

$$G(A) = \int_A g dm$$

(this uses the Orlicz–Pettis Theorem (see 2.12, 2.9) as was noted earlier). Let $A \in \Sigma$ and $\|x'\| \leq 1$ and set

$$A_k = \{t \in S : |f(t) - f_k(t)| \geq \epsilon\}.$$

Note f is $x'm$ integrable by the Dominated Convergence Theorem for scalar measures. Then

$$\left| \int_A (f - f_k) dx'm \right| \leq \left| \int_{A \backslash A_k} (f - f_k) dx'm \right| + \left| \int_{A \cap A_k} (f - f_k) dx'm \right|$$
$$\leq \epsilon |x'm| (A \backslash A_k) + 2 \int_{A \cap A_k} |g| \, d |x'm|$$
$$\leq \epsilon \, semi - var(m)(S) + 2 semi - var(G)(A_k).$$

Thus,

$$\left\| \int_A (f_j - f_k) dm \right\| \leq 2\epsilon \, semi - var(m)(S) + 2 \, semi - var(G)(A_k)$$
$$+ 2 \, semi - var(G)(A_j)$$

and by the result above

$$\lim_k semi - var(G)(A_k) = 0.$$

This justifies the claim. Thus,

$$\lim_k \int_A f_k dm = F(A)$$

exists for every $A \in \Sigma$. The Dominated Convergence Theorem implies that f is scalarly m integrable and the computation above shows $F(A) = \int_A f dm$. $\square$

Note the Dominated Convergence Theorem implies a Bounded Convergence Theorem for the integral since bounded scalar functions are integrable.

For other treatments of the integration of scalar functions with respect to vector valued measures, see [DS], [KK], [Pa].

There are also theories of integration of vector valued functions with respect to operator valued measures. The definitive development of such theories have been carried out by Bartle ([Bar]) and Dobrakov ([Do]). These developments are quite technical so we do not give descriptions. One may use the properties of these integrals to define and treat abstract triples in the same manner as done above.

Chapter 2

Subseries Convergence

In this chapter we study versions of the Orlicz–Pettis Theorem for subseries convergent series in abstract triples and use these results to establish versions of the theorem in various settings. We also give applications of the results to various topics in analysis. A series $\sum_{j=1}^{\infty} x_j$ in a topological Abelian group (X, τ) is τ *subseries convergent* if for every subsequence $\{n_j\}$ the subseries $\sum_{j=1}^{\infty} x_{n_j}$ is τ convergent in X. The classical version of the Orlicz–Pettis Theorem for normed spaces asserts that a series in a normed space which is subseries convergent in the weak topology of the space is subseries convergent in the norm topology ([Or],[Pe]). The theorem has important applications to many areas in the integration of vector valued functions and vector valued measures. In particular, Pettis used the theorem to establish the countable additivity of the Pettis integral which he defined. The theorem has been extended to locally convex spaces and many other situations including topological groups. See [DU],[K1],[FL] for a discussion of the history of the subject. We refer to any result which asserts that a series which is subseries convergent in some weak topology is subseries convergent in a stronger topology as an *Orlicz–Pettis Theorem*.

Throughout this chapter let

$$(E, F : G)$$

be an abstract triple. Let $w(F, E)$ be the weakest topology on F such that all of the maps $\{b(x, \cdot) : x \in E\}$ from F into G are continuous; $w(E, F)$ is defined similarly. A subset B of a topological space (X, τ) is *sequentially conditionally τ compact* if every sequence $\{x_j\}$ in B has a subsequence $\{x_{n_j}\}$ which is τ Cauchy (this is terminology of Dinculeanu ([Din])). A subset B is *sequentially relatively τ compact* if every sequence $\{x_j\}$ in B has a subsequence $\{x_{n_j}\}$ which is τ convergent to an element of X. Thus, a subset $B \subset F$ is $w(F, E)$ sequentially conditionally compact if every sequence $\{y_j\} \subset B$ has a subsequence $\{y_{n_j}\}$ such that $\lim_j x \cdot y_{n_j}$ exists for every $x \in E$.

The method of proof used in treating our versions of the Orlicz–Pettis Theorem relies on the Antosik–Mikusinskiy Matrix Theorem which we now state for convenience. A proof and other versions of the theorem may be found in Appendix E.

Theorem 2.1. (Antosik–Mikusinski) *Let G be an Abelian topological group and $x_{ij} \in G$ for $i, j \in \mathbb{N}$. Suppose*

(I) $\lim_i x_{ij} = x_j$ *exists for each j and*
(II) *for each increasing sequence of positive integers $\{m_j\}$ there is a subsequence $\{n_j\}$ of $\{m_j\}$ such that $\{\sum_{j=1}^{\infty} x_{in_j}\}$ is Cauchy.*

Then $\lim_i x_{ij} = x_j$ uniformly for $j \in \mathbb{N}$. In particular,

$$\lim_i \lim_j x_{ij} = \lim_j \lim_i x_{ij} = 0 \text{ and } \lim_i x_{ii} = 0.$$

A matrix $M = [x_{ij}]$ which satisfies conditions (I) and (II) of Theorem 2.1 is referred to as a $\mathcal{K}$ matrix.

Orlicz–Pettis Theorems.

We now establish several versions of the Orlicz–Pettis Theorem for abstract triples and then give applications to various topics in measure theory and functional analysis. The conclusion of our first result involves a type of convergence for series. A series $\sum_j x_j$ in an Abelian topological group G is *unordered convergent* if

$$\lim_D \sum_{j \in \sigma} x_j$$

converges, where D is the net

$$D = \{\sigma : \sigma \subset \mathbb{N} \text{ finite}\}$$

directed by set inclusion. A family of series $\sum_j x_{i,j}$, $i \in I$, is *uniformly unordered convergent* if the nets

$$\lim_D \sum_{j \in \sigma} x_{ij}$$

converge uniformly for $i \in I$.

Theorem 2.2. *If the series $\sum x_j$ is $w(E, F)$ subseries convergent, then the series*

$$\sum_{j \in \sigma} x_j \cdot y$$

converge uniformly for $y \in B$ and $\sigma \subset \mathbb{N}$, where B is any sequentially conditionally $w(F, E)$ compact subset $B \subset F$ [that is, for every closed neighborhood of 0, U, in G there exists N such that

$$\sum_{j \in \sigma} x_j \cdot y \in U$$

whenever $y \in B$ and $\min \sigma > N$; a strong form of unordered convergence for the series].

Proof. If the conclusion fails to hold, there exists a closed neighborhood, U, of 0 such that for every k there exist σ_k with $\min \sigma_k > k$ and $y_k \in B$ such that

$$\sum_{j \in \sigma_k} x_j \cdot y_k \notin U.$$

Put $k_1 = 1$ so we have

$$\sum_{j \in \sigma_1} x_j \cdot y_1 \notin U.$$

We may assume that σ_1 is finite since U is closed. Put $k_2 = \max \sigma_1$. Apply the condition above to k_2 to obtain

$$\sum_{j \in \sigma_2} x_j \cdot y_2 \notin U$$

with σ_2 finite, $\min \sigma_2 > k_2$ and $y_2 \in B$. This construction produces finite sequences $\{\sigma_k\}$ with $\min \sigma_{k+1} > \max \sigma_k$ and $\{y_k\} \subset B$ satisfying

$$(\&) \quad \sum_{j \in \sigma_k} x_j \cdot y_k \notin U.$$

There exists a subsequence $\{y_{n_k}\}$ such that $\lim x \cdot y_{n_k}$ exists for every $x \in E$. Consider the matrix

$$M = [m_{ij}] = \left[\sum_{l \in \sigma_j} x_l \cdot y_{n_i} \right].$$

We claim that M is a $\mathcal{K}$ matrix. The columns of M converge and for every subsequence $\{r_j\}$ the subseries

$$\sum_{j=1}^{\infty} \sum_{l \in \sigma_{r_j}} x_l$$

is $w(E, F)$ convergent to $x \in E$ and

$$\lim_i \sum_{j=1}^{\infty} m_{i r_j} = \lim_i \sum_{j=1}^{\infty} \sum_{l \in \sigma_{r_j}} x_l \cdot y_{n_i} = \lim_i x \cdot y_{n_i}$$

exists. Therefore, M is a $\mathcal{K}$ matrix whose diagonal converges to 0 by the Antosik–Mikusinski Theorem. But, this contradicts $(\&)$. $\qquad \square$

The conclusion of the theorem is a strong form of unordered convergence in the sense that in the series

$$\sum_{j \in \sigma} x_j \cdot y$$

the sets σ may be infinite.

As noted earlier a subset $B \subset F$ is sequentially relatively $w(F, E)$ compact if every sequence $\{y_k\} \subset B$ has a subsequence $\{y_{n_k}\}$ and there exists $y \in F$ such that $\lim_k x \cdot y_{n_k} = x \cdot y$ for every $x \in E$. A sequentially relatively $w(F, E)$ compact is obviously sequentially conditionally $w(F, E)$ compact so the result above holds for this family of subsets of F.

The unordered convergence form of the conclusion of Theorem 2.2 obviously implies that the series is subseries convergent in the topology of uniform convergence on sequentially conditionally $w(F, E)$ compact subsets and is useful in treating the Hahn–Schur Theorem considered later.

We consider the theorem for $w(F, E)$ compact subsets.

Theorem 2.3. *Let G be metrizable under the metric ρ. If $\sum x_j$ is $w(E, F)$ subseries convergent, then the series*

$$\sum_{j \in \sigma} x_j \cdot y$$

converge uniformly for $y \in B$ and $\sigma \subset \mathbb{N}$, where B is any $w(F, E)$ compact subset $B \subset F$.

Proof. Let B be $w(F, E)$ compact. Define an equivalence relation on B by $y \sim z$ iff $x_j \cdot y = x_j \cdot z$ for every j. Let $\widehat{B}$ be the collection of equivalence classes and $\widehat{y}$ the equivalence class to which y belongs. Define a metric d on $\widehat{B}$ by

$$d(\widehat{y}, \widehat{z}) = \sum_{j=1}^{\infty} \frac{1}{2^j} \frac{\rho(x_j \cdot y, x_j \cdot z)}{1 + \rho(x_j \cdot y, x_j \cdot z)}$$

so a net $\widehat{y_\alpha} \to \widehat{y}$ with respect to d iff

$$\lim_\alpha x_j \cdot y_\alpha = x_j \cdot y$$

for all j.

Let $\mathcal{S}$ be the set of partial sums of the series $\sum x_j$; i.e.,

$$\mathcal{S} = \left\{ \sum_{j \in \sigma} x_j : \sigma \subset \mathbb{N} \right\},$$

where $\sum_{j\in\sigma} x_j$ is the $w(E,F)$ sum of the series. Note that if $y,z \in B$ and $y \sim z$, then $x_\sigma \cdot y = x_\sigma \cdot z$ for all $\sigma \subset \mathbb{N}$. Thus,

$$(\mathcal{S}, \widehat{B} : G)$$

is an abstract triple under the map $(x_\sigma, \widehat{y}) \to x_\sigma \cdot y$. Since B is $w(F,E)$ compact, $\widehat{B}$ is $w(\widehat{B}, \mathcal{S})$ compact. [If $\{\widehat{y_\alpha}\}$ is a net in $\widehat{B}$, then $\{y_\alpha\}$ is a net in B and so has a subnet $\{y_\beta\}$ which is $w(F,E)$ convergent to some $y \in B$ and then

$$x \cdot y_\beta \to x \cdot y$$

for $x \in E$. In particular,

$$x_\sigma \cdot y_\beta \to x_\sigma \cdot y$$

for every σ so $\{\widehat{y_\beta}\}$ is $w(\widehat{B}, \mathcal{S})$ convergent to $\widehat{y}$.] The inclusion

$$(\widehat{B}, w(\widehat{B}, \mathcal{S})) \to (\widehat{B}, d)$$

is continuous so $d = w(\widehat{B}, \mathcal{S})$ on $\widehat{B}$. Now $\sum x_j$ is $w(\mathcal{S}, \widehat{B})$ subseries convergent and $\widehat{B}$ is $w(\widehat{B}, \mathcal{S})$ sequentially compact since this topology is metrizable so it follows from the previous Orlicz–Pettis Theorem that the series $\sum x_j$ converges uniformly on $\widehat{B}$ and, therefore, on B. $\qquad\square$

If the space G is a locally convex space, the metrizability condition in Theorem 2.3 can be dropped.

Theorem 2.4. *Let (G, τ) be a locally convex space. If $\sum x_j$ is $w(E,F)$ subseries convergent, then the series*

$$\sum_{j\in\sigma} x_j \cdot y$$

converge uniformly for $y \in B$ and $\sigma \subset \mathbb{N}$, where B is any $w(F,E)$ compact subset $B \subset F$.

Proof. Let p be a continuous semi-norm on G. Consider the triple

$$(E, F : (G, p))$$

under the map $(x,y) \to x \cdot y$. Then the series $\sum x_j$ is $w(E,F)$ subseries convergent in this triple. The set $B \subset F$ is $w(F,E)$ compact in the triple $(E, F : (G, p))$. For, if $\{y_\delta\}$ is a net in B, there is a subnet $\{y_{\delta'}\}$ and $y \in B$ such that $x \cdot y_{\delta'} \to x \cdot y$ in τ so $x \cdot y_{\delta'} \to x \cdot y$ in p. By Theorem 2.3 the series $\sum_{j=1}^{\infty} x_j \cdot y$ converge uniformly for $y \in B$. $\qquad\square$

We will discuss applications of Theorems 2.2, 2.3 and 2.4 to locally convex spaces below.

A result of Kalton ([Ka2]) asserts that if τ is a separable polar topology on E from the dual pair E, F, then any series $\sum x_j$ in E which is $\sigma(E, F)$ subseries convergent is τ subseries convergent. If $\tau = \tau_{\mathcal{A}}$ is the polar topology of uniform convergence on the family $\mathcal{A}$ of $\sigma(F, E)$ bounded subsets of F, then for every $x \in E$ and $A \in \mathcal{A}$ the set $\{x \cdot y : y \in A\}$ is sequentially relatively compact in the scalar field. We give an abstraction of this condition and use it to give a generalization of Kalton's result to abstract triples.

Definition 2.5. A subset $B \subset F$ is sequentially conditionally compact at each $x \in E$ if
$$\{x \cdot y : y \in B\} = x \cdot B$$
is sequentially conditionally compact in G for every $x \in E$.

Note that if $B \subset F$ is sequentially conditionally $w(F, E)$ compact, then B is sequentially conditionally compact at each $x \in E$. Under a separability assumption, the converse holds.

Theorem 2.6. *Let G be sequentially complete. Let $\mathcal{F}$ be a family of subsets of F such that each member of $\mathcal{F}$ is sequentially conditionally compact at each $x \in E$ and let τ be the topology on E of uniform convergence on the members of $\mathcal{F}$. If (E, τ) is separable, then each member of $\mathcal{F}$ is sequentially conditionally $w(F, E)$ compact.*

Proof. Let $D = \{d_k : k \in \mathbb{N}\}$ be τ dense in E. Let $B \in \mathcal{F}$ and $\{y_k\} \subset B$. Since B is sequentially conditionally compact at each $x \in E$, by the diagonalization procedure $\{y_k\}$ has a subsequence $\{y_{n_k}\}$ such that the sequence $\{d \cdot y_{n_k}\}$ converges in G for every $d \in D$ ([Ke] p.238). Let $x \in E$. There is a net $\{d^\alpha\} \subset D$ which is τ convergent to x so
$$\lim d^\alpha \cdot y = x \cdot y$$
uniformly for $y \in B$. Let U be a neighborhood of 0 in G and pick a symmetric neighborhood, V, of 0 in G such that $V + V + V \subset U$. There exists β such that
$$d^\beta \cdot y_{n_k} - x \cdot y_{n_k} \in V$$
for all k. Since $\{d^\beta \cdot y_{n_k}\}_k$ converges, there exists N such that $k, j \geq N$ implies
$$d^\beta \cdot y_{n_k} - d^\beta \cdot y_{n_j} \in V.$$

Hence, if $k, j \geq N$, then

$$\begin{aligned}
x \cdot y_{n_k} - x \cdot y_{n_j} &= (x \cdot y_{n_k} - d^\beta \cdot y_{n_k}) + (d^\beta \cdot y_{n_k} - d^\beta \cdot y_{n_j}) \\
&\quad + (d^\beta \cdot y_{n_j} - x \cdot y_{n_j}) \\
&\in V + V + V \subset U
\end{aligned}$$

so $\{x \cdot y_{n_k}\}$ is Cauchy and, therefore, convergent since G is sequentially complete. Hence, B is sequentially conditionally $w(F, E)$ compact. $\quad\square$

Corollary 2.7. *If the conditions of Theorem 2.6 hold and $\sum x_j$ is $w(E, F)$ subseries convergent, then $\sum x_j$ is τ subseries convergent.*

Proof. The result is immediate from Theorem 2.6 and Theorem 2.2. $\quad\square$

Kalton's Orlicz–Pettis Theorem for separable polar topologies follows from Corollary 7; see Theorem 15.

The separability assumption in Theorem 2.6 is important.

Example 2.8. The series $\sum e^j$ is $\sigma(l^\infty, l^1)$ subseries convergent but is not

$$\beta(l^\infty, l^1) = \|\cdot\|_\infty$$

subseries convergent.

Locally Convex Spaces.

We consider applications of the abstract Orlicz–Pettis theorems to locally convex spaces.

Let E be a Hausdorff locally convex space with dual E'. Suppose $\sum x_j$ is subseries convergent with respect to $\sigma(E, E')$, the weak topology of E. Let $\gamma(E, E')$ $(\lambda(E, E'); \tau(E, E'))$ be the topology of uniform convergence on the sequentially conditionally $\sigma(E', E)$ compact subsets of E' $(\sigma(E', E)$ compact subsets; convex $\sigma(E', E)$ compact subsets). It follows from Theorem 2.2 that $\sum x_j$ is $\gamma(E, E')$ subseries convergent. Also, it follows from Theorem 2.2 that $\sum x_j$ is subseries convergent with respect to $\lambda(E, E')$ and, therefore, subseries convergent with respect to $\tau(E, E')$, the Mackey topology. Thus, we have an Orlicz–Pettis Theorem for locally convex spaces.

Theorem 2.9. *If the series $\sum_{j=1}^\infty x_j$ is $\sigma(E, E')$ subseries convergent, then the series is $\gamma(E, E')$, $\tau(E, E')$ and $\lambda(E, E')$ subseries convergent and also subseries convergent in the original topology of E.*

The locally convex version of the Orlicz–Pettis Theorem for the Mackey topology is due to McArthur ([Mc]); the version for the topology $\lambda(E, E')$ is due to Bennett and Kalton ([BK]) and the version for $\gamma(E, E')$ is due to Dierolf ([Die]).

Dierolf has also shown that there is a strongest polar topology which has the same subseries convergent series as the weak topology. We will give a brief description of the Dierolf topology. Let E, F be a pair of vector spaces in duality and let $\mathcal{M}$ be the family of all $M \subset F$ such that M is $\sigma(F, E)$ bounded and for every linear, continuous map

$$T : (F, \sigma(F, E)) \to (l^1, \sigma(l^1, m_0)),$$

TM is relatively compact in $(l^1, \|\cdot\|_1)$. The Dierolf topology is the polar topology of uniform convergence on the elements of $\mathcal{M}$ ([Die],[Sw4]).

Kalton's version of the Orlicz–Pettis Theorem will be considered later.

It should be noted that the result in Theorem 2.9 cannot be improved to subseries convergence in the strong topology.

Example 2.10. The series $\sum_j e^j$ is subseries convergent in the weak topology

$$\sigma(l^\infty, l^1)$$

but is not subseries convergent in the strong topology $\beta(l^\infty, l^1) = \|\cdot\|_\infty$.

The version of Theorem 2.9 for normed spaces is: if a series in a normed space is weak subseries convergent, then the series is subseries convergent in the norm topology. We show how Pettis employed this result in his treatment of the Pettis integral. Let X be a Banach space, Σ a σ algebra of subsets of a set S and λ a σ-finite positive measure on Σ. A function $f : S \to X$ is *scalarly measurable* if $x' \circ f = x'f$ is measurable for every $x' \in X'$ and is scalarly integrable if $x'f$ is λ integrable for every $x' \in X'$. If f is *scalarly integrable*, for every $A \in \Sigma$ there exists $x_A \in X''$ such that

$$x'(x_A) = \int_A x'f d\lambda;$$

x_A is the Dunford integral of f over A and is denoted by $\int_A f d\lambda$. If $\int_A f d\lambda \in X$ for every A, f is Pettis integrable (see Chapter 1 for details). Pettis used the Orlicz–Pettis Theorem to show that the Pettis integral is countably additive. Indeed, if $\{A_k\}$ is a pairwise disjoint sequence from Σ, then for every $x' \in X'$, we have

$$(x')\left(\sum_{k=1}^{\infty} \int_{A_k} f d\lambda\right) = \sum_{k=1}^{\infty} \int_{A_k} x'f d\lambda = \int_{\cup_{k=1}^{\infty} A_k} x'f d\lambda = (x')\left(\int_{\cup_{k=1}^{\infty} A_k} f d\lambda\right)$$

by the countable additivity of the Lebesgue integral. Thus, the series

$$\sum_k \int_{A_k} f d\lambda$$

is weak subseries convergent and by the Orlicz–Pettis Theorem is norm subseries convergent. Hence,

Theorem 2.11. (Pettis) *The indefinite Pettis integral $\int f d\lambda$ is norm countably additive.*

For more information on the Dunford and Pettis integrals see [DU], [BS]. It is interesting to note that a Dunford integrable function is Pettis integrable iff the indefinite integral $\int f d\lambda$ is countably additive ([DU]).

The Orlicz–Pettis Theorem can also be used to establish the countable additivity of the integral of a scalar valued function with respect to a vector valued measure which was developed in Chapter 1. Let X be a Banach space and $m : \Sigma \to X$ be countably additive. A Σ measurable function $f : S \to \mathbb{R}$ is integrable if for every $A \in \Sigma$ there exists $x_A \in X$ such that $\int_A f dx'm = x'(x_A)$. We write $x_A = \int_A f dm$. If $\{A_k\}$ is a pairwise disjoint sequence from Σ, then for every $x' \in X'$,

$$x'\left(\int_{\cup_{k=1}^\infty A_k} f dm\right) = \int_{\cup_{k=1}^\infty A_k} f dx'm = \sum_{k=1}^\infty \int_{A_k} f dx'm = \sum_{k=1}^\infty x'\left(\int_{A_k} f dm\right).$$

Thus,

$$\sum_{k=1}^\infty \int_{A_k} f dm = \int_{\cup_{k=1}^\infty A_k} f dm$$

with respect to $\sigma(X, X')$ and the indefinite integral $\int f dm$ is weakly countably additive and is norm countably additive by the Orlicz–Pettis Theorem.

Theorem 2.12. *The indefinite integral $\int f dm$ is norm countably additive.*

A norm countably additive set function defined on a σ algebra has bounded range (Appendix C, Corollary 2.46) so the indefinite integral is bounded; this fact was used in Theorem 1.30 to define the norm on the spaces of integrable functions.

We next consider a result of Stiles for spaces with a Schauder basis. Stiles' result seems to be the first version of the Orlicz–Pettis Theorem for non-locally convex spaces. A *Schauder basis* for a TVS E is a sequence $\{b_j\}$ from E such that every $x \in E$ has an unique expansion

$$x = \sum_{j=1}^\infty t_j b_j;$$

the functionals $f_j(x) = t_j$ are called the *coordinate functionals* with respect to the basis $\{b_j\}$. When E is a complete metric linear space the coordinate functionals are continuous but not in general ([Sw2],[Wi2]).

Theorem 2.13. (Stiles) *Let E be a topological vector space with a Schauder basis $\{b_k\}$ and coordinate functionals $\{f_k\}$. For each k let $P_k : E \to E$ be the projection*

$$P_k x = \sum_{j=1}^{k} \langle f_j, x \rangle \, b_j.$$

If $\sum x_j$ is subseries convergent with respect to $\sigma(E, \{f_j\})$, then the series is subseries convergent in the original topology of E.

Proof. To see this, set $F = \{P_j : j \in \mathbb{N}\}$ and consider the abstract triple

$$(E, F : E)$$

under the map $(x, P_j) \to P_j x$. Then $\sum x_j$ is $w(E, F)$ subseries convergent and F is sequentially conditionally $w(F, E)$ compact since $P_k x \to x$ so the series $\sum_{j=1}^{\infty} P_k x_j$ is subseries convergent uniformly for $k \in \mathbb{N}$ by Theorem 2.2. To establish the result, let U be a closed neighborhood of 0 in E. Set

$$s^n = \sum_{j=1}^{n} x^j \text{ and } s = \sum_{j=1}^{\infty} x^j,$$

where s is the $\sigma(E, \{f_j\})$ sum of the series. Since

$$\lim_n P_k s^n = P_k s$$

uniformly for $k \in \mathbb{N}$, there exists N such that

$$P_k s^n - P_k s \in U$$

for $n \geq N, k \in \mathbb{N}$. Fixing n and letting $k \to \infty$ gives $s^n - s \in U$ for $n \geq N$. Since the same argument can be applied to any subsequence, the result follows. $\square$

Stiles established this result for metrizable, complete spaces ([Sti]; see also [Bs],[Sw4]); the metrizable and completeness assumptions were later removed ([Sw5]10.4.1).

The result in Theorem 2.13 can be generalized somewhat. Assume that E is a topological vector space and there exist a sequence of linear operators $\{P_k\}$ such that for each $x \in E$ we have

$$x = \sum_{k=1}^{\infty} P_k x$$

with convergence in E. When the $\{P_k\}$ are continuous, then $\{P_k\}$ is called a *Schauder decomposition* ([LT]). Then the proof in Theorem 2.13 shows that if a series is subseries convergent in $w(E, \{P_k\})$, then the series is subseries convergent in the topology of E. Applications of this generalization to sequence spaces are given in Chapter 9 of [Sw4].

We next consider a result of Tweddle. Whereas Dierolf has shown that there is a strongest polar topology with the same subseries convergent series as the weak topology, Tweddle has shown that there is a strongest locally convex topology which has the same subseries convergent series as the weak topology

Remark 2.14. (Tweddle) Let E, F be a pair of vector spaces in duality. Let $\mathcal{E}$ be the family of all $\sigma(E, F)$ subseries convergent series in E and let $E^{\#}$ be all linear functionals x' on E such that

$$\sum_{j=1}^{\infty} \langle x', x_j \rangle = \left\langle x', \sum_{j=1}^{\infty} x_j \right\rangle$$

for all $\{x_j\} \in \mathcal{E}$, where $\sum_{j=1}^{\infty} x_j$ is the $\sigma(E, F)$ sum of the series so $F \subset E^{\#}$. Then $E, E^{\#}$ form a dual pair and each $\{x_j\} \in \mathcal{E}$ is $\sigma(E, E^{\#})$ subseries convergent. It follows from Theorem 2.9 that every $\{x_j\} \in \mathcal{E}$ is subseries convergent in the Mackey topology $\tau(E, E^{\#})$ of uniform convergence on convex $\sigma(E, E^{\#})$ compact subsets of $E^{\#}$. This is the Tweddle topology of E,

$$t(E, F) = \tau(E, E^{\#})$$

and Tweddle has shown that this is the strongest locally convex topology on E which has the same $\sigma(E, F)$ subseries convergent series ([Tw]). To see this, suppose ν is a locally convex topology on E which has the same subseries convergent series as $\sigma(E, F)$. Let $H' = (E, \nu)'$. Then for $\sum_j x_j \in \mathcal{E}$ and $x' \in H'$, we have

$$\sum_{j=1}^{\infty} x'(x_j) = x' \left(\sum_{j=1}^{\infty} x_j \right)$$

so $x' \in E^{\#}$ and $H' \subset E^{\#}$. Therefore, $\tau(E, H')$ is weaker than $t(E, F) = \tau(E, E^{\#})$. But, $\nu \subset \tau(E, H')$ so ν is weaker than $t(E, F)$.

The topology of Tweddle can also be extended to our abstract setting. Let $(E, F : G)$ be an abstract triple and let $\mathcal{E}$ be the family of all $w(E, F)$ subseries convergent series. Let $E^{\#}$ be all functions $f : E \to G$ such that

$$f \left(\sum_{j=1}^{\infty} x_j \right) = \sum_{j=1}^{\infty} f(x_j)$$

for every $\{x_j\} \in \mathcal{E}$. Then

$$(E, E^\# : G)$$

form an abstract triple under the map $(x, f) \to f(x)$ and each $\{x_j\} \in \mathcal{E}$ is $w(E, E^\#)$ subseries convergent. If G is metrizable, it follows from Theorem 2.3 that each $\{x_j\} \in \mathcal{E}$ is subseries convergent in the topology of uniform convergence on $w(E^\#, E)$ compact subsets of $E^\#$.

We indicate an applications of Theorem 2.6 to a result of Kalton ([Ka2]).

Theorem 2.15. (Kalton) *Let E, F be a dual pair of vector spaces and τ a polar topology on E from this duality which is separable. Then any series in E which is $\sigma(E, F)$ subseries convergent is τ subseries convergent.*

Proof. If τ is the polar topology of uniform convergence on the family $\mathcal{A}$ of $\sigma(F, E)$ bounded subsets of F, then every subset A of $\mathcal{A}$ is sequentially conditionally compact at each $x \in E$ so by Theorem 2.6 any series in E which is $\sigma(E, F)$ subseries convergent is τ subseries convergent. □

Partial Sums

Example 2.10 shows that a series which is weakly subseries convergent may fail to be convergent in the strong topology. However, the partial sums of a weakly subseries convergent series may be strongly bounded.

We consider compactness and boundedness for the partial sums of a subseries convergent series. Let $\sum x_j$ be a $w(E, F)$ subseries convergent series. The partial sums of the series is defined to be

$$\mathcal{S} = \left\{ \sum_{j \in \sigma} x_j : \sigma \subset \mathbb{N} \right\}.$$

We first consider compactness for $\mathcal{S}$. Recall that a subseries convergent series is also unordered convergent in the sense that for any neighborhood of 0, U, there exists N such that $\sum_{j \in \sigma} x_j \in U$ whenever $\min \sigma \geq N$ (see [Rob1] or the conclusion of Theorem 2.2).

Lemma 2.16. *Let $\Omega = \{0, 1\}$ and define $\varphi : \Omega^\mathbb{N} \to E$ by*

$$\varphi(\{t_j\}) = \sum_{j \in \sigma} x_j,$$

where $\sigma = \{j : t_j = 1\}$. Then φ is continuous when $\Omega^\mathbb{N}$ has the product topology and E has $w(E, F)$.

Proof. Let $\{t^\delta\}$ be a net in $\Omega^{\mathbb{N}}$ which converges to $t = \{t_j\} \in \Omega^{\mathbb{N}}$. If $\{t_j^\delta\} = t^\delta$ and $t = \{t_j\}$, then $t_j^\delta \to t_j$ for every j so $t_j^\delta = t_j$ eventually. Let U be a neighborhood of 0 in G and pick a symmetric neighborhood of 0, V, such that $V + V \subset U$. Let $y \in F$ and set $\sigma = \{j : t_j = 1\}, \sigma^\delta = \{j : t_j^\delta = 1\}$ and $\sigma(n) = \{j \in \sigma : j \geq n\}$. By the unordered convergence of $\sum x_j$ there exists n such that

$$\sum_{j \in \sigma(n)} x_j \cdot y \in V, \quad \sum_{j \in \sigma^\delta(n)} x_j \cdot y \in V$$

for all δ. There exists δ_0 such that $\delta \geq \delta_0$ implies $t_j^\delta = t_j$ for $1 \leq j \leq n$. Hence, if $\delta \geq \delta_0$, then

$$\varphi(\{t_j^\delta\}) \cdot y - \varphi(\{t_j\}) \cdot y = \sum_{j \in \sigma^\delta(n)} x_j \cdot y - \sum_{j \in \sigma(n)} x_j \cdot y \in V + V \subset U.$$

so $\varphi(t^\delta) \to \varphi(t)$ in $w(E, F)$. $\qquad\square$

Since $\Omega^{\mathbb{N}}$ is compact, sequentially compact and countably compact with respect to the product topology, Lemma 2.16 gives

Theorem 2.17. *S is compact, sequentially compact and countably compact with respect to $w(E, F)$.*

In particular, the set of partial sums of a subseries convergent series in an Abelian topological group is compact, sequentially compact and countably compact ([Rob1],[Rob2]).

There is an interesting "partial" converse to this theorem for TVS.

Theorem 2.18. *Let E be a TVS and $\{x_j\} \subset E$. If*

$$F = \left\{ \sum_{j \in \sigma} x_j : \sigma \text{ finite} \right\}$$

is relatively compact in E, then $\sum_j x_j$ is subseries convergent.

Proof. Since the closure of F is complete, it suffices to show that the partial sums of the series are Cauchy. If this fails to hold, there exist a closed neighborhood of 0, U, and an increasing sequence of intervals $\{I_k\}$ with

$$z_k = \sum_{j \in I_k} x_j \notin U.$$

Pick a symmetric neighborhood of 0, V, with $V + V \subset U$. Since F is bounded, there exists k such that $F \subset kV$. Pick a symmetric neighborhood of 0, W, such that

$$\underbrace{W + \ldots + W}_{k \text{ terms}} \subset V.$$

F is relatively compact so there exist $z_1 + W, ..., z_n + W$ covering

$$Z = \{z_k : k \in \mathbb{N}\}.$$

At least one set $z_1 + W, ..., z_n + W$ contains infinitely many elements of Z, say, $\{z_{n_j}\} \subset z_1 + W$. Then

$$\sum_{j=1}^{k} z_{n_j} \in \underbrace{(z_1 + W) + ... + (z_1 + W)}_{k \text{ terms}} \subset k z_1 + V.$$

Hence,

$$k z_1 \in \sum_{j=1}^{k} z_{n_j} + V \subset F + V \subset kV + V \subset kU$$

which implies $z_1 \in U$. This is a contradiction. The same argument can be applied to any subseries so the result follows. $\square$

- As an aside, we observe that the result above can be used to give an operator theory characterization of subseries convergence in normed spaces. Let X be a normed space and $\{x_j\} \subset X$. Define a summing operator $S : c_{00} \to X$ by

$$St = S(\{t_j\}) = \sum_{j=1}^{\infty} t_j x_j$$

(finite sum). Equip c_{00} with the sup-norm. Note that if S is continuous, then the partial sums $\{\sum_{j \in \sigma} x_j : \sigma \text{ finite}\}$ are bounded since

$$\left\| \sum_{j \in \sigma} x_j \right\| = \left\| S \left(\sum_{j \in \sigma} e^j \right) \right\| \leq \|S\|$$

for all finite σ. Also, it follows from the result above that if S is compact, then $\sum_j x_j$ is subseries convergent since

$$S \left(\left\{ \sum_{j \in \sigma} e^j : \sigma \text{ finite} \right\} \right) = \left\{ \sum_{j \in \sigma} x_j : \sigma \text{ finite} \right\}.$$

We consider the converse of these two statements. For this note that if $t = \sum_{j=0}^{n} t_j e^j \in c_{00}$ is non-negative with

$$0 \leq t_0 \leq t_1 \leq ... \leq t_n \leq 1,$$

then by Abel partial summation we can write

$$t = t_0 \sum_{k=0}^{n} e^k + \sum_{j=0}^{n-1} (t_{j+1} - t_j) \sum_{k=j+1}^{n} e^k$$

so

$$St = t_0 \sum_{k=0}^{n} x_k + \sum_{j=0}^{n-1}(t_{j+1} - t_j) \sum_{k=j+1}^{n} x_k \in co\left(\left\{\sum_{j\in\sigma} x_j : \sigma \text{ finite}\right\}\right)$$

since

$$t_0 + \sum_{j=0}^{n-1}(t_{j+1} - t_j) = t_n \leq 1.$$

Thus, if $\{\sum_{j\in\sigma} x_j : \sigma \text{ finite}\}$ is bounded (relatively compact), then S is bounded (compact).

We next consider the boundedness of the partial sums in a semi-convex topological vector space. A subset U of a topological vector space is *semi-convex* if there exists $a > 0$ such that $U + U \subset aU$; for example, if U is convex we may take $a = 2$. A topological vector space G is *semi-convex* if it has a neighborhood base of semi-convex subsets ([Rob1],[Rob2]). The spaces l^p $(0 < p < 1)$ are semi-convex but not locally convex.

Theorem 2.19. *Let G be a semi-convex space and let $B \subset F$ be pointwise bounded on E, i.e., $\{x \cdot y : y \in B\}$ is bounded in G for every $x \in E$. Then*

$$\{x \cdot y : x \in \mathcal{S}, \; y{\in}B\}$$

is bounded, i.e., B is uniformly bounded on $\mathcal{S}$.

Proof. First, note that if $\sigma \subset \mathbb{N}$ satisfies the condition that the set

$$E_\sigma = \left\{\sum_{j\in\tau} x_j : \tau \subset \sigma\right\}$$

is not absorbed by the semi-convex neighborhood U of G and if V is a symmetric neighborhood such that $V + V \subset U$, then for every $k \in \mathbb{N}$ there exists a partition (α^k, β^k) of σ, $n_k > k$ and $y_k \in B$ such that

$$\sum_{j\in\alpha^k} x_j \cdot y_k \notin n_k V, \quad \sum_{j\in\beta^k} x_j \cdot y_k \notin n_k V.$$

By the bounded hypothesis for each $x = \sum_{j\in\sigma} x_j \in \mathcal{S}$, there is an $n_k \geq k$ such that

$$\left\{\sum_{j\in\sigma} x_j \cdot y : y \in B\right\} \subset n_k V.$$

But, $E_\sigma \not\subset n_k(V + V)$ since $V + V \subset U$. So there exist $\alpha^k \subset \sigma, y_k \in B$ such that

$$\sum_{j\in\alpha^k} x_j \cdot y_k \notin n_k(V + V)$$

and, hence,

$$\sum_{j \in \alpha^k} x_j \cdot y_k \notin n_k V.$$

If $\beta^k = \sigma \setminus \alpha^k$, then $\sum_{j \in \beta^k} x_j \cdot y_k \notin n_k V$ because otherwise

$$\sum_{j \in \alpha^k} x_j \cdot y_k = \sum_{j \in \sigma} x_j \cdot y_k - \sum_{j \in \beta^k} x_j \cdot y_k \in n_k V + n_k V \subset n_k (V + V).$$

If the conclusion fails, there exists a semi-convex neighborhood, U, of 0 which does not absorb $E_{\mathbb{N}}$. Let V be a closed, symmetric neighborhood such that $V + V \subset U$. By the observation above there exists a partition (α^1, β^1) of $\sigma^1 = \mathbb{N}$, n_1 and $y_1 \in B$ such that

$$\sum_{j \in \alpha^1} x_j \cdot y_1 \notin n_1 V, \quad \sum_{j \in \beta^1} x_j \cdot y_1 \notin n_1 V.$$

Either E_{α^1} or E_{β^1} is not absorbed by U [if both are absorbed by U, there is m such that $E_{\alpha^1} + E_{\beta^1} = E_{\sigma^1} \subset m(U + U) \subset m(aU)$; this is where semi-convexity is used]; pick whichever of α^1 or β^1 satisfies this condition and label it A^1 and set $B^1 = \sigma^1 \setminus A^1$. Now treat A^1 as above and obtain a partition (A^2, B^2) of $A^1, n_2 > n_1, y_2 \in B$ such that

$$\sum_{j \in A^2} x_j \cdot y_2 \notin n_2 V, \quad \sum_{j \in B^2} x_j \cdot y_2 \notin n_2 V$$

and E_{A^2} is not absorbed by U. Continuing this construction produces a pairwise disjoint sequence $\{B^k\}$ of subsets of $\mathbb{N}$, increasing $\{n_k\}$ and $y_k \in B$ such that

$$(\#) \quad \sum_{j \in B^k} x_j \cdot y_k \notin n_k V,$$

and since V is closed we may assume that each B^k is finite.

Now consider the matrix

$$M = [m_{ij}] = \left[\frac{1}{n_i} \sum_{l \in B^j} x_l \cdot y_i \right].$$

We claim that M is a $\mathcal{K}$ matrix. First, the columns of M converge to 0 since B is pointwise bounded on E. Suppose $\{k_j\}$ is an increasing sequence of integers and set $\tau = \cup_{j=1}^{\infty} B^{k_j}$. Since the $\{B^j\}$ are pairwise disjoint and finite and $\sum x_j$ is $w(E, F)$ subseries convergent, we have

$$\sum_{j=1}^{\infty} \frac{1}{n_i} \sum_{l \in B^{k_j}} x_l \cdot y_i = \frac{1}{n_i} \sum_{l \in \tau} x_l \cdot y_i = \frac{1}{n_i} x_\tau \cdot y_i \to 0,$$

where x_τ is the $w(E, F)$ sum of the series $\sum_{j \in \tau} x_j$. Hence, M is a $\mathcal{K}$ matrix and by the Antosik–Mikusinski Matrix Theorem the diagonal of M converges to 0. But this contradicts $(\#)$. $\square$

We will make a remark on the semi-convexity assumption later.

Corollary 2.20. *Let E be a semi-convex space with a nontrivial dual E'. If $\sum x_j$ is $\sigma(E, E')$ subseries convergent, then the set of partial sums, S, of $\sum x_j$ is $\beta(E, E')$ bounded. In particular, this holds for locally convex spaces.*

Example 2.10 shows that a series may be weak subseries convergent but fail to be subseries convergent in the strong topology while Corollary 2.20 shows that the partial sums are always bounded in the strong topology.

Continuous Function Spaces

We consider subseries convergence with respect to pointwise convergence in spaces of continuous functions.

Let G be an Abelian topological group and Ω be a sequentially compact topological space and let $SC(\Omega, G)$ be the space of all sequentially continuous functions from Ω into G.

Theorem 2.21. *Suppose $\sum f_j$ is a series in $SC(\Omega, G)$ which is subseries convergent in the topology of pointwise convergence on G. Then the series $\sum f_j(t)$ is subseries convergent uniformly for $t \in \Omega$.*

Proof. To see this consider the abstract triple

$$(SC(\Omega, G), \Omega : G)$$

under the map $(f, t) \to f(t)$. The series $\sum f_j$ is subseries convergent with respect to $w(SC(\Omega, G), \Omega)$ and the set Ω is $w(\Omega, SC(\Omega, G))$ sequentially compact since Ω is sequentially compact so the claim follows from Theorem 2.2. $\square$

If G is either the scalar field or a normed space, $(G, \|\cdot\|)$, the conclusion of the theorem is that if a series is subseries convergent in the topology of pointwise convergence, then the series is subseries convergent in the sup-norm,

$$\|f\|_\infty = \sup\{\|f(t)\| : t \in \Omega\}.$$

We give an improvement of the theorem.

Let D be a dense subset of Ω.

Theorem 2.22. *Suppose $\sum f_j$ is a series in $SC(\Omega, G)$ which is subseries convergent in the topology of pointwise convergence on D. Then the series $\sum f_j(t)$ is subseries convergent uniformly for $t \in \Omega$.*

Proof. Consider the triple

$$(SC(\Omega, G), D : G)$$

under the map $(f, t) \to f(t)$. The series $\sum_j f_j$ is $w(SC(\Omega, G), D)$ subseries convergent. If $\{t_j\} \subset D$, there is a subsequence $\{t_{n_j}\}$ and $t \in \Omega$ such that $t_{n_j} \to t$ so $f(t_{n_j}) \to f(t)$ for $f \in SC(\Omega, G)$. Then D is sequentially conditionally $w(D, SC(\Omega, G))$ compact. By Theorem 2.2 the series $\sum_j f_j(t)$ converge uniformly for $t \in D$. Since D is dense in Ω, the result follows. $\square$

Let Ω be a topological space, G be metrizable and $C(\Omega, G)$ the space of continuous functions from Ω to G.

Theorem 2.23. *Suppose $\sum f_j$ is a series in $C(\Omega, G)$ which is subseries convergent in the topology of pointwise convergence on Ω. Then the series is subseries convergent in the topology of uniform convergence on compact subsets of Ω.*

Proof. To see this consider the abstract triple

$$(C(\Omega, G), \Omega : G)$$

under the map $(f, t) \to f(t)$. The series $\sum f_j$ is $w(C(\Omega, G), \Omega)$ subseries convergent and any compact subset of Ω is $w(\Omega, C(\Omega, G))$ compact so the claim follows from Theorem 2.3. $\square$

Theorems of this type relative to pointwise convergent series in spaces of continuous functions were established in [Th] and [Sw1].

Thomas has observed that the theorem above implies the Orlicz–Pettis Theorem for normed spaces. Let X be a normed space and let B be the unit ball in X' with the weak star topology so B is a compact space. Let $\sum_j x_j$ be series in X which is weak subseries convergent. Then each x_j is a continuous function on B and the series $\sum_j x_j$ is subseries convergent in the topology of pointwise convergence on B. By the result above the series is subseries convergent in the topology of uniform convergence on B, i.e., the series is subseries convergent in norm.

Linear Operators

We now consider subseries convergence in the space of continuous linear operators. Let E, G be topological vector spaces and $L(E, G)$ the space of continuous linear operators from E into G. Let $L_s(E, G)$ be $L(E, G)$ with the topology of pointwise convergence on E, i.e., the strong operator topology. Let $L_c(E,G)$ $(L_b(E,G))$ be $L(E,G)$ with the topology of uniform convergence on compact (bounded) subsets of E.

Let E, G be topological vector spaces and consider the abstract triple

$$(L(E,G), E : G)$$

under the map $(T,x) \to Tx$.

Theorem 2.24. *Let G be metrizable (LCTVS). Suppose that $\sum T_j$ is a series in $L(E,G)$ which is subseries convergent in the strong operator topology. Then the series is subseries convergent in $L_c(E,G)$.*

Proof. Consider the abstract triple

$$(L(E,G), E : G)$$

under the map $(T,x) \to Tx$. Note that any subset $B \subset E$ which is compact in E is $w(E, L(E,G))$ compact so the series $\sum T_j$ which is subseries convergent in $w(L(E,G), E)$ is subseries convergent in $L_c(E,G)$, the topology of uniform convergence on compact subsets of E, by Theorem 2.3 (Theorem 2.4). □

Remark 2.25. A similar result holds for the topology of uniform convergence on sequentially compact subsets of E denoted by $L_{sc}(E,F)$.

It should be noted that even in the case of locally convex spaces the result above cannot be improved to subseries convergence in the topology of uniform convergence on bounded subsets of E, $L_b(E,G)$ (Example 2.10). In fact to obtain subseries convergence in $L_b(E,G)$ it is necessary to consider the space of compact operators as the following example shows.

Example 2.26. Let X be a Banach space with an unconditional Schauder basis $\{b_i\}$, i.e., every $x \in X$ has a unique series representation

$$x = \sum_{i=1}^{\infty} t_i b_i,$$

where the series is subseries convergent. Let $\{f_i\}$ be the coordinate functionals $f_i(x) = t_i$. Each f_i is continuous ([Sw2]10.1.13). Set $P_i x = f_i(x)b_i$. Let Y be a Banach space. If $T \in L(X,Y)$, then

$$Tx = \sum_{i=1}^{\infty} f_i(x)Tb_i = \sum_{i=1}^{\infty} TP_i x,$$

where the series is subseries convergent. Thus, the series

$$\sum_{i=1}^{\infty} TP_i$$

is subseries convergent in $L_s(X, Y)$ to the operator T. If $L(X, Y)$ has the property that any series which is subseries convergent in the strong operator topology is subseries convergent in the norm topology of $L(X, Y)$, it follows that every $T \in L(X, Y)$ is compact being the norm limit of the compact operators $\{\sum_{i=1}^{n} TP_i\}$. That is, if $L(X, Y)$ has this property, then

$$L(X, Y) = K(X, Y),$$

the space of compact operators from X into Y. In particular, if $X = Y$, then the identity operator I on X is compact and X must be finite dimensional.

If the spaces E, G are locally convex spaces, we have the following result concerning the weak and strong operator topologies. The *weak operator topology* on $L(E, G)$ is the topology of pointwise convergence on E when F has the weak topology.

Theorem 2.27. *Let E, G be locally convex spaces. If the series $\sum_j T_j$ is subseries convergent in the weak operator topology, the the series is subseries convergent in the strong operator topology.*

Proof. For each $x \in E$ the series $\sum_j T_j x$ is subseries convergent in the weak topology $\sigma(G, G')$ and is, therefore, subseries convergent in the original topology of G by the Orlicz–Pettis Theorem for locally convex spaces, Theorem 2.9. That is, the series is subseries convergent in the strong operator topology. □

Thus, if E, G are locally convex spaces, the result in Theorem 2.24 can be improved to read that if the series is subseries convergent in the weak operator topology, then the series is subseries convergent in $L_c(E, G)$.

Next, we consider the space of compact operators.

Let E, G be Hausdorff topological vector spaces and $K(E, G)$ the space of all continuous linear operators which carry bounded subsets of E into sequentially conditionally compact subsets of G. If E, G are Banach spaces $K(E, G)$ is the space of compact operators. Then

$$(K(E, G), E : G)$$

is an abstract triple under the map $(T, x) \to Tx$ and if $\mathcal{B}$ is the family of bounded subsets of E, each $B \in \mathcal{B}$ is sequentially conditionally compact at each $T \in K(E, G)$ (Definition 2.5). The topology $w(K(E, G), E)$ is just the strong operator topology. Let $K_b(E, G)$ be the topology of uniform convergence on the members of $\mathcal{B}$; if E, G are normed spaces this is just the uniform operator topology of $K(E, G)$. From Theorem 2.6 we have

Theorem 2.28. *If $\mathcal{F}(E, G)$ is any separable subspace of $K_b(E, G)$, then any series in $\mathcal{F}(E, G)$ which is subseries convergent in the strong operator topology is subseries convergent in $K_b(E, G)$.*

We indicate situations where Theorem 2.28 is applicable. Let E, G be Banach spaces and let $\mathcal{F}(E, G)$ be the space of operators with finite dimensional range. Thus, every operator $T \in \mathcal{F}(E, G)$ has a representation

$$Tx = \sum_{j=1}^{n} \langle x_j', x \rangle y_j$$

with $x_j' \in E', y_j \in G$. If E' and G are separable, then $\mathcal{F}(E, G)$ is a separable subspace of $K_b(E, G)$ so Theorem 2.28 applies. If, in addition, either E' or G has the approximation property, then $\mathcal{F}(E, G)$ is dense in $K_b(E, G)$ ([LT] 1.e.4 or 1.e.5) so Theorem 2.28 also applies in this case to $K_b(E, G)$.

Another situation where Theorem 2.28 applies is given as follows. Let E be a metrizable nuclear space and G be separable. Then any continuous linear operator from E into G is in $K_b(E, G)$ since bounded subsets of E are relatively compact. Now E is separable ([GDS] II.VI.5) and E_b' is separable ([GDS] II.VI.12) so $L_b(E, G)$ is separable ([GDS] III.II.11.b and 13.c) and Theorem 2.28 applies.

Also, if E is dual nuclear (i.e., the strong dual of E is nuclear) and G is nuclear, then $L_b(E, G)$ is nuclear ([Pi] 5.5.1) and, therefore, separable so Theorem 2.28 applies.

Finally, if E, G are normed spaces and E', G are separable, then $\mathcal{F}(E, G)$ is separable under the nuclear norm ν on $\mathcal{F}(E, G)$ (see [Pi] 3.1) so the space of nuclear operators $\mathcal{N}(E, G)$ is separable under the nuclear norm ν([Pi] 3.1.4) and Theorem 2.28 applies. Similar remarks apply to the space of Hilbert–Schmidt (absolutely summing) operators on Hilbert spaces ([Pi] 2.5).

Some of the results of this case were announced without proofs in [LC].

We consider another result related to compact operators and a family of operators introduced by A. Mohsen. Let X, Y, Z be normed spaces and let

$$W^*(Y', Z)$$

be the space of all sequentially weak*-$\|\cdot\|$ continuous linear operators from Y' into Z (these operators were introduced and studied by Mohsen ([Mo]) and were shown to be bounded).

Let $B(Z)$ denote the unit ball of any normed space Z. Assume $B(Y')$ is weak* sequentially compact and that the series $\sum U_j$ is subseries convergent in the strong operator topology of $W^*(Y', Z)$. Consider the abstract triple

$$(W^*(Y', Z), B(Y') : Z)$$

under the map $(U, y') \to Uy'$. Then the series $\sum U_j$ is $w(W^*(Y', Z), B(Y'))$ subseries convergent and $B(Y')$ is $w(B(Y'), W^*(Y', Z))$ sequentially compact since if $\{y'_j\} \subset B(Y')$, then there is a subsequence $\{y'_{n_j}\}$ which is weak* convergent to some $y' \in B(Y')$ and

$$\left\| Uy'_{n_j} - Uy' \right\| \to 0$$

for any $U \in W^*(Y', Z)$ by the definition of $W^*(Y', Z)$. Hence, by Theorem 2.2 the series $\sum_{j=1}^{\infty} U_j y'$ converge uniformly for $y' \in B(Y')$ and similarly for any subseries. That is, the series $\sum U_j$ is subseries convergent in norm so we have

Theorem 2.29. *If the series $\sum U_j$ is subseries convergent in the strong operator topology of $W^*(Y', Z)$, then the series is subseries convergent in norm.*

As a special case of Theorem 2.29, we can obtain a result of Kalton ([Ka3]). We say a normed space Z has the *Diestel–Faires property (DF property)* if any weak* subseries convergent series $\sum z_j$ in Z' is norm subseries convergent ([DF]; Diestel and Faires have characterized the Banach spaces with DF as the spaces Z whose dual does not contain a copy of l^∞).

Theorem 2.30. (Kalton) *Let X, Y be normed spaces. Let $\sum T_j$ be subseries convergent in the weak operator topology of $K(X, Y)$ and assume X has the DF property. Then the series is norm subseries convergent.*

Proof. Since each T_j has separable range we may assume that Y is separable by replacing Y with the union of the ranges of the T_j, if necessary. For each $z' \in Y'$ the series $\sum T'_j z'$ is weak* subseries convergent in X' and by the DF property is norm convergent. Let $K'(X, Y)$ be

$$\{T' \in K(Y', X') : T \in K(X, Y)\}$$

and consider the abstract triple

$$(K'(X,Y), B(Y') : X')$$

under the map $(T', y') \to T'y'$. The series $\sum T'_j$ is $w(K'(X,Y), B(Y'))$ subseries convergent by the observation above and the ball of Y' is weak* sequentially compact since Y is separable. Also,

$$K'(X,Y) \subset W^*(Y', X')$$

([DS] VI.5.6) so Theorem 2.29 implies that the series $\sum T'_j$ is norm subseries convergent and, hence, the series $\sum T_j$ is norm subseries convergent. $\square$

An operator $T \in L(E, G)$ is *completely continuous* if T carries weakly convergent sequences into convergent sequences; denote all such operators by $CC(E, G)$. Note that if T is completely continuous, then T carries weak Cauchy sequences into Cauchy sequences. Now consider the abstract triple

$$(CC(E, G), E : G)$$

under the bilinear map $(CC(E, G), E) \to G$ defined by $(T, z) \to T \cdot z = Tz$. If a subset $K \subset E$ is sequentially conditionally weakly compact, then K is sequentially conditionally $w(CC(E, G), G)$ compact. If CW denotes the set of all sequentially conditionally weakly compact subsets of E and $CC_{CW}(E, G)$ is $CC(E, G)$ with the topology of uniform convergence on CW, then from Theorem 2.2 we have

Theorem 2.31. *If the series $\sum_j T_j$ is subseries convergent in $CC_s(E, G)$, then $\sum_j T_j$ is subseries convergent in $CC_{CW}(E, G)$.*

An operator $T \in L(E, G)$ is *weakly compact* if T carries bounded sets to relatively weakly compact sets; denote all such operators by $W(E, G)$. The space E has the Dunford–Pettis property if every weakly compact operator from E into any locally convex space G carries weak Cauchy sequences into convergent sequences. Consider the abstract triple

$$(W(E, G), E : G)$$

under the bilinear map $(W(E, G), E) \to G$ defined by $(T, z) \to T \cdot z = Tz$. If $K \subset E$ is sequentially conditionally weakly compact and E has the Dunford–Pettis property, then K is sequentially conditionally $w(W(E, G), G)$ compact. If CW denotes the set of all sequentially conditionally weakly compact subsets of E, then from Theorem 2.2 we have

Theorem 2.32. *Assume that E has the Dunford–Pettis property. If the series $\sum_j T_j$ is subseries convergent in $W_s(E, G)$, then $\sum_j T_j$ is subseries convergent in $W_{CW}(E, G)$.*

A space E is *almost reflexive* if every bounded sequence contains a weak Cauchy subsequence ([LaW]). For example, Banach spaces with separable duals, quasi-reflexive Banach spaces and $c_0(S)$ are almost reflexive ([LaW]). If E is almost reflexive and has the Dunford–Pettis property, then every bounded set is sequentially conditionally $w(W(E,G),G)$ compact so from Theorem 2.2, we have

Theorem 2.33. *Assume that E is almost reflexive with the Dunford–Pettis property. If the series $\sum_j T_j$ is subseries convergent in $W_s(E,G)$, then $\sum_j T_j$ is subseries convergent in $W_b(E,G)$.*

Sequence Spaces

We next consider vector valued sequence spaces and Orlicz–Pettis Theorems with respect to the topology of coordinate convergence.

Let X be a topological vector space and assume that E is a vector space of X valued sequences which contains the space $c_{00}(X)$ of all X valued sequences which are eventually 0. If $z \in X$ and $k \in \mathbb{N}$, let $e^k \otimes z$ be the sequence with z in the kth coordinate and 0 in the other coordinates. Assume that E has a locally convex vector topology under which the coordinate mappings

$$Q_k : \{x_k\} \to e^k \otimes x_k$$

from E into E are continuous (i.e., E is a K-space). The space E is an *AK-space* if for every $x = \{x_k\} \in E$ we have

$$x = \sum_{k=1}^{\infty} e^k \otimes x_k,$$

where the series converges in E. We say that a series $\sum_j x^j$ in E is *coordinatewise convergent* if the series

$$\sum_{k=1}^{\infty} e^k \otimes x_k^j$$

converges in E for every j.

Theorem 2.34. *Assume that E is an AK-space. Suppose that $\sum x^j$ is a series in E which is subseries coordinatewise convergent. Then the series $\sum x^j$ is subseries convergent in the original topology of E.*

Proof. To see this consider the following abstract triple. Define $P_k : E \to E$ by

$$P_k x = \sum_{i=1}^{k} e^i \otimes x_i$$

so $P_k x \to x$ in E for every $x \in E$ by the AK assumption. Set $F = \{P_k : k \in \mathbb{N}\}$ and note that

$$(E, F : E)$$

is an abstract triple under the map $(x, P_k) \to P_k x$ and that the series $\sum x^j$ is subseries convergent with respect to $w(E, F)$. The set F is sequentially conditionally $w(F, E)$ compact by the AK hypothesis so it follows from Theorem 2.2 that the series

$$\sum_{j=1}^{\infty} P_k x^j$$

is subseries convergent uniformly for $k \in \mathbb{N}$. To establish the result let U be a closed neighborhood of 0 in E. Set

$$s^n = \sum_{j=1}^{n} x^j \text{ and } s = \sum_{j=1}^{\infty} x^j,$$

where s is the coordinate sum of the series. Since

$$\lim_n P_k s^n = P_k s$$

uniformly for $k \in \mathbb{N}$, there exists N such that

$$P_k s^n - P_k s \in U$$

for $n \geq N, k \in \mathbb{N}$. Fixing n and letting $k \to \infty$ gives $s^n - s \in U$ for $n \geq N$. Since the same argument can be applied to any subsequence, the result follows. □

This result applies to such sequence spaces as $l^p(X)$, $1 < p < \infty$, and $c_0(X)$ (see Appendix B or Appendix C of [Sw4] for these spaces).

Applications

In this section we give a number of applications of the results above to various topics in measure theory and functional analysis.

Measure Theory

We begin by considering the Nikodym Convergence Theorem. In its original form the theorem asserts that the pointwise limit of a sequence of signed measures is countably additive and the countable additivity of the sequence is uniform. We consider a group valued version of the theorem.

Let G be an Abelian topological group and Σ a sigma algebra of subsets of a set S. A sequence of countably additive set functions $m_k : \Sigma \to G$ is *uniformly countably additive* if for every pairwise disjoint sequence $\{A_j\} \subset \Sigma$,

$$\lim_n \sum_{j=n}^{\infty} m_k(A_j) = 0$$

uniformly for $k \in \mathbb{N}$, i.e., the series $\sum_{j=1}^{\infty} m_k(A_j)$ converge uniformly for $k \in \mathbb{N}$. We have the following criteria for uniform countable additivity.

Lemma 2.35. *Let $\{m_k\}$ be a sequence of G valued, countably additive set functions defined on Σ. The following are equivalent:*

(i) *$\{m_k\}$ is uniformly countably additive,*

(ii) *for each decreasing sequence $\{A_j\}$ from Σ with $\cap_{j=1}^{\infty} A_j = \emptyset$, $\lim_j m_k(A_j) = 0$ uniformly for $k \in \mathbb{N}$,*

(iii) *if $\{A_j\}$ is pairwise disjoint, then $\lim_j m_k(A_j) = 0$ uniformly for $k \in \mathbb{N}$.*

Proof. (i) and (ii) are clearly equivalent for countably additive functions and (i) clearly implies (iii).

Suppose (ii) fails. Then we may assume (by passing to a subsequence if necessary) that there exist a decreasing sequence $\{A_j\}$ from Σ with $\cap_{j=1}^{\infty} A_j = \emptyset$ and a symmetric neighborhood, U, of 0 in G such that

$$m_k(A_k) \notin U$$

for every k. Pick a symmetric neighborhood V of 0 such that $V + V \subset U$. There exists k_1 such that $m_1(A_{k_1}) \in V$. There exists $k_2 > k_1$ such that $m_{k_1}(A_{k_2}) \in V$. Continuing this construction produces a subsequence $\{k_j\}$ such that

$$m_{k_j}(A_{k_{j+1}}) \in V.$$

Put $B_j = A_{k_j} \backslash A_{k_{j-1}}$ so $\{B_j\}$ is pairwise disjoint and

$$m_{k_j}(B_j) = m_{k_j}(A_{k_j}) - m_{k_j}(A_{k_{j-1}}) \notin V.$$

Hence, (iii) does not hold. $\qquad\square$

Theorem 2.36. (Nikodym) *Let* $\{m_k\}$ *be a sequence of* G *valued countably additive set functions defined on* Σ. *If*

$$\lim_k m_k(A) = m(A)$$

exists for every $A \in \Sigma$, *then*

(i) m *is countably additive and*
(ii) $\{m_k\}$ *is uniformly countably additive.*

Proof. Let $\{A_j\}$ be a disjoint sequence from Σ and set $\mathcal{M} = \{m_k\}$. Consider the triple

$$(\Sigma, \mathcal{M} : G)$$

under the map $(A, m_k) \to m_k(A)$. The series $\sum_{j=1}^{\infty} A_j$ is $w(\Sigma, \mathcal{M})$ subseries convergent and the sequence $\{m_k\}$ sequentially conditionally $w(\mathcal{M}, \Sigma)$ compact. By Theorem 2.2 the series $\sum_{j=1}^{\infty} m_k(A_j)$ converge uniformly for $k \in \mathbb{N}$. This establishes (ii).

(i) follows from (ii). □

For the case of the Nikodym Theorem for groups see [AS1].

The σ algebra assumption in the theorem is important.

Example 2.37. Let $S = [0,1)$ and $\mathcal{A}$ be the algebra generated by the intervals of the form $[a, b)$, $0 \le a \le b \le 1$. Define μ_n on $\mathcal{A}$ by

$$\mu_n(A) = n\lambda(A \cap [0, 1/n)),$$

where λ is Lebesgue measure. Then each μ_n is countably additive,

$$\lim_n \mu_n(A) = \mu(A)$$

exists for every $A \in \mathcal{A}$ but μ is not countably additive.

Despite this example there are algebras for which the conclusion of the Nikodym Convergence Theorem holds. See Schachermeyer ([Sch]).

Using a result of Drewnowski, the Nikodym Convergence Theorem can be extended to strongly additive set functions. A finitely additive set function $\mu : \Sigma \to G$ is *strongly additive* if $\mu(A_j) \to 0$ for every disjoint sequence $\{A_j\} \subset \Sigma$ and a family of finitely additive set functions, $\mathcal{M}$, is *uniformly strongly additive* if for every disjoint sequence $\{A_j\} \subset \Sigma$, $m(A_j) \to 0$ uniformly for $m \in \mathcal{M}$.

We have a characterization of strongly additive and uniformly strongly additive set functions.

Proposition 2.38. (i) *A finitely additive set function* $m : \Sigma \to G$ *is strongly additive* $\iff$ *the series* $\sum_{j=1}^{\infty} m(A_j)$ *is Cauchy for every pairwise disjoint sequence* $\{A_j\} \subset \Sigma$.

(ii) *The family* $\mathcal{M}$ *of finitely additive set functions from* Σ *into* G *is uniformly strongly additive* $\iff$ *the series* $\sum_{j=1}^{\infty} m(A_j)$, $m \in \mathcal{M}$, *are uniformly Cauchy for every pairwise disjoint sequence* $\{A_j\}$ *from* Σ.

Proof. (i): $\Longleftarrow$ is clear. $\Rightarrow$: If the condition fails to hold, there exist a neighborhood of 0, U, an increasing sequence of finite intervals $\{I_k\}$ such that

$$\sum_{j \in I_k} m(A_j) \notin U.$$

Put $B_k = \cup_{j \in I_k} A_j$. Then $\{B_k\}$ is pairwise disjoint but $m(B_k) \not\to 0$. Hence, m is not strongly additive.

(ii): $\Longleftarrow$ is clear. $\Rightarrow$: If the condition fails to hold, there exist a neighborhood of 0, U, an increasing sequence of finite intervals $\{I_k\}$, and a sequence $\{m_k\} \subset \mathcal{M}$ with

$$\sum_{j \in I_k} m_k(A_j) \notin U.$$

Put $B_k = \cup_{j \in I_k} A_j$. Then $m_k(B_k) \notin U$ so $\mathcal{M}$ is not uniformly strongly additive. $\qquad\square$

Before giving our extension of the Nikodym Convergence Theorem to strongly additive set functions, we pause to prove an important result of Pettis for countably additive set functions with values in a Banach space. First we establish a result which gives a characterization of uniformly strongly additive set functions with values in a Banach space and uses properties of the semi-variation developed in Chapter 1.

Let X be a Banach space. Recall $ba(\Sigma, X)$ is the space of all X valued, bounded, finitely additive set functions defined on Σ.

Proposition 2.39. *Let* $\mathcal{M}$ *be a subset of* $ba(\Sigma, X)$. *The following are equivalent.*

(1) $\mathcal{M}$ *is uniformly strongly additive,*

(2) $\{x'm : m \in \mathcal{M}, \|x'\| \leq 1\}$ *is uniformly strongly additive,*

(3) $\{A_k\} \subset \Sigma$ *pairwise disjoint implies* $\lim_k m(A_k) = 0$ *uniformly for* $m \in \mathcal{M}$,

(4) $\{A_k\} \subset \Sigma$ *pairwise disjoint implies* $\lim_k semi - var(m)(A_k) = 0$
uniformly for $m \in \mathcal{M}$,

(5) $\{|x'm| : m \in \mathcal{M}, \|x'\| \leq 1\}$ *is uniformly strongly additive.*

Proof. That (1) implies (2) implies (3) is clear.

Assume (3) holds but (4) is false. Then there exist pairwise disjoint $\{A_k\}$, $\delta > 0$, $\{m_k\} \subset \mathcal{M}$ such that

$$semi - var(m_k)(A_k) > 2\delta.$$

There exist $B_k \subset A_k$ such that $\|m_k(B_k)\| > \delta$ (see Chapter 1, Theorem 26 for a norm equivalent to the semi-variation norm). That $\{B_k\}$ is pairwise disjoint contradicts (3).

Assume (4) holds but (5) fails. Then there exist pairwise disjoint $\{A_k\} \subset \Sigma$, $\delta > 0$, such that

$$\sup \left\{ \sum_{j=k}^{\infty} |x'm| \, (A_j) : m \in \mathcal{M}, \|x'\| \leq 1 \right\} > \delta.$$

There exists a subsequence $\{p_k\}$ such that $\left\|x'_{p_k}\right\| \leq 1, m_{p_k} \in \mathcal{M}$ and

$$\sum_{j=p_k+1}^{p_{k+1}} \left|x'_{p_k} m_{p_k}\right| (A_j) > \delta.$$

Set $B_k = \cup_{j=p_k+1}^{p_{k+1}} A_j$ so $\{B_k\}$ is pairwise disjoint and $\left|x'_{p_k} m_{p_k}\right| (B_k) > \delta$. This contradicts (4) (see Chapter 1, Theorem 26 and the second formula for the semi-variation). $\qquad \square$

Now we establish Pettis' result about "absolutely continuous" measures which generalizes a well known result for scalar measures. A countably additive set function $m : \Sigma \to G$, an Abelian topological group, is λ *continuous*, where λ is a scalar measure, if

$$\lim_{\lambda(A) \to 0} m(A) = 0.$$

Theorem 2.40. (Pettis) *Let* $m \in ca(\Sigma, X)$ *and* $\lambda : \Sigma \to [0, \infty)$ *be a countably additive measure. Then* m *is* λ *continuous iff* $\lambda(A) = 0$ *implies* $m(A) = 0$.

Proof. Suppose m is not λ continuous. Then there exist $\delta > 0$, $\{A_k\} \subset \Sigma$ such that

$$\|m(A_k)\| > \delta \text{ and } \lambda(A_k) < 1/2^k.$$

Pick $x'_k \in X', \|x'_k\| \leq 1$, such that $|x'_k m(A_k)| > \delta$. Since m is countably additive, $\{x'_k m : k\}$ is uniformly strongly additive. By (5) above, $\{|x'_k m| : k\}$ is uniformly strongly additive. Set $A = \limsup A_k$ and $B_k = \cup_{j=k}^{\infty} A_j$. Then

$$\lambda(A) \leq \sum_{i=k}^{\infty} \lambda(A_i) \leq \sum_{i=k}^{\infty} 1/2^i = 1/2^{k-1}$$

so $\lambda(A) = 0$ and m is 0 on every subset of A. Set

$$C_1 = S \setminus B_1, \ C_{k+1} = B_k \setminus B_{k-1}.$$

Then $\{C_k\}$ is pairwise disjoint and $B_{k-1} \setminus A = \cup_{i=k}^{\infty} C_i$. Now $|x'_j m|(A) = 0$ implies

$$(*) \ \lim_k |x'_j m|(B_{k-1}) = \lim_k |x'_j m|(\cup_{i=k}^{\infty} C_i) = \lim_k \sum_{i=k}^{\infty} |x'_j m|(C_i) = 0$$

uniformly in j by the uniform strong additivity of $\{|x'_j m| : j\}$. But,

$$|x'_{k-1} m|(B_{k-1}) \geq |x'_{k-1} m|(A_{k-1}) \geq |x'_{k-1} m(A_{k-1})| > \delta$$

contradicting $(*)$. The other implication is obvious. $\qquad\square$

It follows from Pettis' Theorem that if $f : S \to X$ is Pettis integrable with respect to the measure λ, then the indefinite Pettis integral,

$$\int_{\cdot} f d\lambda,$$

is λ continuous.

We now give our generalization of the Nikodym Convergence Theorem to strongly additive set functions.

A *quasi-norm* on an Abelian topological group G is a function $|\cdot| : G \to [0, \infty)$ such that $|x| \geq 0$, $|x| = |-x|$, $|x + y| \leq |x| + |y|$ for all $x, y \in G$. Such a quasi-norm induces an invariant metric $d(x, y) = |x - y|$. An Abelian group G with a group topology generated by a quasi-norm is called a *quasi-norm group*. It is interesting that the topology of any Abelian topological group is always generated by a family of quasi-norms ([BM], see Appendix A).

Theorem 2.41. *Assume that G is a quasi-norm group. Let $\{m_k\}$ be a sequence of G valued strongly additive set functions defined on Σ. If*

$$\lim_k m_k(A) = m(A)$$

exists for every $A \in \Sigma$, then

(i) *m is strongly additive and*
(ii) *$\{m_k\}$ is uniformly strongly additive.*

Proof. Let $\{A_j\} \subset \Sigma$ be a pairwise disjoint sequence. To show that

$$\lim_j m_k(A_j) = 0$$

uniformly for $k \in \mathbb{N}$ it suffices to show, by passing to a subsequence if necessary, that $\lim_k m_k(A_k) = 0$. By Drewnowski's Lemma (Appendix D), there exists a subsequence $\{n_k\}$ such that each m_k is countably additive on the sigma algebra Σ_0 generated by $\{A_{n_k}\}$. Since $\lim_k m_k(A) = m(A)$ exists for every $A \in \Sigma_0$, Theorem 2.36 implies that the series $\sum_{j=1}^{\infty} m_k(A_{n_j})$ converge uniformly for $k \in \mathbb{N}$. In particular, $\lim_k m_{n_k}(A_{n_k}) = 0$. Since the same argument can be applied to any subsequence of $\{m_k(A_k)\}$, it follows that $\lim_k m_k(A_k) = 0$ as desired. This establishes (ii). (i) follows from (ii). $\qquad\square$

Using the result of Burzyk and Mikusinski ([BM], see Appendix A), we can extend the result above to arbitrary Abelian topological groups.

Theorem 2.42. *Let $\{m_k\}$ be a sequence of G valued, strongly additive set functions defined on Σ. If*

$$\lim_k m_k(A) = m(A)$$

exists for every $A \in \Sigma$, then

(i) *m is strongly additive and*
(ii) *$\{m_k\}$ is uniformly strongly additive.*

Proof. Since the topology of G is generated by a family of quasi-norms, in order to establish (ii) it suffices to show that $|m_k(A_j)| \to 0$ as $j \to \infty$ uniformly for $k \in \mathbb{N}$ for any continuous quasi-norm $|\cdot|$. But this follows from Theorem 2.41. Then (i) follows from (ii). $\qquad\square$

We note in passing that a version of the Nikodym Boundedness Theorem can be obtained from the Nikodym Convergence Theorem. The original form of the Nikodym Boundedness Theorem asserts that if $\mathcal{M}$ is a family of countably additive scalar valued set functions defined on Σ which is pointwise bounded on Σ, then $\mathcal{M}$ is uniformly bounded on Σ ([DS],[DU]). To show that $\mathcal{M}$ is uniformly bounded on Σ it suffices to show that $\{\mu_n(A_n)\}$ is bounded for every $\{\mu_n\} \subset \mathcal{M}$ and every pairwise disjoint sequence $\{A_n\} \subset \Sigma$ (see Appendix C). $\{\mu_n(A_n)\}$ is bounded if $\frac{1}{n}\mu_n(A_n) \to 0$. Since

$\mathcal{M}$ is pointwise bounded, $\frac{1}{n}\mu_n(A) \to 0$ for every $A \in \Sigma$ so the Nikodym Convergence Theorem implies the series $\sum_{j=1}^{\infty} \frac{1}{n}\mu_n(A_j)$ converge uniformly for $n \in \mathbb{N}$. In particular, $\frac{1}{n}\mu_n(A_n) \to 0$ as desired. The version of the theorem for locally convex valued measures follows from the scalar version and the Uniform Boundedness Theorem. We consider a more general version of the theorem later (Theorem 2.45). In particular, we show the result holds for bounded, finitely additive set functions.

We are now able to use the Nikodym Convergence Theorem to establish another important result from measure theory, the Vitali–Hahn–Saks Theorem. Let Σ be a sigma algebra of subsets of a set S and $\mu : \Sigma \to [0, \infty]$ a countably additive measure. A countably additive set function $m : \Sigma \to G$ is μ *continuous* if

$$\lim_{\mu(A) \to 0} m(A) = 0$$

and a family $\mathcal{M}$ of countably additive set functions from Σ to G is *uniformly* μ *continuous* if $\lim_{\mu(A) \to 0} m(A) = 0$ uniformly for $m \in \mathcal{M}$.

Lemma 2.43. *Let $\{m_k\}$ be a sequence of G valued, countably additive, μ continuous set functions defined on Σ. If $\{m_k\}$ is uniformly countably additive, then $\{m_k\}$ is uniformly μ continuous.*

Proof. If the conclusion fails to hold, then there exists a neighborhood of 0, U, in G such that for every $\delta > 0$ there exist k, $A \in \Sigma$ with

$$m_k(A) \notin U \text{ and } \mu(A) < \delta.$$

Pick a symmetric neighborhood, V, of 0 such that $V + V \subset U$. There exist $E_1 \in \Sigma$, n_1 such that $m_{n_1}(E_1) \notin U$ and $\mu(E_1) < \delta$. There exists $\delta_1 > 0$ such that $m_{n_1}(E) \in V$ when $\mu(E) < \delta_1$. There exist $E_2 \in \Sigma$, $n_2 > n_1$ such that $m_{n_2}(E_2) \notin U$ and $\mu(E_2) < \delta_1/2$. Continuing this construction produces sequences $\{E_k\} \subset \Sigma$, $\delta_{k+1} < \delta_k/2$, $n_k \uparrow$ such that

$$m_{n_k}(E_k) \notin U, \ \mu(E_{k+1}) < \delta_k/2$$

and $m_{n_k}(E) \in V$ when $\mu(E) < \delta_k$. Note that

$$\mu(\cup_{j=k+1}^{\infty} E_j) \leq \sum_{j=k+1}^{\infty} \mu(E_j) < \delta_k/2 + \delta_{k+1}/2 + \dots < \delta_k/2 + \delta_k/2^2 + \dots = \delta_k$$

so that

$$m_{n_k}(E_k \cap \cup_{j=k+1}^{\infty} E_j) \in V.$$

Now set $A_k = E_k \backslash E_k \cap \cup_{j=k+1}^{\infty} E_j$ so the $\{A_k\}$ are disjoint and then

$$m_{n_k}(A_k) = m_{n_k}(E_k) - m_{n_k}(E_k \cap \cup_{j=k+1}^{\infty} E_j) \notin V.$$

However by the uniform countable additivity, we have $\lim_k m_j(A_k) = 0$ uniformly for $j \in \mathbb{N}$ (Lemma 2.35) so we have the desired contradiction. $\square$

Theorem 2.44. (Vitali–Hahn–Saks) *Let $\{m_k\}$ be a sequence of G valued countably additive, μ continuous set functions defined on Σ. If*

$$\lim_k m_k(A) = m(A)$$

exists for every $A \in \Sigma$, then

(i) *m is μ continuous and*
(ii) *$\{m_k\}$ is uniformly μ continuous.*

Proof. By Theorem 2.36 $\{m_k\}$ is uniformly countably additive so (ii) follows from the Lemma above. (i) follows from (ii). $\qquad\square$

We establish a version of the Nikodym Boundedness Theorem for vector valued set functions. As noted above the original version of the theorem asserted that a family of countably additive, scalar valued set functions defined on a σ algebra which is pointwise bounded on a σ algebra is uniformly bounded on the σ algebra. Dunford and Schwartz refer to this remarkable result as a striking improvement of the uniform boundedness principle ([DS] IV.9.8).

Theorem 2.45. (Nikodym Boundedness) *Let Σ be a σ algebra of subsets of a set S, G be a semi-convex space. If $\mathcal{M}$ is a family of countably additive G valued set functions defined on Σ which is pointwise bounded on Σ, then $\mathcal{M}$ is uniformly bounded on Σ, i.e.,*

$$\{\mu(A) : \mu \in \mathcal{M}, A \in \Sigma\}$$

is bounded.

Proof. By the lemma in Appendix C it suffices to show that $\{m_k(A_k)\}$ is bounded for every $\{m_k\} \subset \mathcal{M}$ and every disjoint sequence $\{A_k\} \subset \Sigma$. Consider the triple

$$(\Sigma, \mathcal{M} : G)$$

under the mapping $(A, m) \to m(A)$. The formal series $\sum_k A_k$ is $w(\Sigma, \mathcal{M})$ subseries convergent ($\sum_{k=1}^{\infty} A_{n_k} = \cup_{k=1}^{\infty} A_{n_k}$ with respect to $w(\Sigma, \mathcal{M})$ for every subsequence $\{n_k\}$). We use Theorem 2.19 to establish the result. By Theorem 2.19 the partial sums

$$\left\{ \sum_{j \in \sigma} m_k(A_j) : k \in \mathbb{N}, \sigma \subset \mathbb{N} \right\}$$

are bounded since $\mathcal{M}$ is pointwise bounded on Σ. In particular, $\{m_k(A_k)\}$ is bounded. $\qquad\square$

Theorem 2.45 is applicable to the case when $\mathcal{M}$ is single measure so a countably additive set function with values in a semi-convex space has bounded range; this gives a generalization of 3.6.3 of [Rol].

Corollary 2.46. *Let G be semi-convex. If $m : \Sigma \to G$ is countably additive, then m has bounded range.*

Turpin has given an example of a countably additive set function defined on a σ algebra with values in a non-locally convex space which has unbounded range so the semi-convex assumption cannot be dropped ([Rol] 3.6.4). What conditions on the space which are necessary and sufficient for a vector measure to have bounded range seem to be unknown.

The version of the Nikodym Boundedness Theorem for semi-convex spaces is due to Constantinescu ([Co]) and Weber ([We]). It should be pointed out that the σ algebra assumption in the Nikodym Boundedness Theorem is important.

Example 2.47. Let $\mathcal{A}$ be the algebra of subsets of $\mathbb{N}$ which are either finite or have finite complements. Define $\delta_n : \mathcal{A} \to \mathbb{R}$ by $\delta_n(A) = 1$ if $n \in A$ and 0 otherwise. Define $\mu_n : \mathcal{A} \to \mathbb{R}$ by

$$\mu_n(A) = n(\delta_{n+1}(A) - \delta_n(A))$$

if A is finite and

$$\mu_n(A) = -n(\delta_{n+1}(A) - \delta_n(A))$$

if A has finite complement. Then each μ_n is countably additive, $\{\mu_n\}_n$ is pointwise bounded on $\mathcal{A}$ but not uniformly bounded on $\mathcal{A}$.

Despite this example there are algebras for which the conclusion of the Nikodym Boundedness Theorem do hold. See Schachermeyer ([Sch]) for discussions.

Using Drewnowski's Lemma (Appendix D) we can extend the corollary to strongly additive set functions.

Corollary 2.48. *Let G be semi-convex and quasi-normed. If $m : \Sigma \to G$ is strongly additive, then m has bounded range.*

Proof. Let $\{A_k\}$ be pairwise disjoint from Σ. It suffices to show that $\{m(A_k)\}$ has a convergent subsequence (Appendix C). By Drewnowski's Lemma there is a subsequence $\{A_{n_k}\}$ such that m is countably additive on the σ algebra Σ_0 generated by $\{A_{n_k}\}$. Then $m(A_{n_k}) \to 0$ as desired. $\square$

The converse of this result is false in general even for Banach space valued set functions.

Example 2.49. Let $m : \Sigma \to B(\Sigma)$, the space of bounded, Σ measurable functions, be defined by $m(A) = \chi_A$. If $B(\Sigma)$ has the sup-norm, m is bounded and finitely additive but not strongly additive (for any pairwise disjoint sequence, $\{A_k\}$, of non-empty sets from Σ, $\|m(A_k)\| = 1$).

However, for scalar valued set functions, we have the converse. For this we need an observation.

Lemma 2.50. *Let* $t_j \in \mathbb{R}$ *for every* $j \in \mathbb{N}$. *If there exists* $M \geq 0$ *such that*

$$\left| \sum_{j \in \sigma} t_j \right| \leq M$$

for every finite set σ, *then*

$$\sum_{j=1}^{\infty} |t_j| \leq 2M.$$

Proof. For $\sigma \subset \mathbb{N}$ finite, let $\sigma_+ = \{j \in \sigma : t_j \geq 0\}$ and $\sigma_- = \{j \in \sigma : t_j < 0\}$. Then

$$\sum_{j \in \sigma_+} |t_j| = \sum_{j \in \sigma_+} t_j \leq M$$

and

$$\sum_{j \in \sigma_-} |t_j| = - \sum_{j \in \sigma_-} t_j \leq M$$

so

$$\sum_{j \in \sigma} |t_j| \leq 2M.$$

Since σ is arbitrary $\sum_{j=1}^{\infty} |t_j| \leq 2M$. $\square$

Proposition 2.51. *Let* $\lambda : \Sigma \to \mathbb{R}$ *be finitely additive. Then* λ *is bounded iff* λ *is strongly additive.*

Proof. Let λ be bounded with

$$\sup\{|\lambda(A)| : A \in \Sigma\} = M < \infty.$$

Let $\{A_k\} \subset \Sigma$ be pairwise disjoint. If σ is finite,

$$\left|\sum_{k \in \sigma} \lambda(A_k)\right| = |\lambda(\cup_{k \in \sigma} A_k)| \le M$$

so

$$\sum_{k=1}^{\infty} |\lambda(A_k)| \le 2M$$

(Lemma 50). Then λ is strongly bounded by Proposition 2.38. The other implication follows from the corollary.   □

We give an extension of the Nikodym Boundedness Theorem to strongly additive set functions.

Theorem 2.52. (Nikodym) *Let Σ be a σ algebra of subsets of a set S, G be a semi-convex space. If $\mathcal{M}$ is a family of bounded, strongly additive, G valued set functions defined on Σ which is pointwise bounded on Σ, then $\mathcal{M}$ is uniformly bounded on Σ, i.e.,*

$$\{\mu(A) : \mu \in \mathcal{M}, A \in \Sigma\}$$

is bounded.

Proof. Let $\{m_k\} \subset \mathcal{M}$ and let $\{A_k\} \subset \Sigma$ be a disjoint sequence. By the lemma in Appendix C, it suffices to show $\{m_k(A_k)\}$ is bounded. The sequence $\{(1/k)m_k\}$ converges pointwise to 0 on Σ so by Theorem 2.42 the sequence is uniformly strongly additive. Hence,

$$\lim(1/k)m_k(A_k) = 0$$

and $\{m_k(A_k)\}$ is bounded.   □

The version of the Nikodym Boundedness Theorem above has an interesting duality application. Let $\mathcal{S}(\Sigma)$ be the space of Σ simple functions with the sup norm. Then the dual of $\mathcal{S}(\Sigma)$ is the space $ba(\Sigma)$ of bounded, finitely additive set functions on Σ and the dual norm on $ba(\Sigma)$ is equivalent to the norm

$$\|m\| = \sup\{|m(A)| : A \in \Sigma\}$$

([DS] IV.5.1, [Sw3] 6.3). A scalar consequence of Theorem 2.52 is given below. Note members of $ba(\Sigma)$ are strongly additive (Proposition 2.51). Recall a normed space is barrelled if weak* subsets of the dual are norm bounded in the dual norm.

Corollary 2.53. *In the space* $ba(\Sigma)$, $\sigma(ba(\Sigma), \mathcal{S}(\Sigma))$ *bounded subsets are norm bounded, i.e.,* $\mathcal{S}(\Sigma)$ *is a barrelled space. In particular, if* m_0 *is the subspace of* l^∞ *consisting of the sequences with finite range with the sup-norm,* m_0 *is barrelled.*

The space m_0 is often given as an example of an incomplete normed space which is barrelled.

It follows from the corollary and the Uniform Boundedness Principle that a family of bounded, finitely additive set functions defined on a σ algebra with values in a LCTVS which is pointwise bounded is uniformly bounded on the σ algebra.

The Nikodym Convergence Theorem, the Nikodym Boundedness Theorem and the Vitali–Hahn–Saks Theorem have been extended to certain algebras of subsets. See Schakerrmeyer ([Sch]) for discussions of these extensions.

Finally, we observe that we can obtain a result of Graves and Ruess ([GR] Lemma 6) on the uniform countable additivity of weak compact sets. Let X be a Banach space. We consider the triple

$$(\mathcal{S}(\Sigma), ca(\Sigma, X) : X)$$

under the integration map $(g, \nu) \to \int_S g d\nu$ (see Chapter 1). From Theorem 2.3 we have

Corollary 2.54. *If* K *is* $w(ca(\Sigma, X), \mathcal{S}(\Sigma))$ *compact, then* K *is uniformly countably additive.*

Proof. If $\{A_j\}$ is pairwise disjoint from Σ, then the formal series $\sum_j A_j$ is $w(\mathcal{S}(\Sigma), ca(\Sigma, X))$ subseries convergent so the result follows from Theorem 2.3. $\qquad \Box$

Hahn–Schur Theorems

We next consider a version of the Hahn–Schur Theorem for group valued series. One scalar version of the Hahn–Schur Theorem asserts that a sequence $\{x^k\} = \{x_j^k\}_j$ in l^1 which is weakly convergent is actually norm convergent in l^1. In this statement the series $\sum_{j=1}^\infty x_j^k$ are absolutely convergent. If we are seeking a version of the theorem for normed or locally convex spaces this is a very restrictive assumption. In $\mathbb{R}^n$ a series is absolutely convergent iff the series is subseries convergent so it seems to be reasonable to consider subseries convergent series in any attempted generalization. A weaker version of the Hahn–Schur Theorem than the one given in the statement above reads as follows:

(HS) If $\lim_k \sum_{j\in\sigma} x_j^k$ exists for every $\sigma \subset \mathbb{N}$ and if $x_j = \lim_k x_j^k$ for every j, then $\{x_j\} \in l^1$ and $\lim_k \left\| x^k - x \right\|_1 = 0$.

Using Lemma 50, we can give another version of the last conclusion. This lemma means that the condition

$$\lim_k \left\| x^k - x \right\|_1 = 0$$

above is equivalent to the condition

$$\lim_k \sum_{j\in\sigma} x_j^k = \sum_{j\in\sigma} x_j$$

uniformly for $\sigma \subset \mathbb{N}$. This latter condition makes sense for subseries convergent series in an Abelian topological group and suggests how the generalizations of the Hahn–Schur Theorem should be sought.

Let G be a Hausdorff, Abelian topological group.

Theorem 2.55. (Hahn–Schur Theorem) *Let $\sum_j x_{ij}$ be a subseries convergent series in G for every $i \in \mathbb{N}$ and suppose*

$$\lim_i \sum_{j\in\sigma} x_{ij}$$

exists for every $\sigma \subset \mathbb{N}$. Set $x_j = \lim_i x_{ij}$. Then

(i) $\sum_j x_j$ *is subseries convergent,*

(ii) *the series* $\sum_{j\in\sigma} x_{ij}$ *converge uniformly for $i \in \mathbb{N}, \sigma \subset \mathbb{N}, and$*

(iii) $\lim_i \sum_{j\in\sigma} x_{ij} = \sum_{j\in\sigma} x_j$ *uniformly for $\sigma \subset \mathbb{N}$.*

Proof. We show that (ii) follows directly from Theorem 2.2. Let E be the power set of $\mathbb{N}$, define $f_i : E \to G$ by

$$f_i(\sigma) = \sum_{j\in\sigma} x_{ij}$$

and set $F = \{f_i : i \in \mathbb{N}\}$. Then

$$(E, F : G)$$

is an abstract triple under the map $(\sigma, f_i) \to f_i(\sigma)$ and the (formal) series $\sum_j j$ is $w(E, F)$ subseries convergent with $\sum_j f_i(j) = \sum_j x_{ij}$. By hypothesis, F is sequentially conditionally $w(F, E)$ compact so from Theorem 2.2, it follows that (ii) holds. Conditions (i) and (iii) follow from (ii) by the Iterated Limit Theorem. $\square$

The usual scalar version of the theorem can be obtained easily from Lemma 50.

Corollary 2.56. *Let* $\{x^k\} = \{x_j^k\}_j \in l^1$. *If*

$$\lim_k \sum_{j \in \sigma} x_j^k$$

exists for every $\sigma \subset \mathbb{N}$ *and if* $x_j = \lim_k x_j^k$ *for every* j, *then* $\{x_j\} \in l^1$ *and* $\lim_k \left\| x^k - x \right\|_1 = 0$.

Indeed, if $\epsilon > 0$, it follows from the Hahn–Schur Theorem that there exists N such that

$$\left| \sum_{j \in \sigma} (x_j^k - x_j) \right| \le \epsilon$$

for $k \ge N$, $\sigma \subset \mathbb{N}$. By Lemma 50,

$$\sum_{j=1}^{\infty} \left| x_j^k - x_j \right| \le 2\epsilon$$

for $k \ge N$.

We can also note a vector version of this result follows from the Hahn–Schur Theorem. Let G be a LCTVS whose topology is generated by the sem-norms $\mathcal{P}$. Let

$$ss(G) = m_0^{\beta G}$$

be the space of all G valued subseries convergent series and set

$$\widehat{p}(\{x_j\}) = \sup \left\{ p \left(\sum_{j \in \sigma} x_j \right) : \sigma \subset \mathbb{N} \right\}$$

for $\{x_j\} \in ss(G)$. Note $\widehat{p}(\{x_j\}) < \infty$ by Theorem 17. Give $ss(G)$ the topology, $\tau_{ss}(G)$, generated by the semi-norms $\{\widehat{p} : p \in \mathcal{P}\}$. Consider the triple

$$(ss(G), \{\sigma \subset \mathbb{N}\} : G)$$

under the map $(\{x_j\}, \sigma) \to \sum_{j \in \sigma} x_j$. It follows from the Hahn–Schur Theorem that if the sequence $x^k \to 0$ in $w(ss(G), \{\sigma \subset \mathbb{N}\})$, then $x^k \to 0$ in $\tau_{ss}(G)$. Moreover, if G is sequentially complete, $w(ss(G), \{\sigma \subset \mathbb{N}\})$ is sequentially complete.

Thus, we see that the topology $\sigma(l^1, m_0)$ is sequentially complete and $\sigma(l^1, m_0)$ Cauchy sequences are norm convergent.

We can obtain a generalization of Theorem 17 from the Hahn–Schur Theorem. First, a lemma.

Lemma 2.57. *Let S be a compact Hausdorff space and $g_i : S \to G$ be continuous functions for $i = 0, 1, 2, ...$, with*

$$\lim_i g_i(t) = g_0(t)$$

uniformly for $t \in S$. Then $\mathcal{R} = \cup_{i=0}^\infty \mathcal{R}g_i$ is compact, where $\mathcal{R}g_i$ is the range of g_i.

Proof. Let $\mathcal{G}$ be an open cover of $\mathcal{R}$. For each $x \in \mathcal{R}$ there exists $U_x \in \mathcal{G}$ such that $x \in U_x$. Then $-x + U_x$ is an open neighborhood of 0 so there is an open neighborhood of 0, V_x, such that $V_x + V_x \subset -x + U_x$. Then

$$\mathcal{G}' = \{x + V_x : x \in \mathcal{R}\}$$

is an open cover of $\mathcal{R}$.

Since $\mathcal{R}g_0$ is compact, there exist finite $x_1 + V_{x_1}, ..., x_k + V_{x_k}$ covering $\mathcal{R}g_0$. Put $V = \cap_{j=1}^k V_{x_j}$ so V is an open neighborhood of 0. There exists n such that $g_i(t) - g_0(t) \in V$ for $i \geq n, t \in S$. For $t \in S$ there exists j such that $g_0(t) \in x_j + V_{x_j}$ so

$$g_i(t) \in g_0(t) + V \subset x_j + V_{x_j} + V_{x_j} \subset U_{x_j}$$

for $i \geq n$. Hence, $U_{x_1}, ..., U_{x_k}$ covers $\cup_{i=n}^\infty \mathcal{R}g_i$.

Since $\mathcal{R}g_i$, $i = 0, ..., n-1$, are compact, a finite subcover of $\mathcal{G}$ covers the union of these sets and, hence, $\mathcal{G}$ has a finite subcover which covers $\mathcal{R}$. $\quad\square$

Theorem 2.58. *Let $\sum_j x_{ij}$ be a subseries convergent series in G for every $i \in \mathbb{N}$ and suppose*

$$\lim_i \sum_{j \in \sigma} x_{ij}$$

exists for every $\sigma \subset \mathbb{N}$. Set $x_j = \lim_i x_{ij}$. Then

$$B = \left\{ \sum_{j \in \sigma} x_{ij} : i \in \mathbb{N}, \sigma \subset \mathbb{N} \right\} \cup \left\{ \sum_{j \in \sigma} x_j : \sigma \subset \mathbb{N} \right\}$$

is compact.

Proof. Let $\Lambda = \operatorname{span}\{\chi_\sigma : \sigma \subset \mathbb{N}\}$ and let p be the topology of pointwise convergence on Λ. Let $S_i : \Lambda \to G$ $(S : \Lambda \to G)$ be the summing operator

$$S_i(\sigma) = \sum_{j \in \sigma} x_{ij} \quad \left(S(\sigma) = \sum_{j \in \sigma} x_j \right).$$

Each S_i, S is continuous with respect to p by Lemma 16 and by the Hahn–Schur Theorem $S_i \to S$ uniformly on Λ. Since (Λ, p) is compact, the result follows from the lemma. $\quad\square$

We have a partial converse for the Hahn–Schur Theorem.

Theorem 2.59. *Let G be sequentially complete. Assume that $\sum_j x_{ij}$ is uniformly unordered convergent for $i \in \mathbb{N}$ and such that $\lim_i x_{ij} = x_j$ exists for each j. Then $\sum_j x_j$ is subseries convergent and*

$$\lim_i \sum_{j \in \sigma} x_{ij} = \sum_{j \in \sigma} x_j$$

exists for every $\sigma \subset \mathbb{N}$. [The limit is uniform for $\sigma \subset \mathbb{N}$ by the Hahn–Schur Theorem.]

Proof. Let $\sigma \subset \mathbb{N}$ and U be a neighborhood of 0. Pick a closed, symmetric neighborhood of 0, V, with $V + V + V \subset U$. There exists n such that

$$\sum_{j \in \tau} x_{ij} \in V$$

for all i and τ such that $\min \tau \geq n$. Let $\sigma^n = \{j \in \sigma : j \geq n\}$ and $\sigma_n = \{j \in \sigma : j < n\}$. There exists m such that $i, k \geq m$ implies

$$\sum_{j \in \sigma_n} (x_{ij} - x_{kj}) \in V.$$

If $i, k \geq m$, we have

$$\sum_{j \in \sigma} x_{ij} - \sum_{j \in \sigma} x_{kj} = \sum_{j \in \sigma_n} (x_{ij} - x_{kj}) + \sum_{j \in \sigma^n} x_{ij} - \sum_{j \in \sigma^n} x_{kj}$$
$$\in V + V + V \subset U.$$

Hence, $\{\sum_{j \in \sigma} x_{ij}\}_i$ is Cauchy so $\lim_i \sum_{j \in \sigma} x_{ij}$ exists. The conclusion follows from the Hahn–Schur Theorem. $\qquad\square$

We use the Hahn–Schur Theorem and the Nikodym Convergence Theorem to prove another important result from measure theory, the Phillips' lemma. We first establish a group theory version of the result and then consider the scalar version.

Theorem 2.60. (Phillips) *Let G be quasi-normed and sequentially complete. Let $m_k : \Sigma \to G$ be strongly additive for $k \in \mathbb{N}$ and suppose*

$$\lim_k m_k(A) = 0$$

for every $A \in \Sigma$. Then for every pairwise disjoint sequence $\{A_j\} \subset \Sigma$,

$$\lim_k \sum_{j \in \sigma} m_k(A_j) = 0$$

uniformly for $\sigma \subset \mathbb{N}$.

Proof. By the Hahn–Schur Theorem it suffice to show

$$\lim_k \sum_{j \in \sigma} m_k(A_j) = 0$$

for every $\sigma \subset \mathbb{N}$. If this fails we may assume, by passing to a subsequence if necessary, that there exists a closed neighborhood of 0, U, such that

$$\sum_{j=1}^{\infty} m_k(A_j) \notin U$$

for every k. Pick a symmetric neighborhood of 0, V, such that $V + V \subset U$. There exists n_1 such that

$$\sum_{j=1}^{n_1} m_1(A_j) \notin U.$$

There exists p_1 such that

$$\sum_{j=1}^{n_1} m_k(A_j) \in V$$

for $k \geq p_1$. There exists $n_2 > n_1$ such that $\sum_{j=1}^{n_2} m_{p_1}(A_j) \notin U$. Therefore,

$$\sum_{j=n_1+1}^{n_2} m_{p_1}(A_j) = \sum_{j=1}^{n_2} m_{p_1}(A_j) - \sum_{j=1}^{n_1} m_{p_1}(A_j) \notin V.$$

Continuing this construction produces increasing sequences $\{n_i\}, \{p_i\}$ with

$$\sum_{j=n_i+1}^{n_{i+1}} m_{p_i}(A_j) \notin V.$$

Put $B_i = \cup_{j=n_i+1}^{n_{i+1}} A_j$ so $\{B_i\}$ are pairwise disjoint and

$$(\#) \quad m_{p_i}(B_i) \notin V.$$

By Drewnowski's Lemma (Appendix D), there is a subsequence $\{q_i\}$ such that each m_k is countably additive on the σ algebra Σ_0 generated by $\{B_{q_i}\}$. We have

$$\lim_i m_{p_i}(A) = 0$$

for every $A \in \Sigma_0$. By Theorem 2.36 the series

$$\sum_{j=1}^{\infty} m_{p_i}(B_{q_j})$$

converge uniformly for $i \in \mathbb{N}$. In particular, $\lim_i m_{p_{q_i}}(B_{q_i}) = 0$ contradicting $(\#)$. $\qquad \square$

Corollary 2.61. *Let G be quasi-normed and sequentially complete. Let $m_k : \Sigma \to G$ be strongly additive for $k \in \mathbb{N}$ and suppose*

$$\lim_k m_k(A) = m(A)$$

exists for every $A \in \Sigma$. Then for every pairwise disjoint sequence $\{A_j\} \subset \Sigma$,

$$\lim_k \sum_{j \in \sigma} m_k(A_j) = \sum_{j \in \sigma} m(A_j)$$

uniformly for $\sigma \subset \mathbb{N}$. In particular, m is strongly additive.

Proof. It suffices to show $\lim_k \sum_{j \in \sigma} m_k(A_j)$ exists for every $\sigma \subset \mathbb{N}$ by the Hahn–Schur Theorem. This is trivial if σ is finite so assume $\sigma = \{n_1 < n_2 < ...\}$. We claim that $\{\sum_{j=1}^{\infty} m_i(A_{n_j})\}_i$ is Cauchy. For this assume $p_i \uparrow, q_i \uparrow$ with $p_i < q_i < p_{i+1}$. Then $\lim_i(m_{p_i}(A) - m_{q_i}(A)) = 0$ so Phillips' Theorem implies

$$\lim_i \sum_{j=1}^{\infty} [m_{p_i}(A_{n_j}) - m_{q_i}(A_{n_j})] = 0$$

so $\{\sum_{j=1}^{\infty} m_i(A_{n_j})\}_i$ is Cauchy. The result follows from the completeness assumption. $\qquad\square$

We now observe that the usual scalar version of Phillips' Lemma follows from the corollary. Let $\mathcal{P}$ be the power set of $\mathbb{N}$ and recall that ba is the space of all bounded, finitely additive set functions defined on $\mathcal{P}$.

Theorem 2.62. (Phillips) *Let $m_k \in ba$. If*

$$\lim_k m_k(A) = m(A)$$

exists for every $A \subset \mathbb{N}$, then $m \in ba$ and

$$\lim_k \sum_{j=1}^{\infty} |m_k(j) - m(j)| = 0.$$

Proof. By the corollary, $m \in ba$. Given $\epsilon > 0$ by the corollary there exists N such that

$$\left| \sum_{j \in \sigma} [m_k(j) - m(j)] \right| < \epsilon$$

for $k \geq N, \sigma \subset \mathbb{N}$. By Lemma 50,

$$\sum_{j=1}^{\infty} |m_k(j) - m(j)| \leq 2\epsilon$$

for $k \geq N$. $\qquad\square$

Phillips' Lemma has an interesting duality interpretation. Let J be the canonical imbedding of c_0 into its bidual l^∞. Then the transpose operator $J' : (l^\infty)' = ba \to l^1$ is given by $J'm = \{m(i)\}_i$. Phillips' Lemma implies that if $\{m_k\}$ converges to 0 in the weak topology $\sigma(ba, m_0)$, then $\{J'(m_k)\}$ converges to 0 in the norm topology of l^1. In particular, if $\{m_k\}$ converges to 0 in $\sigma(ba, l^\infty)$, then $\|J'm_k\|_1 \to 0$. Phillips used this result to show that there is no continuous projection from l^∞ onto c_0.

Antosik Interchange Theorem

A problem often encountered in analysis is the interchange of two limiting processes. For example, the Lebesgue Dominated Convergence Theorem gives sufficient conditions to interchange the pointwise limit of a sequence of integrable functions with the Lebesgue integral, i.e., to take the "limit under the integral sign". We consider sufficient conditions for the equality of two iterated series. For real valued series one of the most useful criterion for interchanging the limit of an iterated series

$$\sum_{i=1}^{\infty} \sum_{j=1}^{\infty} t_{ij}$$

is the absolute convergence of the iterated series. However, absolute convergence for series with values in a LCTVS is a very strong condition and is, therefore, not appropriate. Antosik has given a sufficient condition involving subseries convergence of an iterated series with values in a topological group which has proven to be useful in a number of applications ([A]). We use the Hahn–Schur Theorem to give a proof of Antosik's Theorem.

Let X be a Hausdorff, Abelian topological group. Let $x_{ij} \in X$ for $i, j \in \mathbb{N}$. The double series $\sum_{i,j} x_{ij}$ converges to $x \in X$ if for every neighborhood, U, of 0 in X, there exists N such that

$$\sum_{i=1}^{p} \sum_{j=1}^{q} x_{ij} - x \in U$$

for $p, q \geq N$. We have the following familiar properties of double series.

Proposition 2.63. *Let $\sum_{i,j} x_{ij}$ be a double series.*

(i) *If the double series $\sum_{i,j} x_{ij}$ converges to $x \in X$ and if the series $\sum_{j=1}^{\infty} x_{ij}$ converge for each i, then the iterated series $\sum_{i=1}^{\infty} \sum_{j=1}^{\infty} x_{ij}$ converges to x.*

(ii) *If the series $\{\sum_{i=1}^{m} \sum_{j=1}^{\infty} x_{ij} : m \in \mathbb{N}\}$ converge uniformly and if the iterated series $\sum_{i=1}^{\infty} \sum_{j=1}^{\infty} x_{ij}$ converges to x, then the double series $\sum_{i,j} x_{ij}$ converges to x.*

Proof. (i): Let U be a neighborhood of 0 in X and let V be a symmetric neighborhood such that $V + V \subset U$. There exists N_1 such that $p, q \geq N_1$ implies that

$$\sum_{i=1}^{p} \sum_{j=1}^{q} x_{ij} - x \in V.$$

For each p there exists $N_2(p)$ such that

$$\sum_{i=1}^{p} \sum_{j=1}^{\infty} x_{ij} - \sum_{i=1}^{p} \sum_{j=1}^{q} x_{ij} \in V$$

for $q \geq N_2(p)$. Let $p \geq N_1$ and fix $q \geq \max\{N_1, N_2(p)\}$. Then

$$\sum_{i=1}^{p} \sum_{j=1}^{\infty} x_{ij} - x = \sum_{i=1}^{p} \sum_{j=1}^{\infty} x_{ij} - \sum_{i=1}^{p} \sum_{j=1}^{q} x_{ij} + \sum_{i=1}^{p} \sum_{j=1}^{q} x_{ij} - x \in V + V \subset U.$$

(ii): There exists N such that

$$\sum_{i=1}^{p} \sum_{j=q+1}^{\infty} x_{ij} \in V$$

for $q > N$ and for every $p \in \mathbb{N}$. There exists $M > N$ such that

$$\sum_{i=1}^{p} \sum_{j=1}^{\infty} x_{ij} - x \in V$$

for $p \geq M$. If $p, q \geq M$, then

$$\sum_{i=1}^{p} \sum_{j=1}^{q} x_{ij} - x = \sum_{i=1}^{p} \sum_{j=1}^{\infty} x_{ij} - x - \sum_{i=1}^{p} \sum_{j=q+1}^{\infty} x_{ij} \in V + V \subset U.$$

$\square$

Theorem 2.64. (Antosik) *Let $\{x_{ij}\} \subset X$. Suppose the series $\sum_{i=1}^{\infty} \sum_{j=1}^{\infty} x_{im_j}$ converges for every increasing sequence $\{m_j\}$. Then the double series $\sum_{i,j} x_{ij}$ converges and*

$$(*) \quad \sum_{i,j} x_{ij} = \sum_{i=1}^{\infty} \sum_{j=1}^{\infty} x_{ij} = \sum_{j=1}^{\infty} \sum_{i=1}^{\infty} x_{ij}.$$

Proof. Note that the series $\sum_{i=1}^{\infty} x_{ik}$ converges for every k [consider the difference between the two series $\sum_{i=1}^{\infty}\sum_{j=1}^{\infty} x_{in_j}$ and $\sum_{i=1}^{\infty}\sum_{j=1}^{\infty} x_{im_j}$, where $n_j = j$ for every j and $\{m_j\}$ is the sequence $\{1, ..., k-1, k+1, ...\}$]. Set

$$z_{mj} = \sum_{i=1}^{m} x_{ij}.$$

Then for $\sigma \subset \mathbb{N}$,

$$\sum_{j \in \sigma} z_{mj} = \sum_{i=1}^{m} \sum_{j \in \sigma} x_{ij}$$

converges to $\sum_{i=1}^{\infty} \sum_{j \in \sigma} x_{ij}$ as $m \to \infty$ by hypothesis. By the Hahn–Schur Theorem, the series $\sum_{j=1}^{\infty}(\sum_{i=1}^{\infty} x_{ij})$ is subseries convergent and

$$\lim_m \sum_{i=1}^{m} \sum_{j \in \sigma} x_{ij} = \sum_{j \in \sigma} \sum_{i=1}^{\infty} x_{ij}$$

uniformly for $\sigma \subset \mathbb{N}$. In particular,

$$\sum_{i=1}^{\infty} \sum_{j=1}^{\infty} x_{ij} = \sum_{j=1}^{\infty} \sum_{i=1}^{\infty} x_{ij}.$$

By the proposition above, the uniform convergence implies that the double series $\sum_{i,j} x_{ij}$ converges and $(*)$ holds. $\qquad\square$

Although Antosik's Theorem is easy to apply in many concrete situations, it is only a necessary condition for the equality of the two iterated series. For example, suppose that $a_j, b_j \in \mathbb{R}$ and $x_{ij} = a_i b_j$. If both series $\sum_j a_j$ and $\sum_j b_j$ converge, then

$$\sum_{i=1}^{\infty} \sum_{j=1}^{\infty} x_{ij} = \sum_{j=1}^{\infty} \sum_{i=1}^{\infty} x_{ij} = \sum_{i=1}^{\infty} a_i \sum_{j=1}^{\infty} b_j.$$

However, if the "inner" series, $\sum_{j=1}^{\infty} b_j$ is conditionally convergent, the hypothesis in Antosik's theorem is not satisfied.

Antosik's theorem has found applications to various topics in analysis. We show how the result can be used to give another proof of Stiles' version of the Orlicz–Pettis Theorem for spaces with a Schauder basis (Theorem 2.13). Let X be a TVS with a Schauder basis $\{b_j\}$ and coordinate functionals $\{f_j\}$ so

$$x = \sum_{j=1}^{\infty} f_j(x) b_j.$$

for $x \in X$. We do not assume the coordinate functionals are continuous although this is the case when X is complete and metrizable. Let $F = \{f_j\}$ and consider the weak topology $\sigma(X, F)$.

Theorem 2.65. (Stiles) *If $\sum x_j$ is $\sigma(X, F)$ subseries convergent, then the series is subseries convergent in the original topology of X.*

Proof. Let $\{n_j\}$ be a subsequence. Let $x = \sum_{j=1}^{\infty} x_{n_j}$ be the $\sigma(X, F)$ sum of the series. By the $\sigma(X, F)$ convergence,

$$f_i(x) = \sum_{j=1}^{\infty} f_i(x_{n_j})$$

for each i. The series $\sum_{j=1}^{\infty} f_i(x_{n_j}) b_i$ converges in the original topology of X to $f_i(x) b_i$. Then

$$\sum_{i=1}^{\infty} \sum_{j=1}^{\infty} f_i(x_{n_j}) b_i = \sum_{i=1}^{\infty} f_i(x) b_i = x.$$

We may apply Antosik's Theorem and obtain

$$\sum_{j=1}^{\infty} \sum_{i=1}^{\infty} f_i(x_{n_j}) b_i = \sum_{j=1}^{\infty} x_{n_j} = \sum_{i,j} f_i(x_{n_j}) b_i,$$

with convergence in the original topology of X. $\qquad\square$

Uniform Boundedness Principle

We next consider the Uniform Boundedness Principle. The classical version of this result for normed spaces asserts that a family of continuous linear operators, Γ, from a Banach space X into a normed space Y which is pointwise bounded on X is norm bounded, i.e.,

$$\sup\{\|Tx\| : T \in \Gamma, \|x\| \leq 1\} = \{\|T\| : T \in \Gamma\} < \infty.$$

There are two interpretations of the conclusion to this statement. One is that the family Γ is uniformly bounded on bounded subsets of X and the other is that Γ is equicontinuous. We will consider general versions of both of these conclusions later but for now we consider the first version.

Theorem 2.66. (Uniform Boundedness Principle) *Let G be a locally convex space and let E be a sequentially complete locally convex space. Let Γ be a subset of $L(E, G)$ which is pointwise bounded on E. Then Γ is uniformly bounded on bounded subsets of E.*

Proof. Suppose there exists a bounded subset B of E such that $\Gamma(B)$ is not bounded. Then there exists a continuous semi-norm p on G such that

$$\sup\{p(Tx) : T \in \Gamma, x \in B\} = \infty.$$

Pick $T_k \in \Gamma$ and $x_k \in B$ such that

$$(*) \quad p(T_k x_k) > 2^{2k}.$$

Since $\{x_k\}$ is bounded, the series $\sum_k x_k/2^k$ is absolutely convergent and, therefore, subseries convergent in E by the sequential completeness hypothesis. Set

$$F = \{T_k/2^k : k \in \mathbb{N}\}$$

and consider the abstract triple

$$(E, F : G)$$

under the map $(x, T) \to Tx$. Since the series $\sum_k x_k/2^k$ is subseries convergent in E, the series is $w(E, F)$ subseries convergent. For each $x \in E$ the sequence $\{T_k x\}$ is bounded by hypothesis so $T_k x/2^k \to 0$ which implies that that the sequence $\{T_k/2^k\}$ is sequentially relatively $w(F, E)$ compact; i.e., F is sequentially relatively $w(F, E)$ compact. Then Theorem 2.2 implies that the series

$$\sum_{j=1}^{\infty} (T_k/2^k)(x_j/2^j)$$

converge uniformly for $k \in \mathbb{N}$. In particular, $T_k x_k/2^{2k} \to 0$ in G. But, this contradicts $(*)$. $\qquad\square$

The completeness assumption in the Uniform Boundedness Theorem is important.

Example 2.67. Let $f_i(t) = f_i(\{t_j\}) = \sum_{j=1}^{i} t_j$ for $t = \{t_j\} \in c_{00}$. Then each f_i is a continuous linear functional on c_{00} with the sup-norm. The sequence $\{f_i\}$ is pointwise bounded on c_{00}. But $\|f_i\| = i$ so $\{f_i\}$ is not uniformly bounded on bounded subsets of c_{00}.

The method of proof in the theorem above also can be employed to obtain an operator version of the Banach–Mackey Theorem ([Wi2] 10.4.8; [Sw1] 4.2.7).

Theorem 2.68. (Banach–Mackey) *If $\Gamma \subset L(E, G)$ is pointwise bounded on E and $B \subset E$ is bounded, absolutely convex and sequentially complete, then $\Gamma(B)$ is bounded.*

Proof. Suppose that $\Gamma(B)$ is not bounded. Then there exists a continuous semi-norm p on G such that

$$\sup\{p(Tx) : T \in \Gamma, x \in B\} = \infty.$$

Pick $T_k \in \Gamma$ and $x_k \in B$ such that

$$(*) \quad p(T_k x_k) > 2^{2k}.$$

Since $\{x_k\}$ is bounded and B is absolutely convex, the partial sums of the series $\sum_k x_k/2^k$ belong to B and the series is absolutely convergent in B and, therefore, subseries convergent in E by the sequential completeness hypothesis. Set

$$F = \{T_k/2^k : k \in \mathbb{N}\}$$

and consider the abstract triple

$$(E, F : G)$$

under the map $(x, T) \to Tx$. Since the series $\sum_k x_k/2^k$ is subseries convergent in E, the series is $w(E, F)$ subseries convergent. For each $x \in E$ the sequence $\{T_k x\}$ is bounded by hypothesis so $T_k x/2^k \to 0$ which implies that that the sequence $\{T_k/2^k\}$ is sequentially relatively $w(F, E)$ compact; i.e., F is sequentially relatively $w(F, E)$ compact. Then Theorem 2.2 implies that the series

$$\sum_{j=1}^{\infty} (T_k/2^k)(x_j/2^j)$$

converge uniformly for $k \in \mathbb{N}$. In particular, $T_k x_k/2^{2k} \to 0$ in G. But, this contradicts $(*)$. $\qquad\square$

Recall that a subset B of a locally convex space E is strongly bounded if

$$\sup\{|x'(x)| : x \in B, x' \in A\} < \infty$$

for every $\sigma(E', E)$ bounded set A. Thus, if E is a sequentially complete, locally convex space, then weak* bounded subsets of E' are strongly bounded; this is a version of the Banach–Mackey Theorem ([Wi2] 10.4.8; [Sw1] 4.2.7). This means in the terminology of Wilansky that E, E' is a Banach–Mackey pair ([Wi2] 10.4.3) when E is sequentially complete.

We will consider more general versions of the Uniform Boundedness Principle later in Chapter 5.

Bilinear Operators

A theorem of Mazur and Orlicz asserts that a separately continuous bilinear mapping from the product of two metric linear spaces one of which is complete is (jointly) continuous ([MO]). We show that the Orlicz–Pettis Theorem can be used to derive a similar result. First, we consider a boundedness result. If E, F, G are TVS's and $b : E \times F \to G$ is bilinear, then b is *bounded* if

$$b(A, B) = \{b(x, y) : x \in A, y \in B\}$$

is bounded when $A \subset E, B \subset F$ are bounded.

Theorem 2.69. *Let E, G be locally convex spaces and F be a topological vector space. Assume that E is sequentially complete. Let $b : E \times F \to G$ be a bilinear, separately continuous map. If $A \subset E, B \subset F$ are bounded, then $b(A, B)$ is bounded (i.e., b is a bounded bilinear map).*

Proof. If the conclusion fails to hold, there is a continuous semi-norm p on G such that

$$\sup\{p(x, y) : x \in A, y \in B\} = \infty.$$

Pick $x_k \in A, y_k \in B$ such that

$$(\#) \quad p(x_k, y_k) > 2^{2k}$$

and consider the abstract triple

$$(E, F : G)$$

under the map b. The series $\sum x_k / 2^k$ is absolutely convergent in E since A is bounded and, therefore, the series is subseries convergent by the sequential completeness assumption. Since $b(\cdot, y)$ is continuous, the series $\sum x_k / 2^k$ is $w(E, F)$ subseries convergent. Also, $y_k / 2^k \to 0$ in F since B is bounded and $y_k / 2^k \to 0$ in $w(F, E)$ since $b(x, \cdot)$ is continuous. Therefore, $\{y_k / 2^k\}$ is $w(F, E)$ sequentially relatively compact. By Theorem 2.2 the series

$$\sum_{j=1}^{\infty} b(x_j / 2^j, y_k / 2^k)$$

converges uniformly for $k \in \mathbb{N}$. In particular, $b(x_k / 2^k, y_k / 2^k) \to 0$ contradicting $(\#)$. $\qquad\square$

We can use this result to establish a hypocontinuity result. A bilinear map b is (left) *hypocontinuous* whenever B is a fixed bounded set in F for every neighborhood of 0, W, in G there exists a neighborhood U in E such that $b(U, B) \subset W$. Thus, a bilinear map is hypocontinuous if whenever $\{x_\delta\}$ is a net in E which converges to 0 and $B \subset F$ is bounded, then $b(x_\delta, y) \to 0$ uniformly for $y \in B$. Hypocontinuity is a property between separate continuity and joint continuity (we will discuss hypocontinuity in more detail later in Chapter 7).

The space E is a *braked space* if whenever $x_k \to 0$ there exists a sequence $t_k \to \infty$ such that $t_k x_k \to 0$ ([Kh], Appendix A). For example metric linear spaces are braked; see Appendix A.

Theorem 2.70. *Let E, G be locally convex spaces and F be a topological vector space. Assume that E is a sequentially complete braked space. Let $b : E \times F \to G$ be a bilinear, separately continuous map. Then b is sequentially hypocontinuous in the sense that if $x_k \to 0$ in E and $B \subset F$ is bounded, then $b(x_k, y) \to 0$ uniformly for $y \in B$.*

Proof. To see this it suffices to show that $b(x_k, y_k) \to 0$ for $y_k \in B$. Now there exists $t_k \to \infty$ such that $t_k x_k \to 0$ and the result above implies that $\{b(t_k x_k, y_k)\}$ is bounded so

$$\frac{1}{t_k} b(t_k x_k, y_k) = b(x_k, y_k) \to 0$$

as desired. □

This implies the result of Mazur and Orlicz ([MO]).

Corollary 2.71. *If E is braked and sequentially complete, then b is jointly sequentially continuous. In particular, if E is a complete metrizable space, then b is jointly sequentially continuous.*

The completeness assumption in Corollary 2.71 is important.

Example 2.72. Define $b : c_{00} \times c_{00} \to \mathbb{R}$ by

$$b(s, t) = \sum_{j=1}^{\infty} s_j t_j,$$

where $s = \{s_j\}, t = \{t_j\}$ (finite sum). Then b is separately continuous when c_{00} has the sup-norm. However, b is not jointly continuous. Consider $s^k = \sum_{j=1}^{k} e^j / \sqrt{k}$. Then $s^k \to 0$ but $b(s^k, s^k) = 1$ for all k.

We finally consider a uniform boundedness result for bilinear mappings.

Theorem 2.73. *Let E, F, G be locally convex spaces with E, F sequentially complete and E braked. Let Γ be a family of separately continuous bilinear* *mappings from $E \times F$ into G which is pointwise bounded on $E \times F$. If $A \subset E, B \subset F$ are bounded, then $\Gamma(A, B)$ is bounded, i.e., Γ is uniformly bounded on bounded subsets of $E \times F$.*

Proof. If the conclusion fails to hold, there exist a continuous semi-norm p on G, $x_k \in A, y_k \in B, b_k \in \Gamma$ such that

$$(*) \quad p(b_k(x_k, y_k)) > 2^{2k}.$$

Consider the triple

$$(E \times F, \{b_k\} : G)$$

under the map $((x, y), b_k) \to b_k(x, y)$. The series $\sum_k (x_k, y_k)/2^k$ is absolutely convergent in $E \times F$ and, therefore, subseries convergent by the sequential completeness assumption. By Corollary 2.71, each b_k is jointly sequentially continuous so the series is $w(E \times F, \{b_k\})$ subseries convergent. Also, by the pointwise boundedness assumption the sequence $b_k/2^k \to 0$ in $w(\{b_k\}, E \times F)$ so the sequence is sequentially relatively $w(\{b_k\}, E \times F)$ compact. By Theorem 2.2 the series

$$\sum_{j=1}^{\infty} b_k((x_j, y_j)/2^j)/2^k$$

converge uniformly for $k \in \mathbb{N}$. In particular, $b_k(x_k, y_k)/2^{2k} \to 0$ contradicting $(*)$. $\qquad \square$

We say that Γ is left sequentially hypocontinuous if the family

$$\{b(\cdot, y) : b \in \Gamma, y \in B\}$$

is sequentially equicontinuous when $B \subset F$ is bounded. The result above implies that Γ is left sequentially hypocontinuous with respect to the bounded subsets of F.

Corollary 2.74. *Let the notation be as in the corollary above. Then Γ is left sequentially hypocontinuous.*

For suppose $x_k \to 0$ in E and $B \subset F$ is bounded in F and $t_k \to \infty$ with $t_k x_k \to 0$. Let p be a continuous semi-norm on G and set

$$M = \sup\{p(b(t_k x_k, y)) : k \in \mathbb{N}, y \in B, b \in \Gamma\};$$

$M < \infty$ by the result above. Then

$$p(b(x_k, y)) = \frac{1}{t_k} p(b(t_k x_k, y)) \leq M/t_k \to 0$$

uniformly for $b \in \Gamma, y \in B$. In particular, this implies that Γ is sequentially equicontinuous.

A similar uniform boundedness result for bilinear maps is given in 6.3.1 of [Sw1].

We can use the result above to obtain an equicontinuity result analogous to Corollary 2.71.

Theorem 2.75. *Let E, F, G be locally convex spaces with E, F sequentially complete and E braked. Let $\Gamma = \{b_i\}$ be a sequence of separately continuous bilinear mappings from $E \times F$ into G which is pointwise bounded on $E \times F$. Then Γ is sequentially equicontinuous.*

Proof. Let $x_j \to 0$ in E and $y_j \to 0$ in F. If $\lim_k b(x_j, y_j) = 0$ uniformly for $b \in \Gamma$ fails to hold, then note that there exists a neighborhood of 0, U, in G such that for every i there exist $b_{m_i} \in \Gamma, n_i > i$ with

$$b_{m_i}(x_{n_i}, y_{n_i}) \notin U.$$

Applying this condition to $i_1 = 1$, there exist $b_{m_1} \in \Gamma$, $n_1 > 1$ with

$$b_{m_1}(x_{n_1}, y_{n_1}) \notin U.$$

By Corollary 2.71, there exists $i_2 > n_1$ such that $j \geq i_2$ implies

$$b_i(x_j, y_j) \in U$$

for $1 \leq i \leq m_1$. By the observation above there exist $b_{m_2} \in \Gamma, n_2 > n_1$ such that

$$b_{m_2}(x_{n_2}, y_{n_2}) \notin U.$$

Note $m_2 > m_1, n_2 > n_1$. Continuing this construction produces increasing sequences $\{m_k\}, \{n_k\}$ with

$$(\#) \quad b_{m_k}(x_{n_k}, y_{n_k}) \notin U.$$

Let $t_k \to \infty$ such that $t_k x_{n_k} \to 0$. By the theorem above, $\{b_{m_k}(t_k x_{n_k}, y_{n_k})\}$ is bounded. Therefore,

$$\frac{1}{t_k} b_{m_k}(t_k x_{n_k}, y_{n_k}) = b_{m_k}(x_{n_k}, y_{n_k}) \to 0$$

contradicting $(\#)$. $\qquad\square$

As in the case of a single bilinear operator the sequential completeness assumptions in the results above are important.

Example 2.76. Define $b_i : l^\infty \times c_{00} \to \mathbb{R}$ by

$$b_i(s,t) = \sum_{j=1}^{i} s_j t_j$$

and let l^∞, c_{00} have the sup-norm. Then each b_i is separately continuous and the sequence $\{b_i\}$ is pointwise bounded on $l^\infty \times c_{00}$. Let e be the sequence with 1 in each coordinate and $s^i = \sum_{j=1}^{i} e^j$. Then $b_i(e, s^i) = i$ so $\{b_i\}$ is not bounded on the product $\{(e, s^i)\}$. Also, $b_i(e/\sqrt{i}, s^i/\sqrt{i}) = 1$ so $\{b_i\}$ is not sequentially equicontinuous.

We can also obtain a Banach–Steinhaus result for bilinear operators from the results above. The normed space version of the Banach–Steinhaus Theorem asserts that if X is a Banach space, Y is a normed space and $\{T_i\}$ is sequence in $L(X, Y)$ such that

$$Tx = \lim_i T_i x$$

exists for every $x \in X$, then $T \in L(X, Y)$ and $\{T_i\}$ is equicontinuous. Analogously, for bilinear operators we have

Corollary 2.77. *Let E, F, G be locally convex spaces with E, F sequentially complete and E braked. Let $\{b_i\}$ be a sequence of separately continuous bilinear mappings from $E \times F$ into G such that*

$$\lim_i b_i(x, y) = b(x, y)$$

exists for every $(x, y) \in E \times F$. Then $\{b_i\}$ is sequentially equicontinuous and b is bilinear and jointly sequentially continuous.

Proof. The equicontinuity follows from the result above. b is obviously bilinear and if $x_j \to 0, y_j \to 0$, then

$$\lim_j b(x_j, y_j) = \lim_j \lim_i b_i(x_j, y_j) = \lim_i \lim_j b_i(x_j, y_j) = 0$$

by the uniform convergence and the Iterated Limit Theorem. $\square$

Again the completeness assumptions are important.

Example 2.78. Let $\{b_i\}$ be as in the example above. Then

$$\lim_i b_i(s,t) = b(s,t) = \sum_{j=1}^{\infty} s_j t_j$$

and b is not even separately continuous. For if

$$t^i = \sum_{j=1}^{i} e^j / i,$$

then $b(e, t^i) = 1$ while $t^i \to 0$.

We will discuss bilinear operators more extensively later in Chapter 7 and also consider versions of the Uniform Boundedness Principle and the Banach–Steinhaus Theorem in Chapters 5 and 6.

Chapter 3

Bounded Multiplier Convergent Series

A series $\sum_j x_j$ in a TVS is subseries convergent iff the series

$$\sum_{j=1}^{\infty} t_j x_j$$

converges for every $t = \{t_j\} \in m_0 = \text{span}\{\chi_\sigma : \sigma \subset \mathbb{N}\}$ since the series is subseries convergent iff the series $\sum_{j \in \sigma} x_j$ converges for every $\sigma \subset \mathbb{N}$. This suggests that one could consider the convergence of series when the $t = \{t_j\} \in m_0$ are replaced by elements of other sequence spaces. If λ is a scalar sequence space containing the subspace c_{00} of sequences which are eventually 0 and $\sum_j x_j$ is a formal series in a TVS, the series is said to be λ *multiplier convergent* if the series $\sum_{j=1}^{\infty} t_j x_j$ converge for every $t = \{t_j\} \in \lambda$; the elements of λ are referred to as *multipliers*. If the space of multipliers is l^∞, an l^∞ multiplier convergent series is often referred to as a *bounded multiplier convergent* series. If the partial sums of the series $\sum_{j=1}^{\infty} t_j x_j$ are Cauchy for $t \in \lambda$, we say the series is λ *multiplier Cauchy*; similarly, for bounded multiplier convergent series. A bounded multiplier convergent series is obviously subseries convergent since $m_0 \subset l^\infty$; the example below gives an example of a subseries convergent series which is not bounded multiplier convergent. We will study multiplier convergent series in the next chapter. Since many results for subseries convergent series carry over to bounded multiplier convergent series by using an interesting lemma of Rutherford and McArthur, we will consider bounded multiplier convergent series in this brief chapter. This will also serve as motivation for other results about multiplier convergent series and for the general case of multiplier convergent series considered later in Chapter 4.

Lemma 3.1. ([MR]) *Let p be a semi-norm on the vector space V, let σ be* *finite, $x_j \in V$, $t_j \in \mathbb{R}$ for $j \in \sigma$. Then*

$$p\left(\sum_{j\in\sigma} t_j x_j\right) \leq 2\sup_{j\in\sigma}|t_j|\,\sup_{\sigma'\subset\sigma} p\left(\sum_{j\in\sigma'} x_j\right).$$

Proof. First assume all $t_j \geq 0$ with $t_1 \geq t_2 \geq ... \geq t_n \geq 0$. Then

$p(\sum_{j\in\sigma} t_j x_j) = p(\sum_{j=1}^{n-1}(t_j - t_{j+1})(x_1 + ... + x_j) + t_n(x_1 + ... + x_n))$
$\leq \sum_{j=1}^{n-1}(t_j - t_{j+1})p(x_1 + ... + x_j) + t_n p(x_1 + ... + x_n)$
$\leq (\sum_{j=1}^{n-1}(t_j - t_{j+1}) + t_n)\sup_{\sigma'\subset\sigma} p(\sum_{j\in\sigma'} x_j)$
$= \sup_{j\in\sigma}|t_j|\sup_{\sigma'\subset\sigma} p(\sum_{j\in\sigma'} x_j).$

For the general case, apply the inequality above to the positive and negative scalars. $\qquad\Box$

We may apply this inequality to results for subseries convergent series to obtain results for bounded multiplier convergent series.

Theorem 3.2. *Let $\sum_j x_j$ be subseries convergent in the LCTVS E. Then the series $\sum_j x_j$ is bounded multiplier Cauchy. If E is sequentially complete, the series $\sum_j x_j$ is bounded multiplier convergent. Moreover, the series*

$$\sum_{j=1}^{\infty} t_j x_j$$

converge uniformly for $\|\{t_j\}\|_\infty \leq 1$.

Proof. Let p be a continuous semi-norm on E. Let $\epsilon > 0$. By Theorem 2.9 there exists N such that $p(\sum_{j\in\sigma} x_j) \leq \epsilon$ when $\min\sigma \geq N$. Applying the lemma to this inequality gives the desired conclusions. $\qquad\Box$

Without the completeness assumptions the conclusions of the theorem may fail to hold.

Example 3.3. Let $E = m_0$ and let $s = \{s_j\} \in l^1$ with $s_j > 0$ for all j. Define a norm on m_0 by

$$\|t\|_s = \sum_{j=1}^{\infty} s_j |t_j|.$$

Then the series $\sum_j e^j$ is subseries convergent in $(m_0, \|\cdot\|_s)$, is bounded multiplier Cauchy but is not bounded multiplier convergent since, for example, the series $\sum_{j=1}^{\infty} e^j/j$ does not converge to an element of m_0.

The theorem also has an interesting consequence for Banach spaces with an *unconditional Schauder basis*. A Schauder basis $\{b_j\}$ for a Banach space X is said to be unconditional if the series

$$\sum_{j=1}^{\infty} f_j(x)b_j$$

is unconditionally or subseries convergent for every $x \in X$, where $\{f_j\}$ are the coordinate functionals with respect to $\{b_j\}$.

Corollary 3.4. *Let $\{b_j\}$ be an unconditional Schauder basis for the Banach space X. If $x \in X$, the series $\sum_{j=1}^{\infty} t_j x_j$ converge uniformly for $|t_j| \leq |f_j(x)|$.*

Proof. Since X is complete, the series $\sum_{j=1}^{\infty} f_j(x)b_j$ is bounded multiplier convergent by the theorem so the result follows from the lemma. $\square$

We can also use the lemma to obtain a sharper conclusion of the Hahn–Schur Theorem for sequentially complete LCTVS given in Theorem 2.55.

Theorem 3.5. *Let E be a sequentially complete LCTVS. Let $\sum_j x_{ij}$ be subseries convergent for every i and assume that*

$$\lim_i \sum_{j \in \sigma} x_{ij}$$

exists for every $\sigma \subset \mathbb{N}$. Set $x_j = \lim_i x_{ij}$. Then each $\sum_j x_{ij}$ is bounded multiplier convergent, $\sum_j x_j$ is bounded multiplier convergent and

$$\lim_i \sum_{j=1}^{\infty} t_j x_{ij} = \sum_{j=1}^{\infty} t_j x_j$$

uniformly for $\|\{t_j\}\|_\infty \leq 1$.

Proof. It follows from the theorem above that each series $\sum_j x_{ij}$ is bounded multiplier convergent. The series $\sum_j x_j$ is subseries convergent by the Hahn–Schur Theorem 2.55 and then multiplier convergent by the theorem above. Let p be a continuous semi-norm on E and $\epsilon > 0$. By the Hahn–Schur Theorem, there exists N such that

$$p\left(\sum_{j \in \sigma} x_{ij} - \sum_{j \in \sigma} x_j\right) \leq \epsilon \text{ for } i \geq N, \sigma \subset \mathbb{N}.$$

The last conclusion in the statement of the theorem follows from the lemma. $\square$

We can establish a similar result for bounded multiplier convergent series in a TVS without the sequential completeness assumption. For this we need an observation.

Lemma 3.6. *Let E be a TVS. If $\lim_j x_j = 0$ in E, then $\lim_j tx_j = 0$ uniformly for $|t| \leq 1$.*

Proof. Let U be a balanced neighborhood of 0 in E. There exists N such that $j \geq N$ implies $x_j \in U$. Therefore, if $j \geq N$, $tx_j \in U$ for $|t| \leq 1$ since U is balanced. $\qquad\square$

We first prove a special case of the Hahn–Schur Theorem for bounded multiplier convergent series.

Lemma 3.7. *Let E be a Hausdorff TVS and $x_{ij} \in E$ for $i, j \in \mathbb{N}$ be such that $\sum_j x_{ij}$ is bounded multiplier convergent for every i. If*

$$\lim_i \sum_{j=1}^{\infty} t_j x_{ij} = 0$$

for every $t = \{t_j\} \in l^{\infty}$, then $\lim_i \sum_{j=1}^{\infty} t_j x_{ij} = 0$ uniformly for $\|\{t_j\}\|_{\infty} \leq 1$.

Proof. It suffices to show

$$\lim_i \sum_{j=1}^{\infty} t_j^i x_{ij} = 0$$

for every $t^i = \{t_j^i\}_j \in l^{\infty}$ with $\left\|\{t_j^i\}_j\right\|_{\infty} \leq 1$. Let U be a neighborhood of 0 and pick a symmetric neighborhood of 0, V, such that $V + V + V \subset U$. Set $n_1 = 1$. Pick N_1 such that

$$\sum_{j=N_1}^{\infty} t_j^{n_1} x_{n_1 j} \in V.$$

By hypothesis, $\lim_i x_{ij} = 0$ for every j so by the lemma $\lim_i t_j^i x_{ij} = 0$ for every j. Hence, there exists $n_2 > n_1$ with

$$\sum_{j=1}^{N_1-1} t_j^i x_{ij} \in V$$

for $i \geq n_2$. Pick $N_2 > N_1$ such that

$$\sum_{j=N_2}^{\infty} t_j^{n_2} x_{n_2 j} \in V.$$

Continuing this construction produces two increasing sequences $\{n_k\}, \{N_k\}$ with

$$\sum_{k=N_j}^{\infty} t_k^{n_j} x_{n_j k} \in V, \quad \sum_{k=1}^{N_j-1} t_k^i x_{ik} \in V$$

for $i \geq n_{j+1}$. Set

$$I_j = \{k : N_{j-1} \leq k < N_j\}$$

and consider the matrix

$$M = \left[\sum_{k \in I_j} t_k^{n_j} x_{n_i k} \right] = [m_{ij}].$$

The columns of M converge to 0 by hypothesis. If $\{p_j\}$ is an increasing sequence of integers, define $t = \{t_k\} \in l^\infty$ by $t_k = t_k^{n_{p_j}}$ if $k \in I_{p_j}$ and $t_k = 0$ otherwise. Then

$$\sum_{j=1}^{\infty} m_{i p_j} = \sum_{k=1}^{\infty} t_k x_{n_i k} \to 0$$

by hypothesis. Hence, M is a $\mathcal{K}$ matrix and the diagonal of M converges to 0 by the Antosik–Mikusinski Theorem. Hence,

$$m_{ii} = \sum_{k \in I_i} t_k^{n_i} x_{n_i k} \in V$$

for large i. Then

$$\sum_{k=1}^{\infty} t_k^{n_i} x_{n_i k} = \sum_{k=1}^{N_{i-1}-1} t_k^{n_i} x_{n_i k} + \sum_{k \in I_i} t_k^{n_i} x_{n_i k} + \sum_{k=N_i}^{\infty} t_k^{n_i} x_{n_i k}$$

$$\in V + V + V \subset U$$

for large i. Since the same argument can be applied to any subsequence the result follows. $\qquad \square$

We can now establish the general case.

Theorem 3.8. (Hahn–Schur) *Let E be a Hausdorff TVS and $x_{ij} \in E$ for $i, j \in \mathbb{N}$ be such that $\sum_j x_{ij}$ is bounded multiplier convergent for every i. If*

$$\lim_i \sum_{j=1}^{\infty} t_j x_{ij}$$

exists for every $t = \{t_j\} \in l^\infty$ and if $\lim_i x_{ij} = x_j$ for every j, then

(i) $\sum_j x_j$ *is bounded multiplier convergent,*

(ii) $\lim_i \sum_{j=1}^{\infty} t_j x_{ij} = \sum_{j=1}^{\infty} t_j x_j$ *uniformly for $\|\{t_j\}\|_\infty \leq 1$.*

Proof. First note that the sequence $\{\sum_{j=1}^{\infty} t_j x_{ij}\}$ satisfies a Cauchy condition uniformly for $\|\{t_j\}\|_{\infty} \leq 1$. For, if $\{n_k\}, \{m_k\}$ are two arbitrary increasing sequences with $m_k < n_k < m_{k+1}$, the sequence

$$\left\{\sum_{j=1}^{\infty} t_j(x_{m_ij} - x_{n_ij})\right\}$$

converges uniformly to 0 for $\|\{t_j\}\|_{\infty} \leq 1$ by the lemma.

For (i) let $t = \{t_j\} \in l^{\infty}$ with $\|t\|_{\infty} \leq 1$ and set

$$z = \lim_i \sum_{j=1}^{\infty} t_j x_{ij}.$$

We claim

$$\sum_{j=1}^{\infty} t_j x_j = z.$$

Let U be a neighborhood of 0 and pick a closed, symmetric neighborhood of 0, V, such that $V + V + V \subset U$. By the uniform Cauchy condition, there exists N such that $i, k \geq N$ implies

$$z - \sum_{j=1}^{\infty} t_j x_{ij} \in V \text{ and } \sum_{j=1}^{\infty} s_j(x_{ij} - x_{kj}) \in V$$

for $\|\{s_j\}\|_{\infty} \leq 1$. Thus, for every m,

$$\sum_{j=1}^{m} t_j(x_{ij} - x_j) \in V$$

if $i \geq N$. There exists M such that $\sum_{j=m}^{\infty} t_j x_{Nj} \in V$ when $m \geq M$. If $m \geq M$, we have

$$z - \sum_{j=1}^{m} t_j x_j = z - \sum_{j=1}^{\infty} t_j x_{Nj} + \sum_{j=1}^{m} t_j(x_{Nj} - x_j) + \sum_{j=m+1}^{\infty} t_j x_{Nj}$$

$$\in V + V + V \subset U$$

as required.

By what was established above,

$$\lim_i \left(\sum_{j=1}^{\infty} t_j x_{ij} - \sum_{j=1}^{\infty} t_j x_j\right) = 0$$

for every $t \in l^{\infty}$ so conclusion (ii) follows from the lemma. $\qquad\square$

For a single bounded multiplier convergent series, we have

Corollary 3.9. *Let $\sum_j x_j$ be bounded multiplier convergent in the TVS E. Then the series*

$$\sum_{j=1}^{\infty} t_j x_j$$

converge uniformly for $\|\{t_j\}\|_{\infty} \leq 1$.

Proof. Set $x_{ij} = x_j$ for $j \geq i$ and $x_{ij} = 0$ for $j < i$ and apply the theorem. $\square$

A similar conclusion holds for the series in the theorem.

Corollary 3.10. *Let $\{x_{ij}\}$ satisfy the hypothesis of the Hahn–Schur Theorem above. Then the series $\sum_{j=1}^{\infty} t_j x_{ij}$ converge uniformly for $i \in \mathbb{N}$ and $\|\{t_j\}\|_{\infty} \leq 1$.*

Proof. Let U be a neighborhood of 0 and pick a neighborhood of 0, V, such that $V + V \subset U$. By the theorem, there exists N such that $i \geq N$ implies

$$\sum_{j=1}^{\infty} t_j x_{ij} - \sum_{j=1}^{\infty} t_j x_j \in V$$

for $\|\{t_j\}\|_{\infty} \leq 1$. By the corollary, there exists M such that $m \geq M$ implies

$$\sum_{j=m}^{\infty} t_j x_j \in V \text{ and } \sum_{j=m}^{\infty} t_j x_{ij} \in V$$

for $1 \leq i \leq N, \|\{t_j\}\|_{\infty} \leq 1$. Hence, if $m \geq M$ and $i \geq n$, we have

$$\sum_{j=m}^{\infty} t_j x_{ij} = \sum_{j=m}^{\infty} t_j (x_{ij} - x_j) + \sum_{j=m}^{\infty} t_j x_j \in V + V \subset U$$

for $\|\{t_j\}\|_{\infty} \leq 1$. This establishes the result. $\square$

We can also give an interpretation of this version of the Hahn–Schur Theorem as was done following Corollary 2.56. Let G be a LCTVS whose topology is generated by the family of semi-norms $\mathcal{P}$. Let

$$bmc(G) = (l^{\infty})^{\beta G}$$

be the space of all G valued, bounded multiplier convergent series. For $\{x_j\} \in bmc(G)$, set

$$\widehat{p}(\{x_j\}) = \sup\left\{ p\left(\sum_{j=1}^{\infty} t_j x_j \right) : \|\{t_j\}\|_{\infty} \le 1 \right\}$$

and give $bmc(G)$ the topology, $\tau_{bmc}(G)$, generated by the semi-norms $\{\widehat{p} : p \in \mathcal{P}\}$ ($\widehat{p}(\{x_j\}) < \infty$ by the corollary below). Consider the triple

$$(bmc(G), l^{\infty} : G)$$

under the bilinear map $(\{x_j\}, \{t_j\}) \to \sum_{j=1}^{\infty} t_j x_j$. It follows from the Hahn–Schur Theorem that if $x^k \to 0$ in $w(bmc(G), l^{\infty})$, then $x^k \to 0$ in $\tau_{bmc}(G)$. Moreover, if G is sequentially complete, then $w(bmc(G), l^{\infty})$ is sequentially complete.

We prove an analogue of Theorem 2.17. Let $B(l^{\infty})$ be the unit ball of l^{∞}.

Proposition 3.11. *Let $\sum_j x_j$ be bounded multiplier convergent and $S : B(l^{\infty}) \to E$ be the summing operator defined by*

$$S(\{t_j\}) = \sum_{j=1}^{\infty} t_j x_j.$$

Let p be the topology of coordinatewise convergence on $B(l^{\infty})$. Then S is continuous with respect to p and the topology of E.

Proof. Let $\{t^{\alpha}\}$ be a net in $B(l^{\infty})$ which converges to 0 in p. Let U be a neighborhood of 0 in E and pick a neighborhood of 0, V, such that $V + V \subset U$. By Corollary 9 there is n such that $\sum_{j=n}^{\infty} t_j^{\alpha} x_j \in V$ for all α. There exist α_0 such that $\sum_{j=1}^{n-1} t_j^{\alpha} x_j \in V$ when $\alpha \ge \alpha_0$. If $\alpha \ge \alpha_0$, then

$$\sum_{j=1}^{\infty} t_j^{\alpha} x_j = \sum_{j=1}^{n-1} t_j^{\alpha} x_j + \sum_{j=n}^{\infty} t_j^{\alpha} x_j \in V + V \subset U.$$

$\square$

Corollary 3.12. $B = \{\sum_{j=1}^{\infty} t_j x_j : \|\{t_j\}\|_{\infty} \le 1\}$ *is compact.*

Proof. $B(l^{\infty})$ is compact under p and $S(B(l^{\infty})) = B$ so the result follows from the proposition. $\square$

Using the Hahn–Schur Theorem 8 we can establish an analogue of Theorem 2.58.

Theorem 3.13. *Let E be a Hausdorff TVS and $x_{ij} \in E$ for $i, j \in \mathbb{N}$ be such that $\sum_j x_{ij}$ is bounded multiplier convergent for every i. Suppose*

$$\lim_i \sum_{j=1}^{\infty} t_j x_{ij}$$

exists for every $t = \{t_j\} \in l^{\infty}$ and $\lim_i x_{ij} = x_j$ for every j. Then

$$B = \left\{ \sum_{j=1}^{\infty} t_j x_{ij} : \|\{t_j\}\|_{\infty} \leq 1, i \in \mathbb{N} \right\} \cup \left\{ \sum_{j=1}^{\infty} t_j x_j : \|\{t_j\}\|_{\infty} \leq 1 \right\}$$

is compact.

Proof. Let $S_i : B(l^{\infty}) \to E$ ($S : B(l^{\infty}) \to E$) be the summing operator

$$S_i(\{t_j\}) = \sum_{j=1}^{\infty} t_j x_{ij} \quad \left(S(\{t_j\}) = \sum_{j=1}^{\infty} t_j x_j \right).$$

Each S_i, S is continuous with respect to p and $S_i \to S$ uniformly on $B(l^{\infty})$ by Theorem 8. Since $B(l^{\infty})$ is compact under p, the result follows from Lemma 2.57. $\square$

We consider the converse of Theorem 8 for sequentially complete spaces.

Theorem 3.14. *Let E be a sequentially complete, Hausdorff TVS, $x_{ij} \in E$ for $i, j \in \mathbb{N}$. Assume the series $\sum_j x_{ij}$ are bounded multiplier convergent for each i. If $\lim_i x_{ij} = x_j$ exists for each j and for each $t = \{t_j\} \in l^{\infty}$ the series*

$$\sum_{j=1}^{\infty} t_j x_{ij}$$

converge uniformly for $i \in \mathbb{N}$, then

$$\lim_i \sum_{j=1}^{\infty} t_j x_{ij}$$

exists for each $t = \{t_j\} \in l^{\infty}$ [and the stronger conclusion of the Hahn–Schur Theorem holds].

Proof. Let $t = \{t_j\} \in l^\infty$ and U be a neighborhood of 0. Pick a balanced neighborhood of 0, V, with $V + V + V \subset U$. There exists n such that $\sum_{j=n}^{\infty} t_j x_{ij} \in V$ for all i. There exists m such that $\sum_{j=1}^{n-1} t_j(x_{ij} - x_{kj}) \in V$ for $i, k \geq m$. If $i, k \geq m$, we have

$$\sum_{j=1}^{\infty} t_j x_{ij} - \sum_{j=1}^{\infty} t_j x_{kj}$$

$$= \sum_{j=1}^{n-1} t_j(x_{ij} - x_{kj}) + \sum_{j=n}^{\infty} t_j x_{ij} - \sum_{j=n}^{\infty} t_j x_{kj}$$

$$\in V + V + V \subset U.$$

Hence $\{\sum_{j=1}^{\infty} t_j x_{ij}\}_i$ is Cauchy and the result follows from completeness. $\square$

Combining the results above, we have the following

Theorem 3.15. *Let E be a sequentially complete, Hausdorff TVS, $x_{ij} \in E$ for $i, j \in \mathbb{N}$. Assume the series $\sum_j x_{ij}$ are bounded multiplier convergent for each i and $\lim_i x_{ij} = x_j$ exists for each j. The following are equivalent:*

(i) $\lim\limits_i \sum\limits_{j=1}^{\infty} t_j x_{ij}$ *exists for every* $t = \{t_j\} \in l^\infty$;

(ii) $\sum\limits_j x_j$ *is bounded multiplier convergent and*

$$\lim_i \sum_{j=1}^{\infty} t_j x_{ij} = \sum_{j=1}^{\infty} t_j x_j \text{ uniformly for } \|\{t_j\}\|_\infty \leq 1;$$

(iii) $\sum\limits_{j=1}^{\infty} t_j x_{ij}$ *converge uniformly for* $\|\{t_j\}\|_\infty \leq 1, i \in \mathbb{N}$;

(iv) *For each* $t \in l^\infty$, $\sum\limits_{j=1}^{\infty} t_j x_{ij}$ *converge uniformly for* $i \in \mathbb{N}$.

In the case where E is a sequentially complete Hausdorff LCTVS, if $\sum_j x_{ij}$ are subseries convergent (=bounded multiplier convergent) and $\lim_i \sum_{j \in \sigma} x_{ij}$ exists for each $\sigma \subset \mathbb{N}$, then $\lim_i \sum_{j=1}^{\infty} t_j x_{ij}$ exists for each $t = \{t_j\} \in l^\infty$. This may not happen if the space is not sequentially complete even when each series $\sum_j x_{ij}$ is bounded multiplier convergent.

Example 3.16. Let $\sum_j x_j$ be subseries convergent but not bounded multiplier convergent (see Example 3 above). Set $x_{ij} = x_j$ if $1 \leq j \leq i$ and

$x_{ij} = 0$ if $j > i$. Then $\sum_j x_{ij}$ is bounded multiplier convergent for each i and since $\sum_j x_j$ is subseries convergent $\lim_i \sum_{j \in \sigma} x_{ij}$ exists for each $\sigma \subset \mathbb{N}$. However, $\lim_i \sum_{j=1}^{\infty} t_j x_{ij}$ cannot exist for every $t = \{t_j\} \in l^{\infty}$ since $\sum_j x_j$ is not bounded multiplier convergent.

Li's Lemma

We conclude this chapter with a result of Ronglu Li. The result does not fit into the framework of abstract triples but is very much in the same spirit as it is a simple result which has interesting applications.

Lemma 3.17. ([LS]) *Let $\{E_j\}$ be a sequence of sets. Let G be an Abelian (Hausdorff) topological group and $f_j : E_j \to G$. If the series*

$$\sum_{j=1}^{\infty} f_j(t_j)$$

converges for every sequence $\{t_j\}$ with $t_j \in E_j$, then the series $\sum_{j=1}^{\infty} f_j(t_j)$ converge uniformly for all sequences $\{t_j\}$ with $t_j \in E_j$.

Proof. If the conclusion fails to hold, there exists a neighborhood, U, of 0 in G, sequences $\{t_j^i\}_j, t_j^i \in E_j$, and an increasing sequence $\{n_i\}$ such that

$$\sum_{j=n_i}^{\infty} f_j(t_j^i) \notin U.$$

Pick a symmetric neighborhood of 0, V, such that $V + V \subset U$. Since $\lim_k \sum_{j=k}^{\infty} f_j(t_j^1) = 0$ and $\sum_{j=n_1}^{\infty} f_j(t_j^1) \notin U$, there exists $m_1 > n_1$ such that

$$\sum_{j=n_1}^{m_1} f(t_j^1) \notin V.$$

Put $N_1 = 1$ and pick $n_{i_2} = N_2 > m_1$ such that

$$\sum_{j=N_2}^{\infty} f_j(t_j^{i_2}) \notin U.$$

As before pick $m_2 > N_2$ such that

$$\sum_{j=N_2}^{m_2} f_j(t_j^{i_2}) \notin V.$$

Continuing this construction produces increasing sequences $\{N_k\}, \{m_k\}$ and $\{i_k\}$ such that $N_k < m_k < N_{k+1}$ and

$$\sum_{j=N_k}^{m_k} f_j(t_j^{i_k}) \notin V.$$

Pick an arbitrary sequence $\{u_j\}$ with $u_j \in E_j$ for every j. Define a sequence $\{s_j\}$ with $s_j \in E_j$ by $s_j = t_j^{i_k}$ if $N_k \le j \le m_k$ and $s_j = u_j$ otherwise. If the series $\sum_{j=1}^{\infty} f_j(s_j)$ converges, there exists N such that $\sum_{j=m}^{n} f_j(s_j) \in V$ for $n > m \ge N$. But,

$$\sum_{j=N_k}^{m_k} f_j(s_j) = \sum_{j=N_k}^{m_k} f_j(t_j^{i_k}) \notin V$$

for large k so the series $\sum_{j=1}^{\infty} f_j(s_j)$ doesn't satisfy the Cauchy condition and, therefore, doesn't converge. This contradicts the hypothesis. $\quad\square$

To illustrate the utility of Li's Lemma, we derive a couple of previous results for series which were established earlier.

Corollary 3.18. *Let X be a TVS and $\sum_j x_j$ a series in X which is subseries convergent. Then the series $\sum_{j \in \sigma} x_j$ converge uniformly for $\sigma \subset \mathbb{N}$ (i.e., the series is unordered convergent).*

Proof. Let $E_j = \{0, 1\}$ for every j and define $f_j : E_j \to X$ by $f_j(0) = 0$ and $f_j(1) = x_j$. Then the conclusion follows directly from the lemma. $\quad\square$

Next, we consider an improvement of Corollary 9. The result concerns multiplier convergent series which will be considered in detail in the next chapter.

Corollary 3.19. *Let X be a TVS and let λ be a normal (solid) sequence space and let $\sum_j x_j$ be λ multiplier convergent in the sense that the series*

$$\sum_{j=1}^{\infty} s_j x_j$$

converges for every $\{s_j\} \in \lambda$. If $t \in \lambda$, then the series $\sum_{j=1}^{\infty} s_j x_j$ converge uniformly for $|s_j| \le |t_j|$.

Proof. Let $E_j = \{t \in \mathbb{R} : |t| \le |t_j|\}$ and define $f_j : E_j \to X$ by $f_j(t) = tx_j$. Then the conclusion follows directly from the lemma. $\quad\square$

Note that the space λ in the corollary is not assume to be a K-space. From the corollary we can obtain immediately Corollary 9.

Corollary 3.20. *Let X be a TVS and let $\sum_j x_j$ be bounded multiplier convergent. Then the series $\sum_{j=1}^{\infty} t_j x_j$ converge uniformly for $\|\{t_j\}\|_{\infty} \leq 1$.*

Proof. Let e be the constant sequence with 1 in each coordinate. Then the result follows immediately from the corollary. $\qquad\qquad\square$

Li's Lemma can also be used to treat operator valued series with bounded vector valued multipliers. Let X, Y be normed spaces and $\sum_j T_j$ a (formal) series in $L(X, Y)$. The series $\sum_j T_j$ is *bounded multiplier convergent* if the series

$$\sum_{j=1}^{\infty} T_j x_j$$

converges for every bounded sequence $\{x_j\} \subset X$ (this is not in agreement with our previous use of the term, bounded multiplier convergence, but will only be used in this one case).

We have the analogue of Corollary 9 for these series.

Theorem 3.21. *If the series $\sum_j T_j$ is bounded multiplier convergent, then the series*

$$\sum_{j=1}^{\infty} T_j x_j$$

converge uniformly for $\|\{x_j\}\|_{\infty} \leq 1$.

Proof. Set $\Omega = \{\{x_j\} : \|\{x_j\}\|_{\infty} \leq 1\}$ and define $f_j : \Omega \to Y$ by $f_j(x) = T_j x$. The conclusion follows directly from Li's Lemma. $\qquad\square$

We also have a version of Corollary 12.

Corollary 3.22. *Assume the series $\sum_j T_j$ is bounded multiplier convergent. Then*

$$B = \left\{ \sum_{j=1}^{\infty} T_j x_j : \|\{x_j\}\|_{\infty} \leq 1 \right\}$$

is bounded.

Proof. Let $t_k \to 0$ with $|t_k| \leq 1$ and $\left\|\{x_j^k\}\right\|_\infty \leq 1$. Let $\epsilon > 0$. There exists n such that

$$\left\|\sum_{j=n}^\infty T_j x_j^k\right\| \leq \epsilon$$

for all k by the theorem. There exists K such that

$$\left\|t_k \sum_{j=1}^{n-1} T_j x_j^k\right\| < \epsilon$$

for $k \geq K$. Then if $k \geq K$,

$$\left\|\sum_{j=1}^\infty t_k T_j x_j\right\| \leq \left\|t_k \sum_{j=1}^{n-1} T_j x_j^k\right\| + \left\|t_k \sum_{j=n}^\infty T_j x_j\right\| \leq 2\epsilon.$$

Hence B is bounded. $\qquad\square$

Note that in this case there is no hope of obtaining the analogue of Corollary 12; just take a single non-compact operator.

Chapter 4

Multiplier Convergent Series

As noted earlier in Chapter 3 a series $\sum_j x_j$ in a TVS E is subseries convergent iff the series

$$\sum_{j=1}^{\infty} t_j x_j$$

converges in E for every $t = \{t_j\} \in m_0$, the space of sequences with finite range. This suggests one might consider replacing the sequence space m_0 by other sequence spaces. This was what was done in Chapter 3 where we considered bounded multiplier convergent series with the space m_0 replaced by the sequence space l^{∞}. In this chapter we will consider replacing m_0 with other sequence spaces. Let λ be a scalar valued sequence space which contains the space c_{00}, the space of sequences which are eventually 0, and let $\Lambda \subset \lambda$. Let E be a TVS and $\sum_j x_j$ a (formal) series in E. The series $\sum_j x_j$ is Λ *multiplier convergent* if the series $\sum_{j=1}^{\infty} t_j x_j$ converges in E for every $t = \{t_j\} \in \Lambda$; the elements of the sequence space Λ are called *multipliers*. We will begin by considering Orlicz–Pettis type theorems for multiplier convergent series. First, we observe that Orlicz–Pettis type theorems for multiplier convergent series must require some sort of assumptions on the space of multipliers as the following example shows.

Example 4.1. Let c_c be the space of scalar sequences which are eventually constant. Then if $\sum_j x_j$ is a (formal) series in any TVS E, the series $\sum_{j=1}^{\infty} t_j x_j$ is c_c multiplier convergent iff the series $\sum_{j=1}^{\infty} x_j$ converges in E. The series

$$\sum_{j=1}^{\infty} (e^{j+1} - e^j)$$

is $\sigma(c_0, l^1)$ convergent in c_0 (to $-e^1$) but is not convergent in the norm or Mackey topology. Therefore, there is no Orlicz–Pettis Theorem possible for c_c multiplier convergent series with respect to the weak and norm topologies.

We now consider Orlicz–Pettis type theorems for multiplier convergent series in abstract triples. The first assumption on the multiplier space λ which will be considered is the *signed weak gliding hump property* (signed-WGHP). Let $\Lambda \subset \lambda$. The space Λ has the signed-WGHP if for every $t = \{t_j\} \in \Lambda$ and every increasing sequence of intervals $\{I_j\}$ in $\mathbb{N}$, there exist a subsequence $\{n_j\}$ and a sequence of signs $\{s_j\}$ such that the coordinate sum of the series

$$\sum_{j=1}^{\infty} s_j \chi_{I_{n_j}} t \in \Lambda.$$

If all of the signs $\{s_j\}$ can be chosen to be equal to 1, then Λ has the *weak gliding hump property* (WGHP). The spaces $c_0, l^p, 0 < p \leq \infty, m_0$ have WGHP while the space c does not have WGHP. The space bs of bounded series has signed-WGHP but not WGHP. See Appendix B and Appendix B of [Sw4] for further examples and references.

Let E be a vector space, F be a set and G be a TVS which form an abstract triple

$$(E, F : G)$$

under the map $b : E \times F \to G$, where the maps $b(\cdot, y) : E \to G$ are linear for every $y \in F$. As before, let $w(E, F)$ $[w(F, E)]$ be the weakest topology on E $[F]$ such that all of the maps $x \to b(x, y) = x \cdot y$ from E into G $[y \to x \cdot y$ from F into $G]$ are continuous for all $y \in F$ $[x \in E]$. A subset $K \subset F$ is *sequentially conditionally compact* if every sequence $\{y_k\} \subset K$ has a subsequence $\{y_{n_k}\}$ such that $\lim_k x \cdot y_{n_k}$ exists for every $x \in E$ (see Chapter 2). We have an analogue of Theorem 2.2 for multiplier convergent series.

Theorem 4.2. *Let $\Lambda \subset \lambda$ have signed-WGHP. If the series $\sum_j x_j$ is Λ multiplier convergent in E with respect to $w(E, F)$, then for each $t \in \Lambda$ and each sequentially conditionally $w(F, E)$ compact subset $K \subset F$, the series*

$$\sum_{j=1}^{\infty} t_j x_j \cdot y$$

converge uniformly for $y \in K$.

Proof. If the conclusion fails to hold, there exists a neighborhood of 0, W, in G, $y_k \in K$ and an increasing sequence of intervals $\{I_k\}$ such that

$$(\#) \quad \sum_{l \in I_k} t_l x_l \cdot y_k \notin W$$

for every k. We may assume, by passing to a subsequence if necessary, that $\lim_k x \cdot y_k$ exists for every $x \in E$. Consider the matrix

$$M = [m_{ij}] = \left[\sum_{l \in I_j} t_l x_l \cdot y_i \right].$$

We claim that M is a signed $\mathcal{K}$-matrix (Appendix E). First, the columns of M converge. Next, given an increasing sequence of positive integers there is a subsequence $\{n_j\}$ and a sequence of signs $\{s_j\}$ such that

$$u = \sum_{j=1}^{\infty} s_j \chi_{I_{n_j}} t \in \Lambda.$$

If $\sum_{l=1}^{\infty} u_l x_l$ denotes the $w(E, F)$ sum of the series, then

$$\left\{ \sum_{j=1}^{\infty} s_j m_{in_j} \right\}_i = \left\{ \sum_{j=1}^{\infty} s_j \sum_{l \in I_{n_j}} t_l x_l \cdot y_i \right\}_i = \left\{ \sum_{l=1}^{\infty} u_l x_l \cdot y_i \right\}_i$$

converges. Hence, M is a signed $\mathcal{K}$-matrix so the diagonal of M converges to 0 by the signed version of the Antosik–Mikusinski Matrix Theorem (Appendix E). But, this contradicts $(\#)$. $\qquad \square$

We also have the analogue of Theorem 2.3.

Theorem 4.3. *Let $\Lambda \subset \lambda$ have signed-WGHP and G metrizable under the metric ρ. If the series $\sum_j x_j$ is Λ multiplier convergent in E with respect to $w(E, F)$, then for each $w(F, E)$ compact (countably compact) subset $K \subset F$ and each $t \in \Lambda$, the series*

$$\sum_{j=1}^{\infty} t_j x_j \cdot y$$

are convergent uniformly for $y \in K$.

Proof. We need to show that the series $\sum_j t_j x_j \cdot y$ converge uniformly for $y \in K$ with respect to ρ. Define an equivalence relation $\sim$ on F by $y \sim z$ iff $x_j \cdot y = x_j \cdot z$ for all j. If

$$E_0 = \left\{ \sum_{j=1}^{\infty} s_j x_j : s \in \Lambda \right\},$$

where $\sum_{j=1}^{\infty} s_j x_j$ is the $w(E, F)$ sum of the series, then $x \cdot y = x \cdot z$ for every $x \in E_0$ when $y \sim z$. Let y^- be the equivalence class of $y \in F$ and set

$$F^- = \{f^- : f \in F\}.$$

Define a metric d on F^- by

$$d(y^-, z^-) = \sum_{j=1}^{\infty} \rho(x_j \cdot (y - z))/2^j (1 + \rho(x_j \cdot (y - z)));$$

note that d is a metric. Define a mapping

$$\cdot : E_0 \times F^- \to (G, \rho)$$

by $x \cdot y^- = x \cdot y$ so we may consider the triple

$$(E_0, F^- : (G, \rho))$$

as above. The quotient map $F \to F^-$ is $w(F, E) - w(F^-, E_0)$ continuous and the inclusion $(F^-, w(F^-, E_0)) \subset (F^-, \rho)$ is continuous so K^- is compact (countably compact) with respect to $w(F^-, E_0)$ and ρ and, therefore, $w(F^-, E_0) = d$ on K^- and K^- is $w(F^-, E_0)$ sequentially compact. Since the series $\sum_j x_j$ is Λ multiplier convergent with respect to $w(E, F)$, the series $\sum_j x_j$ is Λ multiplier convergent with respect to $w(E_0, F^-)$ in the abstract triple $(E_0, F^- : (X, \rho))$. Since K^- is sequentially compact in $w(F^-, E_0)$, by the theorem above the series

$$\sum_{j=1}^{\infty} t_j x_j \cdot y^- = \sum_{j=1}^{\infty} t_j x_j \cdot y$$

converge uniformly for $y^- \in K^-$ with respect to ρ. $\qquad \square$

We also have the analogue of Theorem 2.4 for multiplier convergent series.

Theorem 4.4. *Let $\Lambda \subset \lambda$ have signed-WGHP and G be a LCTVS. If the series $\sum_j x_j$ is Λ multiplier convergent in E with respect to $w(E, F)$, then for each $w(F, E)$ compact (countably compact) subset $K \subset F$ and each $t \in \Lambda$, the series*

$$\sum_{j=1}^{\infty} t_j x_j \cdot y$$

are convergent uniformly for $y \in K$.

Proof. Let p be a continuous semi-norm on G. By considering the quotient space G/p, we may assume that p is a norm. Consider the triple

$$(E, F : (G, p))$$

under the map $(x, y) \to x \cdot y$. The set K is $w(E, F)$ compact (countably compact) in the triple $(E, F : (G, p))$ so by the result above the series $\sum_{j=1}^{\infty} t_j x_j \cdot y$ are convergent uniformly for $y \in K$. $\qquad \square$

Note that there is a significant difference in the conclusions of the Orlicz–Pettis theorems above and the conclusions of the Orlicz–Pettis theorems for subseries convergent series in Chapter 2. In the subseries convergent theorems the uniform convergence of the series $\sum_{j \in \sigma} x_j$ was over compact subsets K of F and also for the multipliers χ_σ, $\sigma \subset \mathbb{N}$. In the theorems above the multiplier $t = \{t_j\}$ is fixed. The following example shows that to obtain uniform convergence for some bounded subsets of the multipliers this will require additional assumptions on the multiplier space.

Example 4.5. Consider the series $\sum_j e^j$ in $(c_0, \|\cdot\|_\infty)$. This series is c_0 multiplier convergent. However, the set of multipliers $K = \{t^k = \sum_{j=1}^{k} e^j : k \in \mathbb{N}\}$ is bounded, but the series $\sum_{j=1}^{\infty} t_j^k e^j = \sum_{j=1}^{k} e^j$ do not converge uniformly for $k \in \mathbb{N}$.

One gliding hump assumption which will allow uniform convergence over certain bounded subsets of the multiplier space is the *signed strong gliding hump* (signed-SGHP) property. The signed-SGHP property requires a topology on the multiplier space λ in contrast to the signed-WGHP which is an algebraic property. Let λ be a K-space. Let $\Lambda \subset \lambda$. The set Λ has the *signed-SGHP* if for every bounded sequence $\{t^j\} \subset \Lambda$ and every increasing sequence of intervals $\{I_j\}$, there exist a subsequence $\{n_j\}$ and a sequence of signs $\{s_j\}$ such that the coordinate sum of the series

$$\sum_{j=1}^{\infty} s_j \chi_{I_{n_j}} t^{n_j} \in \Lambda;$$

if the signs $\{s_j\}$ can all be chosen to be equal to 1, Λ has the *strong gliding hump property* (SGHP). The space l^∞ has SGHP while the subset $\Lambda = \{\chi_\sigma : \sigma \subset \mathbb{N}\}$ of m_0 has SGHP but m_0 does not have SGHP. The space bs of bounded series has signed-SGHP but not SGHP (see Appendix B of [Sw4]). Clearly signed-SGHP implies signed-WGHP for K-spaces but $c_0, l^p, 0 < p < \infty$, have WGHP but not SGHP. Further examples can be found in Appendix B.

Theorem 4.6. *Let* $\Lambda \subset \lambda$ *have signed-SGHP. If* $\sum_j x_j$ *is* Λ *multiplier convergent with respect to* $w(E, F)$, *then for each sequentially conditionally* $w(F, E)$ *compact subset* $K \subset F$ *and each bounded subset* $B \subset \Lambda$, *the series*

$$\sum_{j=1}^{\infty} t_j x_j \cdot y$$

converge uniformly for $y \in K, t \in B$.

Proof. If the conclusion fails to hold, there exist a neighborhood, W, in G, $y_k \in K, t^k \in B$ and an increasing sequence of intervals $\{I_k\}$ such that

$$(\#) \quad \sum_{l \in I_k} t_l^k x_l \cdot y_k \notin W$$

for every k. We may assume, by passing to a subsequence if necessary, that $\lim_k x \cdot y_k$ exists for every $x \in E$. Consider the matrix

$$M = [m_{ij}] = \left[\sum_{l \in I_j} t_l^j x_l \cdot y_i \right].$$

We claim that M is a signed $\mathcal{K}$ matrix. First, the columns of M converge. Next given an increasing sequence of positive integers, there exist a sequence of signs $\{s_j\}$ and a subsequence $\{n_j\}$ such that

$$u = \sum_{k=1}^{\infty} s_k \chi_{I_{n_k}} t^{n_k} \in \Lambda.$$

If $\sum_{l=1}^{\infty} u_l x_l$ denotes the $w(E, F)$ sum of the series, then

$$\left\{ \sum_{j=1}^{\infty} s_j m_{in_j} \right\}_i = \left\{ \sum_{j=1}^{\infty} s_j \sum_{l \in I_{n_j}} t_l^{n_j} x_l \cdot y_i \right\}_i = \left\{ \sum_{l=1}^{\infty} u_l x_l \cdot y_i \right\}_i$$

converges. Hence, M is a signed $\mathcal{K}$ matrix so the diagonal of M converges to 0 by the signed version of the Antosik–Mikusinski Matrix Theorem (Appendix E). But, this contradicts $(\#)$. $\qquad \square$

Results analogous to those in Theorems 4.3 and 4.4 also hold. We give a statement of these results.

Theorem 4.7. *Let* G *be metrizable or a LCTVS. Let* $\Lambda \subset \lambda$ *have signed-SGHP. If* $\sum_j x_j$ *is* Λ *multiplier convergent with respect to* $w(E, F)$, *then the series*

$$\sum_{j=1}^{\infty} t_j x_j \cdot y$$

converge uniformly for y *belonging to any* $w(F, E)$ *compact subset of* F *and for* t *belonging to any bounded subset of* Λ.

Note the example above points out the difference in the conclusions of the results for the WGHP and the SGHP.

Locally Convex Spaces

From Theorems 4.2-4.4 applied to dual pairs we can obtain results similar to those in Theorem 2.9 for LCTVS. Let E be a Hausdorff LCTVS with dual E'. Let $\gamma(E, E')$ $[\lambda(E, E'), \tau(E, E')]$ be the polar topology on E of uniform convergence on sequentially conditionally $\sigma(E', E)$ compact subsets of E' $[\sigma(E', E)$ compact sets; absolutely convex $\sigma(E', E)$ compact sets].

From Theorems 4.2-4.4, we have a version of the Orlicz–Pettis Theorem for multiplier convergent series.

Theorem 4.8. *Assume that $\Lambda \subset \lambda$ has signed-WGHP and that the series $\sum_j x_j$ is Λ multiplier convergent with respect to $\sigma(E, E')$. Then the series is Λ multiplier convergent with respect to $\gamma(E, E')$ $[\lambda(E, E'), \tau(E, E')]$.*

From Theorems 4.6, 4.7 we have a stronger version of the Orlicz–Pettis Theorem.

Theorem 4.9. *Assume that $\Lambda \subset \lambda$ has signed-SGHP and that the series $\sum_j x_j$ is Λ multiplier convergent with respect to $\sigma(E, E')$. Then the series*

$$\sum_{j=1}^{\infty} t_j x_j$$

converge with respect to $\gamma(E, E')$ $[\lambda(E, E'), \tau(E, E')]$ uniformly for t belonging to bounded subsets of Λ.

We can employ Theorem 9 to establish generalizations of results about summing operators for subseries and bounded multiplier convergent series. We establish continuity results for summing operators.

Lemma 4.10. *Let $\Lambda \subset \lambda$. If $\sum_j x_j$ is Λ multiplier convergent and the series $\sum_{j=1}^{\infty} t_j x_j$ converge uniformly for $t \in \Lambda$, then the summing operator $S : \Lambda \to X$,*

$$S(\{t_j\}) = \sum_{j=1}^{\infty} t_j x_j,$$

is continuous with respect to the topology p of coordinatewise convergence on Λ and the topology of X.

Proof. Let $t^\delta = \{t_j^\delta\}$ be a net in Λ which converges to $t \in \Lambda$ with respect to p. Let U be a neighborhood of 0 in E and pick a symmetric neighborhood V such that $V + V + V \subset U$. There exists n such that

$$\sum_{j=n}^{\infty} t_j x_j \in V$$

for every $t \in \Lambda$. There exists δ such that $\alpha \geq \delta$ implies

$$\sum_{j<n} (t_j^\alpha - t_j) x_j \in V.$$

Thus, for $\alpha \geq \delta$,

$$S(t^\alpha) - S(t) = \sum_{j<n} (t_j^\alpha - t_j) x_j + \sum_{j=n}^{\infty} t_j^\alpha x_j - \sum_{j=n}^{\infty} t_j x_j \in V + V + V \subset U$$

so S is continuous with respect to p.   □

From Theorem 9 and the lemma we have

Corollary 4.11. *Let λ be a K-space and $\Lambda \subset \lambda$ have signed-SGHP and let $\sum_j x_j$ be Λ multiplier convergent. If $\Lambda \subset \lambda$ is bounded, then the summing operator $S : \Lambda \to X$ is continuous with respect to p.*

Without some assumption on the multiplier space, the conclusion of the corollary may fail.

Example 4.12. Let $\Lambda = \{t \in l^p : \|t\|_p \leq 1\}$ for $1 \leq p < \infty$. The series $\sum e^j$ is l^p multiplier convergent in l^p but the summing operator $S = I : \Lambda \to l^p$ is not continuous with respect to p $[e^j \to 0$ in p but $e^j \not\to 0$ in $l^p]$.

From the corollary we have the following compactness result.

Theorem 4.13. *Let λ be a K-space with signed-SGHP and let $\sum_j x_j$ be λ multiplier convergent. If $\Lambda \subset \lambda$ is bounded in λ and compact with respect to p, then*

$$S\Lambda = \left\{ \sum_{j=1}^{\infty} t_j x_j : t \in \Lambda \right\}$$

is compact in X.

This theorem applies to the subseries case with $\Lambda = \{\chi_\sigma : \sigma \subset \mathbb{N}\}$, $\lambda = m_0$ and the bounded multiplier case with $\Lambda = \{\{t_j\} \in l^\infty : \|\{t_j\}\|_\infty \leq 1\}$, $\lambda = l^\infty$. See Theorem 2.17 and Corollary 3.12.

It was pointed out in Example 2.10 that in general an Orlicz–Pettis Theorem does not hold for weakly subseries convergent series and the strong topology. However, if the multiplier space has stronger properties, an Orlicz–Pettis Theorem for the strong topology is possible. To show this we take a detour through a result which characterizes Orlicz–Pettis theorems for LCTVS.

A locally convex topology $W(E, E')$ defined for dual pairs E, E' is called a *Hellinger–Toeplitz topology* if whenever a linear map $T : E \to F$ is $\sigma(E, E') - \sigma(F, F')$ continuous, then T is $W(E, E') - W(F, F')$ continuous ([Wi2] 11.1.5). Note that Hellinger–Toeplitz topologies must be defined for dual pairs. For example, the polar topologies $\beta(E, E'), \gamma(E, E'), \lambda(E, E'), \tau(E, E')$ are Hellinger–Toeplitz topologies.

We have the following characterization of Orlicz–Pettis theorems for Hellinger–Toeplitz topologies.

Theorem 4.14. *Let $W(E, E')$ be a Hellinger–Toeplitz topology for dual pairs. The following are equivalent.*

(1) *For every dual pair E, E' a series which is λ multiplier convergent with respect to $\sigma(E, E')$ is λ multiplier convergent with respect to $W(E, E')$.*
(2) *$(\lambda, W(\lambda, \lambda^\beta))$ is an AK-space.*

Proof. Assume (1): Then $\sum_j e^j$ is λ multiplier convergent with respect to $\sigma(\lambda, \lambda^\beta)$ so by (1) the series is λ multiplier convergent with respect to $W(\lambda, \lambda^\beta)$. But, this means that if $t \in \lambda$, then $t = \sum_{j=1}^\infty t_j e^j$, where the series is $W(\lambda, \lambda^\beta)$ convergent. That is, $(\lambda, W(\lambda, \lambda^\beta))$ is an AK-space.

Assume (2): Let $\sum_j x_j$ be λ multiplier convergent with respect to $\sigma(E, E')$. Consider the summing operator $S : \lambda \to E$ defined by

$$St = \sum_{j=1}^\infty t_j x_j$$

$[\sigma(E, E')$ sum]. If $x' \in E'$, we have

$$x'(St) = x' \left(\sum_{j=1}^\infty t_j x_j \right) = \sum_{j=1}^\infty t_j x'(x_j).$$

Thus, $\{x'(x_j)\} \in \lambda^\beta$ by the convergence of the series and

$$x'(St) = t \cdot \{x'(x_j)\}.$$

This implies that S is $\sigma(\lambda, \lambda^\beta) - \sigma(E, E')$ continuous and, therefore, $W(\lambda, \lambda^\beta) - W(E, E')$ continuous. If $t \in \lambda$, then

$$t = W(\lambda, \lambda^\beta) - \lim_n \sum_{j=1}^n t_j e^j$$

so

$$St = W(E, E') - \lim_n \sum_{j=1}^n t_j Se^j = \lim_n \sum_{j=1}^n t_j x_j.$$

Hence, (1) holds. □

We give an application of this characterization to the strong topology. First, a lemma.

Lemma 4.15. *If λ is a barrelled AK-space, then $\lambda' = \lambda^\beta$.*

Proof. Let $t \in \lambda^\beta$. For each n define a linear functional f_n on λ by

$$f_n(s) = \sum_{j=1}^n s_j t_j, \ s = \{s_j\} \in \lambda.$$

Each f_n is continuous since λ is a K-space. Now

$$\lim_n f_n(s) = \sum_{j=1}^\infty s_j t_j = s \cdot t = f(s)$$

defines a continuous linear functional f by the Banach–Steinhaus Theorem for barrelled spaces. Therefore, $t \in \lambda'$ and $\lambda^\beta \subset \lambda'$. Let $f \in \lambda'$ and set $t_j = f(e^j)$. By the AK assumption, if $s \in \lambda$,

$$s = \sum_{j=1}^\infty s_j e^j$$

so

$$f(s) = \sum_{j=1}^\infty s_j f(e^j) = \sum_{j=1}^\infty s_j t_j = s \cdot t.$$

Hence, $f = t$ and $\lambda' \subset \lambda^\beta$. □

Theorem 4.16. *Let λ be a barrelled AK-space. If $\sum_j x_j$ is λ multiplier convergent with respect to $\sigma(E, E')$, then the series is λ multiplier convergent with respect to the strong topology $\beta(E, E')$.*

Proof. Since λ is barrelled, λ carries the strong topology $\beta(\lambda, \lambda')$. But,

$$\beta(\lambda, \lambda') = \beta(\lambda, \lambda^\beta)$$

by the lemma. Since the strong topology is a Hellinger–Toeplitz topology, the result follows from the theorem. $\square$

There are other gliding hump assumptions which guarantee convergence in the strong topology; see Chapter 5 of [Sw4].

From Theorem 6 of Chapter 2, we also have a result for polar topologies which applies to the strong topology. See Kalton's Theorem 2.15.

Theorem 4.17. *Let E, E' be a pair of vector spaces in duality and assume $\tau = \tau_{\mathcal{A}}$ is a polar topology of uniform convergence on the family $\mathcal{A}$ of $\sigma(E', E)$ bounded sets. If (E, τ) is separable, then any series $\sum_j x_j$ in E which is λ multiplier convergent with respect $\sigma(E, E')$ is λ multiplier convergent with respect to $\beta(E, E')$.*

Proof. Each $\sigma(E', E)$ bounded set is sequentially conditionally compact at each point $x \in E$ so by the separability assumption and Theorem 2.6 every $\sigma(E', E)$ bounded set is sequentially conditionally $\sigma(E', E)$ compact and Theorem 2 applies. $\square$

Continuous Function Spaces

We will restate some of the results for spaces of continuous functions given in Chapter 2 for multiplier convergent series.

Let Ω be a sequentially compact topological space and let $SC(\Omega, G)$ be the space of all sequentially continuous functions from Ω into G.

Theorem 4.18. *Let $\Lambda \subset \lambda$. Suppose $\sum f_j$ is a series in $SC(\Omega, G)$ which is Λ multiplier convergent in the topology of pointwise convergence on Ω. If Λ has signed-WGHP, then for each $t \in \Lambda$ the series*

$$\sum_{j=1}^{\infty} t_j f_j(s)$$

is Λ multiplier convergent uniformly for $s \in \Omega$. If Λ has signed-SGHP, then the series $\sum_{j=1}^{\infty} t_j f_j(s)$ converge uniformly for $s \in \Omega$ and t running through any bounded subset of Λ.

Proof. To see this consider the abstract triple

$$(SC(\Omega, G), \Omega : G)$$

under the map $(f, t) \to f(t)$. The series $\sum f_j$ is Λ multiplier convergent with respect to $w(SC(\Omega, G), \Omega)$ and the set Ω is $w(\Omega, SC(\Omega, G))$ sequentially compact since Ω is sequentially compact so the claims follows from Theorems 4.2 and 4.6. $\square$

If G is either the scalar field or a normed space, $(G, \|\cdot\|)$, the conclusion of the theorem is that if a series is λ multiplier convergent in the topology of pointwise convergence, then the series is λ multiplier convergent in the sup-norm,

$$\|f\|_\infty = \sup\{\|f(t)\| : t \in \Omega\}.$$

Let Ω be a topological space, G be metrizable and $C(\Omega, G)$ the space of continuous functions from Ω to G.

Theorem 4.19. *Let* $\Lambda \subset \lambda$. *Suppose* $\sum f_j$ *is a series in* $C(\Omega, G)$ *which is* Λ *multiplier convergent in the topology of pointwise convergence on* Ω. *If* Λ *has signed-WGHP, then for each* $t \in \Omega$ *the series*

$$\sum_{j=1}^{\infty} t_j f_j(s)$$

converge uniformly for s running through any compact subset of Ω. *If* Λ *has signed-SGHP, then the series* $\sum_{j=1}^{\infty} t_j f_j(s)$ *converge uniformly for s running through any compact subset of* Ω *and t running through any bounded subset of* Λ.

Proof. To see this consider the abstract triple

$$(C(\Omega, G), \Omega : G)$$

under the map $(f, t) \to f(t)$. The series $\sum f_j$ is $w(C(\Omega, G), \Omega)$ Λ multiplier convergent and any compact subset of Ω is $w(\Omega, C(\Omega, G))$ compact so the claims follow from Theorems 4.3 and 4.7. $\square$

We can obtain an improvement for sequentially compact spaces.

Theorem 4.20. *Let* $\Lambda \subset \lambda$. *Let* Ω *be a compact metric space and* D *a dense subset. Suppose* $\sum f_j$ *is a series in* $C(\Omega, G)$ *which is* Λ *multiplier* *convergent in the topology of pointwise convergence on* D. *If* Λ *has signed-WGHP, then for each* $t \in \Lambda$ *the series*

$$\sum_{j=1}^{\infty} t_j f_j(s)$$

converge uniformly for $s \in \Omega$. *If* Λ *has signed-SGHP, then the series* $\sum_{j=1}^{\infty} t_j f_j(s)$ *converge uniformly for* $s \in \Omega$ *and* t *running through any bounded subset of* Λ.

Proof. Consider the triple

$$(C(\Omega, G), D : G)$$

under the map $(f, t) \to f(t)$. The series $\sum_j f_j$ is Λ multiplier convergent in $w(C(\Omega, G), D)$. If $\{t_j\} \subset D$, then there is a subsequence $\{t_{n_j}\}$ which converges to a point $t \in \Omega$ so if $f \in C(\Omega, G)$, then $f(t_{n_j}) \to f(t)$. Hence, D is $w(D, C(\Omega, G))$ sequentially conditionally compact. Theorems 4.2 and 4.6 give the conclusions of the theorem for $s \in D$ and from the denseness of D in Ω the result holds for $s \in \Omega$. $\qquad\square$

Theorems of this type relative to pointwise convergent series in spaces of continuous functions were established in [Sw4].

Linear Operators

We consider Orlicz–Pettis Theorems for linear operators.

Let X, Y be TVS and $L(X, Y)$ the space of all continuous linear operators from X into Y. Let $L_s(X, Y)$ be the space $L(X, Y)$ with the topology of pointwise convergence on X, the *strong operator topology*. Let $L_c(X, Y)$ $[L_b(X, Y)]$ be $L(X, Y)$ with the topology on $L(X, Y)$ of uniform convergence on compact [bounded] subsets of X.

Theorem 4.21. *Let* $\Lambda \subset \lambda$. *Assume* Y *is either metrizable or a LCTVS. Let* $\sum_j T_j$ *be* Λ *multiplier convergent in* $L_s(X, Y)$. *If* Λ *has signed-WGHP, then for each* $t \in \Lambda$ *the series*

$$\sum_{j=1}^{\infty} t_j T_j$$

is λ *multiplier convergent in* $L_c(X, Y)$. *If* Λ *has signed-SGHP, then the series* $\sum_{j=1}^{\infty} t_j T_j$ *converge in* $L_c(X, Y)$ *uniformly for* t *running through bounded subsets of* Λ.

Proof. Consider the triple

$$(L(X,Y), X : Y)$$

under the map $(T, x) \to Tx$. Any compact subset of E is $w(X, L(X,Y))$ compact so the result follows from Theorems 4.3 and 4.4. $\square$

As noted in Chapter 2 Orlicz–Pettis Theorems for $L_b(X, Y)$ do not hold in general. Later we will establish an Orlicz–Pettis Theorem for $L_b(X, Y)$ with strong assumptions on the multiplier space λ.

For LCTVS we have the analogue of a result for subseries convergent series.

Let X, Y be LCTVS. The *weak operator topology* on $L(X,Y)$ is the topology of pointwise topology on $L(X,Y)$ when Y has the weak topology, i.e., the topology generated by the semi-norms

$$T \to |y'Tx|, x \in X, y' \in Y'.$$

Theorem 4.22. *Let $\Lambda \subset \lambda$. Let Λ have signed-WGHP. If $\sum_j T_j$ is Λ multiplier convergent in the weak operator topology, then the series is Λ multiplier convergent in the strong operator topology.*

Proof. For each $x \in X$ the series $\sum_{j=1}^{\infty} T_j x$ is Λ multiplier convergent with respect to $\sigma(Y, Y')$. By the Orlicz–Pettis Theorem for LCTVS (Theorem 4.8) the series is Λ multiplier convergent for the original topology of Y. That is, $\sum_j T_j$ is Λ multiplier convergent in $L_s(X, Y)$. $\square$

Without an assumption on the multiplier space, the conclusion of the theorem may fail.

Example 4.23. Let $\lambda = c_c$. Then a series $\sum_j x_j$ is λ multiplier convergent in a TVS iff the series $\sum_j x_j$ is convergent in the TVS. Define a continuous linear operator $T_j : \mathbb{R} \to c_0$ by

$$T_j s = s(e^{j+1} - e^j).$$

Then for every $s \in \mathbb{R}$, the series $\sum_{j=1}^{\infty} T_j s$ converges in $\sigma(c_0, l^1)$ but does not converge in $(c_0, \|\cdot\|_{\infty})$. That is, the series converges in the weak operator topology but not in the strong operator topology.

For our first Orlicz–Pettis Theorem for $L_b(X, Y)$, we establish the analogue of Theorem 16 for operators.

Theorem 4.24. *Let λ be a barrelled AK space and X be barrelled. If the series $\sum_j T_j$ is λ multiplier convergent in the weak operator topology of $L(X,Y)$, then $\sum_j T_j$ is λ multiplier convergent in $L_b(X, Y)$.*

Proof. If $x \in X$ and $y' \in Y'$, define a continuous linear functional $x \otimes y'$ on $L(X, Y)$ by $\langle x \otimes y', T \rangle = \langle y', Tx \rangle$. Let

$$X \otimes Y' = \text{span}\{x \otimes y' : x \in X, y' \in Y'\}.$$

Note that the weak operator topology on $L(X, Y)$ is just $\sigma(L(X, Y), X \otimes Y')$. From Theorem 16 it follows that the series $\sum_j T_j$ is λ multiplier convergent in the strong topology $\beta(L(X, Y), X \otimes Y')$. Thus, it suffices to show that the strong topology $\beta(L(X, Y), X \otimes Y')$ is stronger than $L_b(X, Y)$. Let $\{S_\delta\}$ be a net in $L(X, Y)$ which converges to 0 in $\beta(L(X, Y), X \otimes Y')$. Let $A \subset X$ be bounded, $B \subset Y'$ be equicontinuous and set $C = \{x \otimes y' : x \in A, y' \in B\}$. Since A is $\beta(X, X')$ bounded from the barrelledness assumption,

$$\sup\{|\langle x \otimes y', T \rangle| : x \in A, y' \in B\} < \infty$$

for every $T \in L(X, Y)$; that is, C is $\sigma(X \otimes Y', L(X, Y))$ bounded. Thus,

$$\sup\{|\langle x \otimes y', S_\delta \rangle| : x \in A, y' \in B\} \to 0$$

so $S_\delta \to 0$ in Y uniformly for $x \in A$ or $S_\delta \to 0$ in $L_b(X, Y)$. Thus, $\beta(L(X, Y), X \otimes Y')$ is stronger than $L_b(X, Y)$ as desired. $\qquad\square$

Example 2.26 suggests that if one wishes to establish Orlicz–Pettis theorems for λ multiplier convergent series with respect to $L_b(X, Y)$ without imposing strong conditions on the multiplier space λ, one should consider the space $K(X, Y)$ of compact operators. The major result in this area is a result of Kalton. Kalton has shown that if X has the DF property (or, if X' contains no subspace isomorphic to l^∞) and if $\sum_j T_j$ is a series of compact operators from a Banach space X into a Banach space Y which is subseries convergent in the weak operator topology of $K(X, Y)$, then the series $\sum_j T_j$ is subseries convergent in the uniform operator topology of $K(X, Y)$ (2.30).

We first establish a result of Wu and Lu which characterizes the Orlicz–Pettis property for the space of compact operators ([WL]). Their result contains Kalton's result as a special case (see Theorem 66). Let $K_b(X, Y)$ be the topology on $K(X, Y)$ induced by $L_b(X, Y)$.

For this theorem we require some preliminary results. Let $\Lambda \subset \lambda$. The β dual of Λ with respect to X is defined to be

$$\Lambda^{\beta X} = \left\{\{x_j\} \subset X : \sum_{j=1}^\infty t_j x_j \text{ converges for every } t \in \Lambda\right\}.$$

Write

$$x \cdot t = \sum_{j=1}^{\infty} t_j x_j$$

for $x = \{x_j\} \in \Lambda^{\beta X}, t = \{t_j\} \in \Lambda.$

Lemma 4.25. *Let $\Lambda \subset \lambda$ have signed-WGHP. If $\{x^k\} \subset \Lambda^{\beta X}$ is such that*

$$\lim_k t \cdot x^k$$

exists for every $t \in \Lambda$ and $\lim_k x_j^k$ exists for every j, then for every $t \in \Lambda$ the series

$$\sum_{j=1}^{\infty} t_j x_j^k$$

converge uniformly for $k \in \mathbb{N}$.

Proof. If the conclusion fails, then

($*$) *there exists a neighborhood of 0, U, in X such that for every n there exist $k_n, n_n > m_n > n$ such that*

$$\sum_{j=m_n}^{n_n} t_j x_j^{k_n} \notin U.$$

By ($*$) for $n = 1$, there exist $k_1, n_1 > m_1 > 1$ such that

$$\sum_{j=m_1}^{n_1} t_j x_j^{k_1} \notin U.$$

There exists $m' > n_1$ such that

$$\sum_{j=m}^{n} t_j x_j^k \in U$$

for $n > m > m', 1 \leq k \leq k_1$. By ($*$) there exist $k_2, n_2 > m_2 > m'$ such that

$$\sum_{j=m_2}^{n_2} t_j x_j^{k_2} \notin U.$$

Hence, $k_2 > k_1$. Continuing this construction produces increasing sequences $\{k_i\}, \{m_i\}, \{n_i\}$ with $m_i < n_i < m_{i+1}$ and

(#) $x^{k_i} \cdot \chi_{I_i} t \notin U$, where $I_i = [m_i, n_i].$

Define the matrix M by

$$M = [m_{ij}] = [x^{k_i} \cdot \chi_{I_j} t].$$

We show that M is a signed $\mathcal{K}$-matrix. First, the columns of M converge by hypothesis. Second, given any increasing sequence of integers, there is a subsequence $\{p_k\}$ and a sequence of signs $\{s_k\}$ such that

$$z = \{z_j\} = \sum_{j=1}^{\infty} s_j \chi_{I_{p_j}} t \in \Lambda.$$

Then

$$\sum_{j=1}^{\infty} s_j m_{ip_j} = \sum_{j=1}^{\infty} s_j x^{k_i} \cdot \chi_{I_{p_j}} t = x^{k_i} \cdot z$$

and $\lim x^{k_i} \cdot z$ exists. Hence, M is a signed $\mathcal{K}$-matrix. By the signed version of the Antosik–Mikusinski Matrix Theorem (Appendix E), the diagonal of M converges to 0. But, this contradicts (#). $\qquad\square$

It follows from the lemma that if $F \subset \lambda^{\beta X}$ is sequentially conditionally compact and $t \in \lambda$, then the series $\sum_{j=1}^{\infty} t_j x_j$ converge uniformly for $\{x_j\} \in F$.

Lemma 4.26. *Let $\Lambda \subset \lambda$. If $\{x^k\} \subset \Lambda^{\beta X}$ is such that*

$$\lim_k x^k \cdot t$$

exists for every $t \in \Lambda$, $\lim_k x_j^k = x_j$ exists for each j and for each $t \in \Lambda$ the series

$$\sum_{j=1}^{\infty} t_j x_j^k$$

converge uniformly for $k \in \mathbb{N}$, then $x = \{x_j\} \in \Lambda^{\beta X}$ is such that $x^k \cdot t \to x \cdot t$ for every $t \in \Lambda$.

Proof. Set $x = \{x_j\}$. We claim that $x \in \Lambda^{\beta X}$ and $x^k \cdot t \to x \cdot t$ for every $t \in \Lambda$. Put $u = \lim x^k \cdot t$. It suffices to show that $u = \sum_{j=1}^{\infty} t_j x_j$. Let U be a balanced neighborhood of 0 in X and pick a balanced neighborhood V such that $V + V + V \subset U$. There exists p such that

$$\sum_{j=n}^{\infty} t_j x_j^k \in V$$

for $n \geq p, k \in \mathbb{N}$. Fix $n \geq p$. Pick $k = k_n$ such that

$$\sum_{j=1}^{\infty} t_j x_j^k - u \in V \text{ and } \sum_{j=1}^{n} t_j(x_j^k - x_j) \in V.$$

Then

$$\sum_{j=1}^{n} t_j x_j - u = \left(\sum_{j=1}^{\infty} t_j x_j^k - u \right) + \sum_{j=1}^{n} t_j (x_j - x_j^k) - \sum_{j=n+1}^{\infty} t_j x_j^k \in V + V + V \subset U$$

and the result follows. $\square$

From the lemmas we have a completeness result due to Stuart ([St1],[St2]).

Corollary 4.27. (Stuart) *Let λ have signed-WGHP and let X be sequentially complete. Then $(\lambda^{\beta X}, w(\lambda^{\beta X}, \lambda))$ is sequentially complete.*

Proof. If $\{x^k\}$ is $w(\lambda^{\beta X}, \lambda)$ Cauchy and X is sequentially complete, then $\lim_k x^k \cdot t$ exists for every $t \in \lambda$ so the results above apply. $\square$

We can now establish the result of Wu and Lu.

Theorem 4.28. *Let λ have signed-WGHP. The following are equivalent:*

(i) *Every series $\sum_j T_j$ which is λ multiplier convergent in the weak operator topology of $K(X, Y)$ is λ multiplier convergent in $K_b(X, Y)$.*

(ii) *Every continuous linear operator $S : X \to (\lambda^\beta, \sigma(\lambda^\beta, \lambda))$ is sequentially compact (an operator is sequentially compact if it carries bounded sets into relatively sequentially compact sets).*

Proof. Suppose (ii) holds. Let $\sum_j T_j$ be λ multiplier convergent in the weak operator topology of $K(X, Y)$. By Theorem 22 the series is λ multiplier convergent in the strong operator topology of $K(X, Y)$. Suppose there exists $t \in \lambda$ such that the series $\sum_j t_j T_j$ is not convergent in $K_b(X, Y)$. Then there exist $T \in K(X, Y)$ and a bounded set $A \subset X$ such that

$$\sum_{j=1}^{\infty} t_j T_j x = T x$$

for every $x \in X$ but the series do not converge uniformly for $x \in A$. Thus, there exist a continuous semi-norm p on Y, increasing sequences $\{m_k\}$ and $\{n_k\}$ with $m_k < n_k < m_{k+1}$, $x_k \in A$ and $\epsilon > 0$ such that

$$p\left(\sum_{l=m_k}^{n_k} t_l T_l x_k \right) > \epsilon$$

for all k. By the Hahn–Banach Theorem there is a sequence $\{y_k'\} \subset Y'$ such that

$$(*) \qquad \left\langle y_k', \sum_{l=m_k}^{n_k} t_l T_l x_k \right\rangle > \epsilon$$

and

$$\sup\{|\langle y'_k, y\rangle| : p(y) \le 1\} \le 1.$$

Let Y_0 be the closure in Y of span$\{T_i x_j : i, j \in \mathbb{N}\}$. Then (Y_0, p) is a separable semi-norm space. By the Banach–Alaoglu Theorem for separable semi-norm spaces, $\{y'_k\}$ has a subsequence $\{y'_{n_k}\}$ and $y' \in Y'$ such that

$$\lim \langle y'_{n_k}, y\rangle = \langle y', y\rangle$$

for every $y \in Y_0$ and

$$\sup\{|\langle y', y\rangle| : p(y) \le 1\} \le 1.$$

For notational convenience, assume that $n_k = k$.

Define a semi-norm q on X' by

$$q(x') = \sup\{|\langle x', x_k\rangle| : k \in \mathbb{N}\}.$$

We claim that if $U \in K(X, Y)$ satisfies $Ux_k \in Y_0$, then

$$(**) \quad \lim q(U'y'_k - U'y') = 0.$$

If $(**)$ fails to hold, there exist $\delta > 0$, a subsequence $\{y'_{n_k}\}$ and a subsequence $\{x_{n_k}\}$ such that

$$(***) \quad |\langle U'y'_{n_k} - U'y', x_{n_k}\rangle| > \delta.$$

Since U is compact, $\{Ux_{n_k}\}$ is a relatively compact subset of Y and, therefore, a relatively compact subset of (Y_0, p). Without loss of generality, we may assume that there exists $y \in Y_0$ such that $p(Ux_{n_k} - y) \to 0$. Then

$$|\langle y'_{n_k} - y', Ux_{n_k}\rangle| \le |\langle y'_{n_k} - y', Ux_{n_k} - y\rangle| + |\langle y'_{n_k} - y', y\rangle|$$
$$\le \sup\{|\langle y'_{n_k} - y', z\rangle| : p(z) \le 1\} p(Ux_{n_k} - y)$$
$$+ |\langle y'_{n_k} - y', y\rangle|$$
$$\le 2p(Ux_{n_k} - y) + |\langle y'_{n_k} - y', y\rangle| \to 0.$$

This contradicts $(***)$ and establishes the claim.

If $s \in \lambda$, $x \in X$ and $z' \in Y'$, the series

$$\sum_{j=1}^{\infty} s_j \langle z', T_j x\rangle$$

converges so we define a linear operator

$$S(= S_{z'}) : X \to (\lambda^{\beta}, \sigma(\lambda^{\beta}, \lambda))$$

by $Sx = \{\langle z', T_j x\rangle\}$. Since S is obviously continuous, S is sequentially compact by condition (ii). Thus, SA is sequentially compact with respect to $\sigma(\lambda^\beta, \lambda)$. By the lemma above, if $s \in \lambda$, then the series $\sum_{j=1}^\infty s_j \langle z', T_j x \rangle$ converge uniformly for $x \in A$ or, equivalently, the series $\sum_{j=1}^\infty s_j \langle T_j' z', x \rangle$ converge uniformly for $x \in A$.

Now consider the matrix

$$M = [m_{ij}] = \left[\sum_{l=m_j}^{n_j} t_l T_l' y_i' \right].$$

We show that M is a signed $\mathcal{K}$-matrix with values in the semi-norm space (X', q). First, the columns of M converge by condition $(**)$. Next, given any subsequence $\{p_j\}$ there is a further subsequence $\{q_j\}$ and a sequence of signs $\{\epsilon_j\}$ such that

$$s = \{s_j\} = \sum_{l=1}^\infty \epsilon_l \sum_{j=m_{q_l}}^{n_{q_l}} t_j \in \lambda.$$

There exist $U \in K(X, Y)$ such that

$$\sum_{j=1}^\infty s_j T_j = \sum_{l=1}^\infty \epsilon_l \sum_{j=m_{q_l}}^{n_{q_l}} t_j T_j$$

converges to U in the strong operator topology. By the paragraph above

$$\sum_{l=1}^\infty \epsilon_l \sum_{j=m_{q_l}}^{n_{q_l}} t_j T_j' y_k'$$

converges to $U' y_k'$ uniformly for $x \in A$. In particular,

$$q \left(\sum_{l=1}^\infty \epsilon_l \sum_{j=m_{q_l}}^{n_{q_l}} t_j T_j' y_k' - U' y_k' \right) \to 0.$$

Thus, M is a signed $\mathcal{K}$-matrix (with respect to (X', q)). By the signed version of the Antosik–Mikusinski Matrix Theorem (Appendix E), the diagonal of M converges to 0 in (X', q). This contradicts $(*)$ and establishes that (ii) implies (i).

Suppose that (i) holds. Let

$$S : X \to (\lambda^\beta, \sigma(\lambda^\beta, \lambda))$$

be linear and continuous. So $Sx = \{Sx \cdot e^j\}$. Let $y \in Y, y \neq 0$. Define $T_j \in K(X, Y)$ by

$$T_j x = (Sx \cdot e^j) y.$$

Let $t \in \lambda$. Define $T \ (= T_t) \in K(X,Y)$ by

$$Tx = (Sx \cdot t)y.$$

Then $\sum_{j=1}^{\infty} t_j T_j x = Tx$ for every $x \in X$, i.e., the series $\sum_j t_j T_j$ converges to T in the strong operator topology of $K(X,Y)$. By (i) the series

$$\sum_{j=1}^{\infty} t_j T_j x = \sum_{j=1}^{\infty} t_j (Sx \cdot e^j) y = Tx = (Sx \cdot t)y$$

converge uniformly for x belonging to bounded subsets of X or the series

$$\sum_{j=1}^{\infty} t_j (Sx \cdot e^j) = Sx \cdot t$$

converge uniformly for x belonging to bounded subsets of X. Now to show S is sequentially compact, let $\{x_k\}$ be a bounded sequence in X. Then $\{Sx_k\}$ is coordinatewise bounded in λ^β since S is bounded. By the diagonal method ([Ke] p.238, [DeS] 26.10), there is a subsequence $\{n_k\}$ such that $\lim_k Sx_{n_k} \cdot e^j$ exists for every j and since the series

$$\sum_{j=1}^{\infty} t_j (Sx_{n_k} \cdot e^j)$$

converge uniformly for $k \in \mathbb{N}$, $\lim_k Sx_{n_k} \cdot t$ exists. Thus, $\{Sx_{n_k}\}$ is $\sigma(\lambda^\beta, \lambda)$ Cauchy. By Stuart's Corollary above, there exists $u \in \lambda^\beta$ such that $Sx_{n_k} \to u$ in $\sigma(\lambda^\beta, \lambda)$. Therefore, $\{Sx_{n_k}\}$ is relatively sequentially compact in $(\lambda^\beta, \sigma(\lambda^\beta, \lambda))$ and (ii) holds. $\qquad \square$

Remark 4.29. Wu Junde has shown that subsets of λ^β are $\sigma(\lambda^\beta, \lambda)$ sequentially compact iff they are $\sigma(\lambda^\beta, \lambda)$ compact so condition (ii) of the theorem above can be replaced with the hypothesis that the operator S is compact ([Wu]).

For the case of subseries convergent series, that is, when $\lambda = m_0$, we have

Theorem 4.30. *Let X be a barrelled LCTVS. The following are equivalent:*

(i) *Every series $\sum_j T_j$ which is subseries convergent in the weak operator topology of $K(X,Y)$ is subseries convergent in $K_b(X,Y)$.*

(ii) *Every continuous linear operator $S : X \to (l^1, \sigma(l^1, m_0))$ is compact.*

(iii) *Every continuous linear operator $S : X \to (l^1, \|\cdot\|_1)$ is compact.*

Proof. Since subsets of l^1 are $\sigma(l^1, m_0)$ $[\|.\|_1]$ sequentially compact iff they are compact ([Kö1] 22.4), (i), (ii) and (iii) are equivalent by the theorem above. □

Hahn–Schur Theorems

We next consider Hahn–Schur Theorems for multiplier convergent series. Let G be a TVS.

Theorem 4.31. *Assume* $\Lambda \subset \lambda$ *has signed-WGHP and that* $\sum_j x_{ij}$ *is* Λ *multiplier convergent for every* $i \in \mathbb{N}$. *If*

$$\lim_i \sum_{j=1}^{\infty} t_j x_{ij}$$

exists for every $t \in \Lambda$ *and* $\lim_i x_{ij} = x_j$, *then*

(1) *for every* $t \in \Lambda$ *the series* $\displaystyle\sum_{j=1}^{\infty} t_j x_{ij}$ *converge uniformly for* $i \in \mathbb{N}$,

(2) *the series* $\displaystyle\sum_j x_j$ *is* Λ *multiplier convergent,*

(3) *for every* $t \in \Lambda$, $\displaystyle\lim_i \sum_{j=1}^{\infty} t_j x_{ij} = \sum_{j=1}^{\infty} t_j x_j$.

Proof. For each i define $f_i : \Lambda \to G$ by

$$f_i(t) = \sum_{j=1}^{\infty} t_j x_{ij}.$$

Consider the triple

$$(\Lambda, \{f_i\} : G)$$

under the map $(t, f_i) \to f_i(t)$. The series $\sum_j e^j$ is Λ multiplier convergent with respect to $w(\Lambda, \{f_i\})$. For if $t \in \Lambda$,

$$\sum_{j=1}^{\infty} t_j e^j \cdot f_i = \sum_{j=1}^{\infty} t_j f_i(e^j) = \sum_{j=1}^{\infty} t_j x_{ij}.$$

Now $\{f_i\}$ is $w(\{f_i\}, \Lambda)$ sequentially conditionally compact since $\lim_i \sum_{j=1}^{\infty} t_j x_{ij}$ exists for every t. By Theorem 4.2 the series

$$\sum_{j=1}^{\infty} t_j f_i(e^j) = \sum_{j=1}^{\infty} t_j x_{ij}$$

converge uniformly for $i \in \mathbb{N}$. Thus (1) holds.

For (2) and (3) fix $t \in \Lambda$ and set

$$z = \lim_i \sum_{j=1}^{\infty} t_j x_{ij}.$$

We claim that $z = \sum_{j=1}^{\infty} t_j x_j$; this will establish (2) and (3). Let U be a neighborhood of 0 in G and pick a symmetric neighborhood of 0, V, such that $V + V + V \subset U$. By (1) there exists n such that

$$\sum_{j=n}^{\infty} t_j x_{ij} \in V$$

for every i. There exists i_0 such that $i \geq i_0$ implies

$$\sum_{j=1}^{n-1} t_j(x_j - x_{ij}) \in V \text{ and } \sum_{j=1}^{\infty} t_j x_{ij} - z \in V.$$

If $i \geq i_0$, then

$$(\#) \sum_{j=1}^{n-1} t_j x_j - z = \sum_{j=1}^{n-1} t_j(x_j - x_{ij}) - \sum_{j=n}^{\infty} t_j x_{ij} + \left(\sum_{j=1}^{\infty} t_j x_{ij} - z \right)$$
$$\in V + V + V \subset U$$

and the claim is established. $\qquad\qquad\qquad\qquad\qquad\qquad\qquad\square$

The Hahn–Schur result above can be used to establish a weak sequential completeness result of Stuart (Corollary 27, [St1], [St2]).

Corollary 4.32. (Stuart) *Let X be sequentially complete and λ have signed-WGHP. Then $w(\lambda^{\beta X}, \lambda)$ is sequentially complete.*

Stuart has actually established a version of this result for vector valued sequence spaces.

Without an assumption on the multiplier space, the result above may fail to hold.

Example 4.33. Let $\lambda = c$. Define $x_{ij} = 1$ if $i = j$ and $x_{ij} = 0$ otherwise. If $t \in c$, then $\lim_i t_j x_{ij} = \lim_i t_i$ and $x_j = \lim_i x_{ij} = 0$. But,

$$\lim_i \sum_{j=1}^{\infty} t_j x_{ij} = \lim_i t_i \neq \sum_{j=1}^{\infty} t_j x_j = 0$$

if $\lim_i t_i \neq 0$.

We can use this version of the Hahn–Schur Theorem to obtain a version of an Orlicz–Pettis result of Kalton for separable spaces ([Ka2]). We give a sketch of the result. A LCTVS E is an *infra-Ptak space* if a $\sigma(E', E)$ dense subspace $M \subset E'$ is $\sigma(E', E)$ closed whenever $M \cap U^0$ is $\sigma(E', E)$ closed for every neighborhood of 0, U, where U^0 is the polar of U. For example, any complete metrizable LCTVS is an infra-Ptak space (see [Kö2] 34.3(5), [Sw2] 23.8). Assume E is a separable infra-Ptak space with $M \subset E'$ a subspace which separates the points of E and λ has signed-WGHP. We claim that if the series $\sum_j x_j$ is λ multiplier convergent with respect to $\sigma(E, M)$, then the series is λ multiplier convergent in E. By the Orlicz–Pettis Theorem 4.8, it suffices to show the series is λ multiplier convergent with respect to $\sigma(E, E')$. If $t \in \lambda$, then $\sum_{j=1}^{\infty} t_j x_j$ will denote the $\sigma(E, M)$ sum of the series. Set

$$M' = \left\{ x' \in E' : x' \left(\sum_{j=1}^{\infty} t_j x_j \right) = \sum_{j=1}^{\infty} x'(t_j x_j) \text{ for all } t \in \lambda \right\}.$$

Now M is $\sigma(E', E)$ dense in E' and since $M \subset M'$, M' is also $\sigma(E', E)$ dense. If M' is $\sigma(E', E)$ closed, we have $E' = M'$ and we are finished. By the infra-Ptak assumption it suffices to show $M' \cap U^0$ is $\sigma(E', E)$ closed when U is a neighborhood of 0. Since E is separable, $(U^0, \sigma(E', E))$ is metrizable ([Kö1] 21.3(4)) so it suffices to show $M' \cap U^0$ is sequentially $\sigma(E', E)$ closed. Suppose $\{x'_k\} \subset M' \cap U^0$ and $x'_k \to x'$ in $\sigma(E', E)$. For $t \in \lambda$,

$$x'_k \left(\sum_{j=1}^{\infty} t_j x_j \right) = \sum_{j=1}^{\infty} x'_k(t_j x_j) \to x' \left(\sum_{j=1}^{\infty} t_j x_j \right)$$

and

$$\lim_k x'_k(x_j) = x'(x'_j)$$

for every j. By the scalar version of the Hahn–Schur Theorem above, the series $\sum_{j=1}^{\infty} x'(x_j)$ is λ multiplier convergent and for every $t \in \lambda$,

$$x'_k \left(\sum_{j=1}^{\infty} t_j x_j \right) = \sum_{j=1}^{\infty} x'_k(t_j x_j) \to \sum_{j=1}^{\infty} x'(t_j x_j) = x' \left(\sum_{j=1}^{\infty} t_j x_j \right).$$

Thus, $x' \in M'$ as desired.

Kalton's result is for F-spaces which may not be locally convex ([Ka2]).

If the multiplier space Λ has the signed-SGHP, we can obtain a strengthened version of the Hahn–Schur Theorem. For this we need an observation (see Lemma 3.6).

Lemma 4.34. *If $x_j \to 0$ in the TVS G, then $\lim_j t x_j = 0$ uniformly for $|t| \leq 1$.*

Proof. Let U be a balanced neighborhood of 0 in G. There exists N such that $k \geq N$ implies $x_k \in U$. If $k \geq N$ and $|t| \leq 1$, $t x_k \in t U \subset U$. $\square$

Theorem 4.35. *Assume $\Lambda \subset \lambda$ has signed-SGHP and that $\sum_j x_{ij}$ is Λ multiplier convergent for every $i \in \mathbb{N}$. If*

$$\lim_i \sum_{j=1}^{\infty} t_j x_{ij}$$

exists for every $t \in \Lambda$ and $\lim_i x_{ij} = x_j$, then

(1′) for every bounded set $B \subset \Lambda$ the series

$$\sum_{j=1}^{\infty} t_j x_{ij} \text{ converge uniformly for } i \in \mathbb{N}, t \in B,$$

(2′) the series $\sum_j x_j$ is Λ multiplier convergent,

(3′) $\lim_i \sum_{j=1}^{\infty} t_j x_{ij} = \sum_{j=1}^{\infty} t_j x_j$ uniformly for t belonging to bounded

subsets of Λ.

Proof. The proof of (1') proceeds as in the theorem above where the stronger conclusion follows by applying the Orlicz–Pettis Theorem 4.6. The proof of (2') and (3') follow as in the proof of the inequality ($\#$) in Theorem 4.31, where the Lemma is used to treat the first term of the right hand side of ($\#$) for t belonging to a bounded subset of Λ (the t_j are bounded since λ is a K-space). $\square$

By applying this result to a single series, it follows that if Λ has signed-SGHP and $\sum_j x_j$ is Λ multiplier convergent, then the series $\sum_{j=1}^{\infty} t_j x_j$ converge uniformly for t belonging to bounded subsets of Λ.

Applying the Hahn–Schur Theorem above to the situation where $\lambda = m_0$ and $\Lambda = \{\chi_\sigma : \sigma \subset \mathbb{N}\}$ gives the Hahn–Schur Theorem for subseries convergent series in Theorem 2.55 since Λ has SGHP.

Applying the Hahn–Schur Theorem above to the case where $\lambda = \Lambda = l^\infty$ gives the Hahn–Schur Theorem for bounded multiplier convergent series in Theorem 3.8 since l^∞ has SGHP.

It is possible for the multiplier space to have WGHP but not SGHP and the stronger conclusions (1') and (3') fail to hold.

Example 4.36. Let $\lambda = \Lambda = l^p = G, 1 \leq p < \infty$. Define $x_{ij} = e^j$ if $1 \leq j \leq i$ and $x_{ij} = 0$ if $j > i$. Then $\sum_{j=1}^{\infty} x_{ij}$ is λ multiplier convergent for every i, $\lim_i x_{ij} = e^j = x_j$ for every j and

$$\lim_i \sum_{j=1}^{\infty} t_j x_{ij} = \lim_i \sum_{j=1}^{i} t_j e^j = \sum_{j=1}^{\infty} t_j e^j = t$$

for every $t \in l^p$. However, both (1') and (3') fail to hold. Consider $t^k = e^k$ so $\{t^k\}$ is bounded in l^p but

$$\sum_{j=1}^{\infty} t_j^k x_{ij} = \sum_{j=1}^{i} t_j^k e^j = e^k$$

if $i \geq k$.

We have a generalization of Theorems 2.58 and 3.13 for multiplier convergent series.

Theorem 4.37. *Assume that $\Lambda \subset \lambda$ is bounded, has signed-SGHP and is compact with respect to the topology p of pointwise convergence on Λ. Suppose that $\sum_j x_{ij}$ is Λ multiplier convergent for every $i \in \mathbb{N}$ and suppose*

$$\lim_i \sum_{j=1}^{\infty} t_j x_{ij}$$

exists for every $t \in \Lambda$ and $\lim_i x_{ij} = x_j$. Then

$$B = \left\{ \sum_{j=1}^{\infty} t_j x_{ij} : i \in \mathbb{N}, t \in \Lambda \right\} \cup \left\{ \sum_{j=1}^{\infty} t_j x_j : t \in \Lambda \right\}$$

is compact.

Proof. Let S_i be the summing operator of $\sum_{j=1}^{\infty} t_j x_{ij}$ and S the summing operator of $\sum_{j=1}^{\infty} t_j x_j$. Each S_i, S is continuous by Lemma 10 and by Theorem 35, $S_i \to S$ uniformly on Λ so the result follows (Lemma 2.57). □

Note Theorems 2.58 and 3.13 follow since $\Lambda = \{\chi_\sigma : \sigma \subset \mathbb{N}\}$ and $\Lambda = \{t \in l^\infty : \|t\|_\infty \leq 1\}$ have SGHP.

We consider a partial converse to the Hahn–Schur Theorems.

Proposition 4.38. *Let $\sum_j x_{ij}$ be Λ multiplier convergent for every $i \in \mathbb{N}$ and assume that $\lim_i x_{ij} = x_j$ exists for every $j \in \mathbb{N}$.*

(1) *If for every $t \in \Lambda$ the series $\sum_{j=1}^{\infty} t_j x_{ij}$ converge uniformly for $i \in \mathbb{N}$, then for every $t \in \Lambda$ the sequence $\{\sum_{j=1}^{\infty} t_j x_{ij}\}_i$ is Cauchy.*

(2) *If the series $\sum_{j=1}^{\infty} t_j x_{ij}$ converge uniformly for $i \in \mathbb{N}$ and t belonging to bounded subsets of Λ, then the sequences $\{\sum_{j=1}^{\infty} t_j x_{ij}\}_i$ satisfy a Cauchy condition uniformly for t belonging to bounded subsets of Λ.*

Proof. (1): Let $t \in \Lambda$. Let U be a neighborhood of 0 in G. Pick a symmetric neighborhood of 0, V, such that $V + V + V \subset U$. There exists N such that $n \geq N$ implies

$$\sum_{j=n}^{\infty} t_j x_{ij} \in V$$

for every $i \in \mathbb{N}$. There exists $n > N$ such that $i, k \geq n$ implies

$$\sum_{j=1}^{N} t_j (x_{ij} - x_{kj}) \in V.$$

If $i, k \geq n$, then

$$(*) \quad \sum_{j=1}^{\infty} t_j x_{kj} - \sum_{j=1}^{\infty} t_j x_{ij} =$$

$$\sum_{j=1}^{N} t_j (x_{kj} - x_{ij}) + \sum_{j=N+1}^{\infty} t_j x_{kj} - \sum_{j=N+1}^{\infty} t_j x_{ij} \in V + V + V \subset U.$$

(2): Let $B \subset \Lambda$ be bounded. Let U be a neighborhood of 0 in G. Pick a symmetric neighborhood of 0, V, such that $V + V + V \subset U$. There exists N such that $n \geq N$ implies

$$\sum_{j=n}^{\infty} t_j x_{ij} \in V$$

for every $i \in \mathbb{N}, t \in B$. By the K-space assumption $\{t_j : t \in B\}$ is bounded for every j. By the lemma above (4.34) there exists $n > N$ such that $i, k \geq n$ implies

$$\sum_{j=1}^{N} t_j (x_{ij} - x_{kj}) \in V$$

for every $t \in B$. If $i, k \geq n$, $t \in B$, then $(*)$ holds. $\qquad\square$

Corollary 4.39. *Assume G is sequentially complete. Let $\sum_j x_{ij}$ be Λ multiplier convergent for every $i \in \mathbb{N}$ and assume that $\lim_i x_{ij} = x_j$ exists for every $j \in \mathbb{N}$.*

(1) *If for every $t \in \Lambda$ the series $\sum_{j=1}^{\infty} t_j x_{ij}$ converge uniformly for $i \in \mathbb{N}$,
then for every $t \in \Lambda$*

$$\lim_i \sum_{j=1}^{\infty} t_j x_{ij} = \sum_{j=1}^{\infty} t_j x_j.$$

(2) *If the series $\sum_{j=1}^{\infty} t_j x_{ij}$ converge uniformly for $i \in \mathbb{N}$ and t belonging
to bounded subsets of Λ, then*

$$\lim_i \sum_{j=1}^{\infty} t_j x_{ij} = \sum_{j=1}^{\infty} t_j x_j$$

uniformly for t belonging to bounded subsets of Λ.

We consider the space of multiplier convergent series. Let

$$\Lambda(G) = \Lambda^{\beta G}$$

be the space of all Λ multiplier convergent G valued series. We define a
topology for $\Lambda(G)$. For this we require some preliminary results.

Theorem 4.40. *Let $\Lambda \subset \lambda$ have signed-SGHP. If the series $\sum_{j=1}^{\infty} x_j$ is
Λ multiplier convergent, then the series $\sum_{j=1}^{\infty} t_j x_j$ converge uniformly for
$\{t_j\}$ belonging to bounded subsets of Λ.*

Proof. Suppose B is a bounded subset of Λ for which the conclusion fails.
Then there exist a balanced neighborhood, U, of 0 in G, $t^k \in B$ and an
increasing sequence of intervals $\{I_k\}$ such that

$$(*) \quad \sum_{j \in I_k} t_j^k x_j \notin U.$$

By signed-SGHP, there is a subsequence $\{n_k\}$ and signs $\{s_k\}$ such that the
pointwise sum of the series

$$t = \sum_{j=1}^{\infty} s_k \chi_{I_k} t^k \in \Lambda.$$

But, then $(*)$ implies that the series $\sum_{j=1}^{\infty} t_j x_j$ does not converge. $\qquad \square$

See also the remark following Theorem 35.

Lemma 4.41. *If $B \subset \Lambda$ is bounded and the series $\sum_{j=1}^{\infty} t_j x_j$ converge
uniformly for $t = \{t_j\} \in B$, then*

$$\left\{ \sum_{j=1}^{\infty} t_j x_j : t = \{t_j\} \in B \right\}$$

is bounded.

Proof. Let U be a balanced neighborhood of 0 in G. There exists N such that

$$\sum_{j=N}^{\infty} t_j x_j \in U$$

for $\{t_j\} \in B$. Since λ is a K-space, $\{t_j x_j : \{t_j\} \in B\}$ is bounded for each j. Thus, there exists $s > 1$ such that

$$\left\{ \sum_{j=1}^{N-1} t_j x_j : \{t_j\} \in B \right\} \subset sU.$$

Then

$$\sum_{j=1}^{\infty} t_j x_j = \sum_{j=1}^{N-1} t_j x_j + \sum_{j=N}^{\infty} t_j x_j \in sU + U \subset s(U + U)$$

and the result follows. $\qquad\qquad\square$

From the theorem and lemma, we have

Corollary 4.42. *Let $\Lambda \subset \lambda$ have signed-SGHP and let $B \subset \Lambda$ be bounded. If $\sum x_j$ is Λ multiplier convergent, then*

$$\left\{ \sum_{j=1}^{\infty} t_j x_j : \{t_j\} \in B \right\}$$

is bounded.

We can now define a natural topology for $\Lambda(G)$. Let Λ have signed-SGHP. Assume G is a LCTVS whose topology is generated by the semi-norms $\mathcal{P}$. If $B \subset \Lambda$ is bounded and $p \in \mathcal{P}$, set

$$\widehat{p_B}(\{x_j\}) = \sup \left\{ p\left(\sum_{j=1}^{\infty} t_j x_j \right) : \{t_j\} \in B \right\}$$

for $\{x_j\} \in \Lambda(G)$ and give $\Lambda(G)$ the locally convex topology, τ_Λ, generated by the semi-norms $\widehat{p_B}$ when p runs through $\mathcal{P}$ and B runs through the bounded subsets B of Λ. Note $\widehat{p_B}(\{x_j\}) < \infty$ by the corollary. Consider the triple

$$(\Lambda(G), \Lambda : G)$$

under the map $(x, t) \to \sum_{j=1}^{\infty} t_j x_j$. It follows from the Hahn–Schur Theorem (Theorem 4.35) that if the sequence $\{x^k\} \subset \Lambda(G)$ converges to x with respect to $w(\Lambda(G), \Lambda)$, then the sequence converges to x with respect to the topology τ_Λ. Moreover, if G is sequentially complete, then $w(\Lambda(G), \Lambda)$ is sequentially complete.

When $\lambda = m_0$ and $\Lambda = \{\chi_\sigma : \sigma \subset \mathbb{N}\}$, this covers the case of the space, $ss(G)$, of subseries convergent series treated in Chapter 2. When $\lambda = \Lambda = l^\infty$, this covers the case of the space, $bmc(G)$, of bounded multiplier convergent series treated in Chapter 3.

Li's Hahn–Schur Theorems ([LS])

We now establish a Hahn–Schur Theorem in the spirit of Li's Lemma 3.17 ([LS]). Again this framework does not fit into abstract triples but is an abstract setup with interesting applications. These theorems are useful in treating subseries convergent series, bounded multiplier convergent series and operator valued series with vector valued multipliers.

Let Ω be a non-empty set and G be an Abelian topological group. Let $f_{ij} : \Omega \to G$ for $i, j \in \mathbb{N}$ and assume that Ω has a distinguished element w_0 such that $f_{ij}(w_0) = 0$ for every i, j.

Theorem 4.43. ([LS]) *Assume that the series $\sum_{j=1}^\infty f_{ij}(w_j)$ converges for every i and every sequence $\{w_j\} \subset \Omega$ and that*

$$\lim_i \sum_{j=1}^\infty f_{ij}(w_j)$$

exists for every sequence $\{w_j\} \subset \Omega$. Then

(1) $\lim_i f_{ij}(w) = f_j(w)$ *exists for every* $w \in \Omega, j \in \mathbb{N}$,
(2) *the series $\sum_{j=1}^\infty f_{ij}(w_j)$ converge uniformly for $i \in \mathbb{N}$ and all sequences $\{w_j\} \subset \Omega$,*
(3) *the series $\sum_{j=1}^\infty f_j(w_j)$ converges and*

$$\lim_i \sum_{j=1}^\infty f_{ij}(w_j) = \sum_{j=1}^\infty f_j(w_j)$$

for every sequence $\{w_j\} \subset \Omega$.

Proof. Let $w \in \Omega$ and $j \in \mathbb{N}$. Define a sequence in Ω by $w_j = w$ and $w_i = w_0$ for $i \neq j$. Then $f_{ij}(w_j) = 0$ if $i \neq j$ so $\lim_i \sum_{j=1}^\infty f_{ij}(w_j) = \lim_i f_{ij}(w)$ exists by hypothesis and (1) holds.

We first show that for each $\{w_j\}$ the series $\sum_{j=1}^{\infty} f_{ij}(w_j)$ converge uniformly for $i \in \mathbb{N}$. If this fails to hold, there exists a neighborhood of 0, U, in G such that

(*) for every k there exist $p > k$ and q such that $\displaystyle\sum_{j=p}^{\infty} f_{qj}(w_j) \notin U$.

Hence, there exist $n_1 > 1, i_1$ such that

$$\sum_{j=n_1}^{\infty} f_{i_1 j}(w_j) \notin U.$$

Pick a neighborhood of 0, V, such that $V + V \subset U$. There exists $m_1 > n_1$ such that

$$\sum_{j=m_1+1}^{\infty} f_{i_1 j}(w_j) \in V.$$

Hence,

$$\sum_{j=n_1}^{m_1} f_{i_1 j}(w_j) \notin V.$$

Since $\sum_{j=1}^{\infty} f_{ij}(w_j)$ converge for $i = 1, ..., i_1$ by (*) there exist $n_2 > m_1, i_2 > i_1$ such that

$$\sum_{j=n_2}^{\infty} f_{i_2 j}(w_j) \notin U$$

and as above there exists $m_2 > n_2$ such that

$$\sum_{j=n_2}^{m_2} f_{i_2 j}(w_j) \notin V.$$

Continuing this construction produces increasing sequences $\{i_p\}, \{m_p\}$ and $\{n_p\}$ with $n_{p+1} > m_p > n_p$ such that

(**) $$\sum_{j=n_p}^{m_p} f_{i_p j}(w_j) \notin V.$$

Now consider the matrix

$$M = [m_{pq}] = \left[\sum_{j=n_q}^{m_q} f_{i_p j}(w_j) \right].$$

We claim that M is a $\mathcal{K}$-matrix. The columns of M converge by (1). If $\{k_q\}$ is an increasing sequence, set $v_j = w_j$ if $n_{k_q} \leq j \leq m_{k_q}$ and $v_j = w_0$ otherwise. Then

$$\lim_p \sum_{q=1}^{\infty} m_{pk_q} = \lim_p \sum_{j=1}^{\infty} f_{i_p j}(v_j)$$

exists by hypothesis. Hence, M is a $\mathcal{K}$-matrix so by the Antosik–Mikusinski Matrix Theorem the diagonal of M converges to 0. But, this contradicts $(**)$.

If (2) fails to hold, then as above there exist increasing sequences $\{i_k\}, \{m_k\}$ and $\{n_k\}$ with $n_k < m_k < n_{k+1}$, a matrix $\{w_{ij}\} \subset \Omega$ and a neighborhood, V, with

$$(***) \qquad \sum_{j=n_k}^{m_k} f_{i_k j}(w_{kj}) \notin V.$$

Now define a sequence $\{w_j\} \subset \Omega$ by $w_j = w_{kj}$ if $n_k \leq j \leq m_k$ and $w_j = w_0$ otherwise. But, then the series

$$\sum_{j=1}^{\infty} f_{ij}(w_j)$$

do not satisfy the Cauchy condition uniformly for $i \in \mathbb{N}$ by $(***)$ and, therefore, violates the condition established above.

For (3), let U be a neighborhood of 0 and $\{w_j\} \subset \Omega$. Pick a neighborhood of 0, V, such that $V + V + V \subset U$. Put

$$g = \lim_i \sum_{j=1}^{\infty} f_{ij}(w_j).$$

We show that the series $\sum_{j=1}^{\infty} f_j(w_j)$ converges to g. By (2) there exists n such that

$$\sum_{j=m}^{\infty} f_{ij}(w_j) \in V$$

for $m \geq n$ and $i \in \mathbb{N}$. Suppose $m > n$. Then by (1) there exists i such that

$$\sum_{j=1}^{n}(f_{ij}(w_j) - f_j(w_j)) \in V \text{ and } g - \sum_{j=1}^{\infty} f_{ij}(w_j) \in V.$$

So if $m > n$,

$$g - \sum_{j=1}^{m} f_j(w_j) = (g - \sum_{j=1}^{\infty} f_{ij}(w_j)) + \sum_{j=m+1}^{\infty} f_{ij}(w_j) + \sum_{j=1}^{m}(f_{ij}(w_j) - f_j(w_j))$$

$$\in V + V + V \subset U.$$

$\square$

Concerning the converse of Theorem 43, we have

Theorem 4.44. *Assume that the series $\sum_{j=1}^{\infty} f_{ij}(w_j)$ converges for every i and every sequence $\{w_j\} \subset \Omega$ and that*

$$\lim_i f_{ij}(w) = f_j(w)$$

exists for every j and $w \in \Omega$. If for every $\{w_j\} \subset \Omega$ the series $\sum_{j=1}^{\infty} f_{ij}(w_j)$ converge uniformly for $i \in \mathbb{N}$, then

$$\left\{ \sum_{j=1}^{\infty} f_{ij}(w_j) \right\}_i$$

is Cauchy. If G is sequentially complete, then the stronger conclusion (3) of Theorem 43 holds.

Proof. Let $\{w_j\} \subset \Omega$ and let U be a neighborhood of 0. Pick a symmetric neighborhood of 0, V, such that $V + V + V \subset U$. By hypothesis there exists n such that

$$\sum_{j=n}^{\infty} f_{ij}(w_j) \in V$$

for all i. Since $\lim_i f_{ij}(w) = f_j(w)$ exists for every j and $w \in \Omega$ there exists m such that

$$\sum_{j=1}^{n-1}(f_{ij}(w_j) - f_{kj}(w_j)) \in V$$

for all $i, k \geq m$. Then for all $i, k \geq m$,

$$\sum_{j=1}^{\infty} f_{ij}(w_j) - \sum_{j=1}^{\infty} f_{kj}(w_j) =$$

$$\sum_{j=1}^{n-1}(f_{ij}(w_j) - f_{kj}(w_j)) + \sum_{j=n}^{\infty} f_{ij}(w_j) - \sum_{j=n}^{\infty} f_{kj}(w_j) \in V + V + V \subset U.$$

The last statement follows from Theorem 43. $\square$

Under stronger assumptions we establish a stronger convergence conclusion than condition (3) in Theorem 43.

Theorem 4.45. *Assume that the series $\sum_{j=1}^{\infty} f_{ij}(w_j)$ converges for every i and every sequence $\{w_j\} \subset \Omega$. If for each $j \in \mathbb{N}$,*

$$\lim_i f_{ij}(w) = f_j(w)$$

converges uniformly for $w \in \Omega$ and if the series $\sum_{j=1}^{\infty} f_{ij}(w_j)$ converge uniformly for all sequences $\{w_j\} \subset \Omega$ and $i \in \mathbb{N}$, then the sequences

$$\left\{ \sum_{j=1}^{\infty} f_{ij}(w_j) \right\}_i$$

satisfy a Cauchy condition uniformly for all sequences $\{w_j\} \subset \Omega$. If G is sequentially complete, then

$$\lim_i \sum_{j=1}^{\infty} f_{ij}(w_j) = \sum_{j=1}^{\infty} f_j(w_j)$$

uniformly for all sequences $\{w_j\} \subset \Omega$.

Proof. Let U be a closed neighborhood of 0 in G and pick a symmetric neighborhood of 0, V, such that $V + V + V \subset U$.

There exists n such that

$$\sum_{j=n}^{\infty} f_{ij}(w_j) \in V$$

for all $\{w_j\} \subset \Omega$ and $i \in \mathbb{N}$. There exists m such that

$$\sum_{j=1}^{n-1} (f_{ij}(w) - f_{kj}(w)) \in V$$

for all $i, k \geq m$ and $w \in \Omega$ by the uniform convergence assumption. Hence, if $i, k \geq m$ and $\{w_j\} \subset \Omega$, we have

$$(*) \quad \sum_{j=1}^{\infty} f_{ij}(w_j) - \sum_{j=1}^{\infty} f_{kj}(w_j) =$$

$$\sum_{j=1}^{n-1} (f_{ij}(w_j) - f_{kj}(w_j)) + \sum_{j=n}^{\infty} f_{ij}(w_j) - \sum_{j=n}^{\infty} f_{kj}(w_j) \in V + V + V \subset U$$

so the first part of the statement is established.

If G is sequentially complete, then $\lim_i \sum_{j=1}^{\infty} f_{ij}(w_j)$ exists by $(*)$. The last statement then follows from (3) of Theorem 43 and $(*)$ above. $\qquad \square$

Corollary 4.46. *Assume that the series $\sum_{j=1}^{\infty} f_{ij}(w_j)$ converges for every i and every sequence $\{w_j\} \subset \Omega$, that*

$$\lim_i \sum_{j=1}^{\infty} f_{ij}(w_j)$$

exists for every sequence $\{w_j\} \subset \Omega$ and for every j,

$$\lim_i f_{ij}(w) = f_j(w)$$

uniformly for $w \in \Omega$. Then

(i) *the series $\sum_{j=1}^{\infty} f_{ij}(w_j)$ converge uniformly for $i \in \mathbb{N}$ and all sequences $\{w_j\} \subset \Omega$,*
(ii) *the series $\sum_{j=1}^{\infty} f_j(w_j)$ converges and*

$$\lim_i \sum_{j=1}^{\infty} f_{ij}(w_j) = \sum_{j=1}^{\infty} f_j(w_j)$$

uniformly for $\{w_j\} \subset \Omega$.

We can apply these results to subseries convergent series and bounded multiplier convergent series to obtain Hahn–Schur Theorems for these series.

First, let G be an Abelian topological group and $x_{ij} \in G$. Assume $\sum_j x_{ij}$ is subseries convergent for every i and

$$\lim_i \sum_{j \in \sigma} x_{ij}$$

exists for every $\sigma \subset \mathbb{N}$ with $\lim_i x_{ij} = x_j$ for every j. Set $\Omega = \{0, 1\}$ and define

$$f_{ij}(0) = 0 \text{ and } f_{ij}(1) = x_{ij}.$$

Then the assumptions of the corollary are satisfied so the series $\sum_j x_j$ is subseries convergent, the series $\sum_{j \in \sigma} x_{ij}$ converge uniformly for $\sigma \subset \mathbb{N}$, and

$$\lim_i \sum_{j \in \sigma} x_{ij} = \sum_{j \in \sigma} x_j$$

uniformly for $\sigma \subset \mathbb{N}$. This is the Hahn–Schur result in 2.55.

Next, let G be a TVS and $x_{ij} \in G$. Assume the series $\sum_j x_{ij}$ is bounded multiplier convergent for every j, that

$$\lim_i \sum_{j=1}^{\infty} t_j x_{ij}$$

exists for every $\{t_j\} \in l^{\infty}$ and $\lim_i x_{ij} = x_j$ for every j. Set $\Omega = [0, 1]$ and define $f_{ij}(t) = t x_{ij}$. Using Lemma 34 we see that the hypothesis of the corollary are satisfied so the series $\sum_{j=1}^{\infty} x_j$ is bounded multiplier convergent, the series $\sum_{j=1}^{\infty} t_j x_{ij}$ converge uniformly for $\|\{t_j\}\|_{\infty} \leq 1$, and

$$\lim_i \sum_{j=1}^{\infty} t_j x_{ij} = \sum_{j=1}^{\infty} t_j x_j$$

uniformly for $\|\{t_j\}\|_{\infty} \leq 1$. This is the Hahn–Schur Theorem 3.8 for bounded multiplier convergent series.

The results above can also be used to treat operator valued series and vector valued multipliers. We will briefly describe this situation. Let X, Y be LCTVS and $T_j \in L(X, Y)$. The series $\sum_j T_j$ is bounded multiplier convergent if the series

$$\sum_{j=1}^{\infty} T_j x_j$$

converges for every bounded sequence $\{x_j\} \subset X$ (this is not in agreement with our previous use of the term bounded multiplier convergence but it will only be used in this one situation as in Chapter 3). Assume $\sum_{ij} T_{ij}$ is bounded multiplier convergent for every i and that

$$\lim_i \sum_{j=1}^{\infty} T_{ij} x_j$$

exists for every bounded sequence $\{x_j\}$. Then $\lim_i T_{ij} x = T_j x$ exists for every $x \in X$ and defines a linear operator T_j on X (T_j may not be continuous). Let p be a continuous semi-norm on X. Set $\Omega = \{x : p(x) \leq 1\}$ and define $f_{ij}(x) = T_{ij} x$. Then the hypothesis of the first Hahn–Schur Theorem (4.43) are satisfied so the series $\sum_j T_j$ is bounded multiplier convergent since p is an arbitrary semi-norm on X, the series $\sum_{j=1}^{\infty} T_{ij} x_j$ converge uniformly for $p(x_j) \leq 1$ and

$$\lim_i \sum_{j=1}^{\infty} T_{ij} x_j = \sum_{j=1}^{\infty} T_j x_j$$

for every bounded sequence $\{x_j\}$. If in addition

$$(\#) \quad \lim_i T_{ij}x = T_jx$$

uniformly for x belonging to bounded subsets of X, then the corollary (4.46) gives that

$$(\#\#) \quad \lim_i \sum_{j=1}^{\infty} T_{ij}x_j = \sum_{j=1}^{\infty} T_jx_j$$

uniformly for $p(x_j) \leq 1$. Note that condition $(\#)$ is a necessary condition for $(\#\#)$ to hold.

We can also obtain a boundedness analogue of Theorem 3.13 for these series. Let the hypotheses of the paragraph above hold. Set

$$B = \left\{ \sum_{j=1}^{\infty} T_{ij}x_j : p(x_j) \leq 1, i \in \mathbb{N} \right\}.$$

Then B is bounded. Let $t_i \to o$ with $|t_i| \leq 1$, $\epsilon > 0$ and $p(x_i) \leq 1$. There exists n such that

$$p\left(t_i \sum_{j=n}^{\infty} T_{ij}x_j \right) < \epsilon$$

for $p(x) \leq 1, i \in \mathbb{N}$, and there exists k such that $i \geq k$ implies

$$p\left(t_i \sum_{j=1}^{n-1} T_{ij}x_j \right) < \epsilon$$

for $p(x_j) \leq 1$. If $i \geq k$, then

$$p\left(t_i \sum_{j=1}^{\infty} T_{ij}x_j \right) \leq 2\epsilon$$

for $p(x_j) \leq 1$. Hence, B is bounded.

We can also obtain a version of Lemma 2.16.

Proposition 4.47. *Let Ω be a topological space with $g_j : \Omega \to G$ continuous and assume that the series $\sum_{j=1}^{\infty} g_j(w_j)$ converges for every $\{w_j\} \subset \Omega$. If $F : \Omega^{\mathbb{N}} \to G$ is defined by*

$$F(\{w_j\}) = \sum_{j=1}^{\infty} g_j(w_j),$$

then F is continuous with respect to the product topology.

Proof. Let $w^k = \{w_j^k\}$ be a net in $\Omega^{\mathbb{N}}$ which converges to $w = \{w_j\}$ in the product topology. Let U be a neighborhood of 0 in G and pick a symmetric neighborhood, V, such that $V + V + V \subset U$. By Lemma 3.17 there exists n such that

$$\sum_{j=n}^{\infty} g_j(v_j) \in V$$

for all $\{v_j\} \subset \Omega$. There exists k_0 such that $k \geq k_0$ implies

$$\sum_{j=1}^{n-1}(g_j(w_j^k) - g_j(w_j)) \in V.$$

If $k \geq k_0$, then

$$F(w^k) - F(w) = \sum_{j=1}^{n-1}(g_j(w_j^k) - g_j(w_j)) + \sum_{j=n}^{\infty} g_j(w_j^k) - \sum_{j=n}^{\infty} g_j(w_j) \in V + V + V \subset U.$$

Thus, F is continuous. $\qquad\qquad\qquad\qquad\qquad\qquad\qquad\qquad\square$

From Proposition 47, we have

Corollary 4.48. *Let Ω be a compact topological space with $g_j : \Omega \to G$ continuous and assume that the series $\sum_{j=1}^{\infty} g_j(w_j)$ converges for every $\{w_j\} \subset \Omega$. Then*

$$S = \left\{ \sum_{j=1}^{\infty} g_j(w_j) : \{w_j\} \subset \Omega \right\}$$

is compact.

From Lemma 2.57, we also obtain

Corollary 4.49. *Let Ω be a compact topological space. Assume that each f_{ij} is continuous, the series $\sum_{j=1}^{\infty} f_{ij}(w_j)$ converge uniformly for $\{w_j\} \subset \Omega$ and $i \in \mathbb{N}$ and for each $j \in \mathbb{N}$, $\lim_i f_{ij}(w) = f_j(w)$ converges uniformly for $w \in \Omega$, then*

$$S = \left\{ \sum_{j=1}^{\infty} f_{ij}(w_j) : \{w_j\} \subset \Omega, i \in \mathbb{N} \right\} \cup \left\{ \sum_{j=1}^{\infty} f_j(w_j) : \{w_j\} \subset \Omega \right\}$$

is compact.

Proof. As in Proposition 47 define $F_i : \Omega^{\mathbb{N}} \to G$ ($F_0 : \Omega^{\mathbb{N}} \to G$) by $F_i(\{w_j\}) = \sum_{j=1}^{\infty} f_{ij}(w_j)$ ($F_0(\{w_j\}) = \sum_{j=1}^{\infty} f_j(w_j)$). By Proposition 47 each F_i is continuous and by Theorem 46, $F_i \to F_0$ uniformly on $\Omega^{\mathbb{N}}$. The result follows from Lemma 2.57. $\qquad\square$

As noted previously the results above cover the cases of subseries convergent series and bounded multiplier convergent series given in Theorems 2.58 and 3.13. If $\sum_j x_{ij}$ are the series in these statements, in the subseries case we take $\Omega = \{0, 1\}$ and define $f_{ij}(0) = 0$ and $f_{ij}(1) = x_{ij}$. In the bounded multiplier convergent case, we take $\Omega = [0, 1]$ and define $f_{ij}(t) = t x_{ij}$. In both cases the distinguished element is $w_0 = 0$. That $\lim_i f_{ij}(w) = f_j(w)$ converges uniformly for $w \in \Omega$ in the bounded multiplier case follows from Lemma 34.

Applications of multiplier Convergent Series

We will indicate several applications of multiplier convergent series to topics in geometric functional analysis.

We first consider results which involve series which are c_0 multiplier Cauchy and c_0 multiplier convergent. These series are often described in a different way which we now consider.

Let X be a Hausdorff LCTVS.

Definition 4.50. A series $\sum_j x_j$ in X is said to be weakly unconditionally Cauchy (wuc) if $\sum_{j=1}^{\infty} |\langle x', x_j \rangle| < \infty$ for every $x' \in X'$.

Note that a series $\sum_j x_j$ is wuc iff the series $\sum_j x_j$ is subseries Cauchy in the weak topology $\sigma(X, X')$. A series which is subseries convergent in the weak topology $\sigma(X, X')$ is wuc, but a wuc series may not be subseries convergent in the weak topology (consider the series $\sum_j e^j$ in c_0). We give several characterizations of wuc series.

Proposition 4.51. *Let $\{x_j\} \subset X$. The following are equivalent:*

 (i) *The series $\sum_j x_j$ is wuc.*
 (ii) *$\{\langle x', x_j \rangle\} \in l^1$ for every $x' \in X'$.*
 (iii) *The series $\sum_j x_j$ is c_0 multiplier Cauchy.*
 (iv) *$\{\sum_{j \in \sigma} x_j : \sigma$ finite$\}$ is bounded in X.*
 (v) *For every continuous semi-norm p on X, there exists $M > 0$ such that $p(\sum_{j \in \sigma} t_j x_j) \leq M \|t\|_{\infty}$ for every $t \in l^{\infty}$ and σ finite.*
 (vi) *The map $T : c_{00} \to X$, $Tt = \sum_{j=1}^{\infty} t_j x_j$, is linear and continuous.*

(vii) *The series* $\sum_j x_j$ *is* c_0 *multiplier Cauchy in* $\sigma(X, X')$.

Proof. Clearly (i) and (ii) are equivalent, and (iii) implies (ii) is clear.
Assume that (i) holds. If $x' \in X'$ and σ is finite, then

$$\left| \left\langle x', \sum_{j \in \sigma} x_j \right\rangle \right| \leq \sum_{j=1}^{\infty} |\langle x', x_j \rangle| < \infty$$

so $\{\sum_{j \in \sigma} x_j : \sigma \ finite\}$ is $\sigma(X, X')$ bounded and, therefore, bounded in X so (iv) holds.

Assume that (iv) holds. Let p be a continuous semi-norm on X. Set

$$M = 2 \sup \left\{ p \left(\sum_{j \in \sigma} x_j \right) : \sigma \text{ finite} \right\}.$$

By the McArthur/Rutherford inequality (3.1), $p(\sum_{j \in \sigma} t_j x_j) \leq M \|t\|_\infty$ for every $t \in l^\infty$ so (v) holds.

That (v) implies (vi) is immediate.

Suppose that (vi) holds. Then the adjoint operator $T' : X' \to c'_{00} = l^1$ is continuous so $T'x' = \{\langle x', x_j \rangle\} \in l^1$. Therefore, $\sum_{j=1}^{\infty} s_j \langle x', x_j \rangle$ converges for every $s \in c_0$ and $\sum_{j=1}^{\infty} s_j x_j$ is $\sigma(X, X')$ Cauchy or $\sum_j x_j$ is c_0 multiplier Cauchy in $\sigma(X, X')$. Thus, (vii) holds.

Assume that (vii) holds. Then $\sum_{j=1}^{\infty} s_j \langle x', x_j \rangle$ converges for every $x' \in X'$ and for every $s \in c_0$. Hence, $\sum_{j=1}^{\infty} |\langle x', x_j \rangle| < \infty$ for every $x' \in X'$ and (i) holds.

Assume (vi). The series $\sum_j e^j$ is c_0 multiplier Cauchy in c_{00} so $\sum_j Te^j = \sum_j x_j$ is c_0 multiplier Cauchy by (vi). Therefore, (iii) holds. $\square$

Note that it follows from this result that a continuous linear operator between LCTVS carries wuc series into wuc series (condition (iv)).

Corollary 4.52. *Let* $\sum_j x_j$ *be* c_0 *multiplier convergent in* X. *Then*

(i) $\sum_j x_j$ *is wuc,*
(ii) *for every continuous semi-norm* p *on* X *there exists* $M > 0$ *such that* $p(\sum_{j=1}^{\infty} t_j x_j) \leq M \|t\|_\infty$ *for every* $t \in c_0$,
(iii) *the linear map* $T : c_0 \to X$, $Tt = \sum_{j=1}^{\infty} t_j x_j$, *is continuous.*

Proof. (i) and (ii) follow from the proposition above; (iii) follows directly from (ii). $\square$

We can now use the notions of wuc series and c_0 multiplier convergent series to give a characterization of a locally complete LCTVS due to Madrigal and Arrese ([MA]). Recall that a LCTVS X is *locally complete* if for every closed, bounded, absolutely convex set $B \subset X$, the space $X_B = \text{span}\, B$ equipped with the Minkowski functional p_B of B in X_B is complete ([Kö2]).

Theorem 4.53. *The LCTVS X is locally complete iff every wuc series in X is c_0 multiplier convergent.*

Proof. Suppose that X is locally complete and let $\sum_j x_j$ be a wuc series in X. Then $S = \{\sum_{j \in \sigma} x_j : \sigma \text{ finite}\}$ is bounded in X by Proposition 4.51. Let B be the closed, absolutely convex hull of S so (X_B, p_B) is complete. Since S is bounded in (X_B, p_B), $\sum_j x_j$ is wuc in (X_B, p_B) by Proposition 4.51. By the completeness of (X_B, p_B) and condition (iii) of Proposition 4.51, $\sum_j x_j$ is c_0 multiplier convergent in (X_B, p_B). Since the inclusion of (X_B, p_B) into X is continuous, $\sum_j x_j$ is c_0 multiplier convergent in X.

Let B be a closed, bounded, absolutely convex subset of X and suppose that $\{x_j\}$ is Cauchy in (X_B, p_B). Pick an increasing sequence $\{n_j\}$ such that

$$p_B\left(x_{n_{j+1}} - x_{n_j}\right) < 1/j2^j$$

for every j and set $y_j = x_{n_{j+1}} - x_{n_j}$. Then $\sum_{j=1}^{\infty} jy_j$ is p_B absolutely convergent ($\sum_{j=1}^{\infty} p_B(jy_j) \leq \sum_{j=1}^{\infty} 1/2^j < \infty$) so by Proposition 4.51, $\sum_{j=1}^{\infty} jy_j$ is wuc in (X_B, p_B) and, therefore, $\sum_{j=1}^{\infty} jy_j$ is wuc in X. By hypothesis $\sum_{j=1}^{\infty} jy_j$ is c_0 multiplier convergent in X so the series $\sum_j y_j$ is convergent to, say, $y \in X$. Thus,

$$\sum_{j=1}^{k} y_j = x_{n_{k+1}} - x_{n_1} \to y$$

or $x_{n_{j+1}} \to y + x_{n_1} = z$ in X. Now, $\{x_{n_j}\}$ is Cauchy in (X_B, p_B), $\{x_{n_j}\}$ converges in X to z and the topology p_B is linked to the relative topology of X_B from X so $\{x_{n_j}\}$ converges to z in X_B ([Wi] 6.1.9). Thus, X_B is complete with respect to p_B. □

Theorem 4.53 has an interesting corollary due to Madrigal and Arrese ([MA]).

Corollary 4.54. *Let X be a locally complete LCTVS. The following are equivalent:*

(i) *every wuc series in X is subseries convergent,*

(ii) *every wuc series in X is l^∞ multiplier convergent,*

(iii) *every continuous linear operator $T : c_0 \to X$ has a compact extension $T : l^\infty \to X$.*

Proof. Suppose that (i) holds. Let $\sum_j x_j$ be wuc and let $t \in l^\infty$. By Theorem 4.53, $\sum_j x_j$ is c_0 multiplier convergent. By Proposition 4.51 and Corollary 4.52, the series $\sum_j t_j x_j$ is wuc. Hence, $\sum_j t_j x_j$ converges by (i) and (ii) holds.

Suppose that (ii) holds. Let $T : c_0 \to X$ be linear and continuous. Since $\sum e^j$ is wuc in c_0, $\sum_j T e^j$ is wuc in X. By (ii), $\sum_j T e^j$ is l^∞ multiplier convergent. By Corollary 3.12, $\{\sum_{j=1}^\infty t_j T e^j : \|t\|_\infty \le 1\}$ is compact. Therefore, $Tt = \sum_j t_j T e^j$, defines a compact operator from l^∞ into X which extends T. Hence, (iii) holds.

Suppose that (iii) holds. Let $\sum_j x_j$ be wuc in X. By Theorem 4.53, $\sum_j x_j$ is c_0 multiplier convergent so $Tt = \sum_{j=1}^\infty t_j x_j$ defines a continuous linear operator from c_0 into X by Corollary 4.52. By (iii) T is compact so

$$S = \left\{ \sum_{j \in \sigma} x_j : \sigma \text{ finite} \right\}$$

is relatively compact. By Theorem 2.18, $\sum_j x_j$ is subseries convergent. $\qquad \square$

Bessaga and Pelczynski have shown that a Banach space X contains no subspace isomorphic to c_0 iff every wuc series in X is subseries convergent ([BP]). We now extend this characterization to LCTVS. For this we require several preliminary lemmas.

Lemma 4.55. *Let $x_{ij} \in \mathbb{R}$, $\varepsilon_{ij} > 0$ for every $i, j \in \mathbb{N}$. If $\lim_i x_{ij} = 0$ for every j and $\lim_j x_{ij} = 0$ for every i, then there exists an increasing sequence $\{m_j\}$ such that $|x_{m_i m_j}| \le \varepsilon_{ij}$ for $i \ne j$.*

Proof. Set $m_1 = 1$. There exists $m_2 > m_1$ such that $|x_{m_1 j}| < \varepsilon_{12}$ and $|x_{i m_1}| < \varepsilon_{21}$ for all $i, j \ge m_2$. There exists $m_3 > m_2$ such that $|x_{m_1 j}| < \varepsilon_{13}, |x_{m_2 j}| < \varepsilon_{23}, |x_{i m_1}| < \varepsilon_{31}, |x_{i m_2}| < \varepsilon_{32}$ for all $i, j \ge m_3$. Now just continue. $\qquad \square$

Lemma 4.56. *Let X be a semi-normed space and $x_{ij} \in X$ for $i, j \in \mathbb{N}$. If $\lim_i x_{ij} = 0$ for every j and $\lim_j x_{ij} = 0$ for every i, then given $\epsilon > 0$ there exists a subsequence $\{m_j\}$ such that*

$$\sum_{i=1}^\infty \sum_{j \ne i} \|x_{m_i m_j}\| < \epsilon.$$

Proof. Pick $\epsilon_{ij} > 0$ such that $\sum_{i=1}^{\infty} \sum_{j=1}^{\infty} \epsilon_{ij} < \epsilon$. Let $\{m_j\}$ be the subsequence from the lemma applied to the double sequence $\|x_{ij}\|$. Then $\|x_{m_i m_j}\| \le \epsilon_{ij}$ for $i \ne j$ so the result follows. $\qquad\square$

Lemma 4.57. *Let X be a semi-normed space that contains a c_0 multiplier convergent series $\sum_j x_j$ with $\|x_j\| \ge \delta > 0$ for every j. Then there exists a subsequence $\{m_j\}$ such that for any subsequence $\{n_j\}$ of $\{m_j\}$,*

$$T\{t_j\} = Tt = \sum_{j=1}^{\infty} t_j x_{n_j}$$

defines a topological isomorphism of c_0 into X.

Proof. By replacing X by the linear subspace spanned by $\{x_j\}$, we may assume that X is separable. For each j pick $x_j' \in X'$, $\|x_j'\| \le 1$, such that $\langle x_j', x_j \rangle = \|x_j\|$. By the Banach–Alaoglu Theorem, $\{x_j'\}$ has a subsequence which is weak* convergent to an element $x' \in X'$; to avoid cumbersome notation later, assume that $\{x_j'\}$ is weak* convergent to x'. Then

$$|\langle x_j' - x', x_j \rangle| \ge \delta - |\langle x', x_j \rangle| > \delta/2$$

for large j since $\langle x', x_j \rangle \to 0$; again to avoid cumbersome notation assume that

$$|\langle x_j' - x', x_j \rangle| \ge \delta/2$$

for all j. The matrix

$$M = [\langle x_i' - x', x_j \rangle]$$

satisfies the assumption of the lemma above so let $\{m_j\}$ be the subsequence from the lemma with $\epsilon = \delta/4$.

Now define a continuous linear operator $T : c_0 \to X$ by $Tt = \sum_{j=1}^{\infty} t_j x_{m_j}$ (Corollary 4.52). If $z_i' = x_{m_i}' - x'$, then by the conclusion of the lemma, we have

$$2\|T\{t_j\}\| \ge |\langle z_i', T\{t_j\} \rangle| \ge |t_i \langle z_i', x_{m_i} \rangle| - \sum_{j \ne i} |t_j \langle z_i', x_{m_j} \rangle|$$

$$\ge |t_i|\,\delta/2 - \|\{t_j\}\|_{\infty}\,\delta/4.$$

Taking the supremum over all i in the inequality above gives

$$\|T\{t_j\}\| \ge (\delta/8)\,\|\{t_j\}\|_{\infty}$$

so T has a bounded inverse.

The same computation applies to any subsequence $\{n_j\}$ of $\{m_j\}$ so the result follows. $\qquad\square$

We now give a characterization of sequentially complete LCTVS which have the property that any wuc series is subseries convergent. In the statement below, if X is a semi-normed space, $B(X)$ denotes the closed unit ball of X.

Theorem 4.58. *Let X be a sequentially complete LCTVS. The following are equivalent:*

(i) *X contains no subspace (topologically) isomorphic to c_0.*

(ii) *If $\sum_j x_j$ is c_0 multiplier convergent in X, then $x_j \to 0$.*

(iii) *If $\sum_j x_j$ is c_0 multiplier convergent in X, then $\sum_j x_j$ is subseries convergent in X.*

(iv) *If $\sum_j x_j$ is c_0 multiplier convergent in X, then $\sum_j x_j$ is bounded multiplier convergent in X.*

(v) *If $\sum_j x_j$ is c_0 multiplier convergent in X, then $\sum_{j=1}^{\infty} t_j x_j$ converges uniformly for $\{t_j\} \in B(l^{\infty})$.*

(vi) *If $\sum_j x_j$ is c_0 multiplier convergent in X, then $\sum_{j=1}^{\infty} t_j x_j$ converges uniformly for $\{t_j\} \in B(c_0)$.*

(vii) *If $\sum_j x_j$ is c_0 multiplier convergent in X, then $\sum_{j=1}^{\infty} t_j x_j$ converges uniformly for $\{t_j\} \in B(l^1)$.*

(viii) *Every continuous linear operator $T : c_0 \to X$ is compact and has a compact extension to l^{∞}.*

Proof. (i) implies (ii): Suppose there exists a c_0 multiplier convergent series $\sum_j x_j$ with $x_j \nrightarrow 0$. Then we may assume there exists a continuous semi-norm p on X and $\delta > 0$ such that $p(x_j) \geq \delta$ for all j. By Lemma 4.57 there is a subsequence $\{m_i\}$ such that $H\{t_j\} = \sum_{j=1}^{\infty} t_j x_{m_j}$ defines a topological isomorphism from c_0 onto (Hc_0, p). Let I be the continuous inclusion operator from X onto (X, p). By Corollary 4.52, $T\{t_j\} = \sum_{j=1}^{\infty} t_j x_{m_j}$ defines a continuous linear operator from c_0 into X, and $T^{-1} = H^{-1}I$ is continuous so T defines a linear homeomorphism from c_0 into X.

(ii) implies (iii): Suppose there exists a c_0 multiplier convergent series $\sum_j x_j$ in X such that $\sum_j x_j$ diverges. Since X is sequentially complete, $\{s_n\} = \{\sum_{j=1}^{n} x_j\}$ is not Cauchy. Hence, there exist a neighborhood of 0, V, in X and an increasing sequence $\{n_j\}$ such that

$$y_j = s_{n_{j+1}} - s_{n_j} \notin V$$

for all j. Since $\sum_j x_j$ is c_0 multiplier convergent, the series $\sum_{j=1}^\infty t_j y_j$ converges for every $\{t_j\} \in c_0$. By (ii), $y_j \to 0$. This contradiction shows that (ii) implies (iii).

That (iii) implies (iv) is given in Theorem 3.2.

That (iv) implies (v) is given in Theorem 3.2.

That (v) implies (vi) and (vi) implies (vii) is clear.

(vii) implies (ii): Suppose there is a c_0 multiplier convergent series $\sum_j x_j$ in X such that the series $\sum_{j=1}^\infty t_j x_j$ converges uniformly for $\{t_j\} \in B(l^1)$ but $x_j \nrightarrow 0$. There exists a neighborhood of 0, V, and a subsequence $\{x_{n_j}\}$ such that $x_{n_j} \notin V$ for every j. Let $t^k = \{t_j^k\} = e^{n_k} \in B(l^1)$. Then $\sum_{j=1}^\infty t_j^k x_j = x_{n_k} \notin V$ so the series $\sum_{j=1}^\infty t_j x_j$ fail to converge uniformly for $\{t_j\} \in B(l^1)$.

(viii) implies (i) since no continuous, linear, 1-1 map from c_0 into X can have a continuous inverse by the compactness of the map.

Finally, (iv) implies (viii): Let $T : c_0 \to X$ be linear and continuous and set $T e^j = x_j$. Then $\sum_j x_j$ is c_0 multiplier convergent and, hence, bounded multiplier convergent by (iv). By Corollary 3.12,

$$\{T\{t_j\} : \|\{t_j\}\|_\infty \le 1\} = \left\{ \sum_{j=1}^\infty t_j x_j : \|\{t_j\}\|_\infty \le 1 \right\}$$

is compact so (viii) holds. $\qquad\square$

Remark 4.59. The equivalence of (i) and (iii) for the case when X is a Banach space is a well known result of Bessaga and Pelczynski ([BP]). Bessaga and Pelczynski derive their result from results on basic sequences in B-spaces; Diestel and Uhl give a proof based on Rosenthal's Lemma ([DU]I.4.5). The equivalence of (i) and (viii) was noted by Li. The conditions (v),(vi) and (vii) are contained in [LB].

Without the sequential completeness assumption, the conclusions in the theorem above may fail.

Example 4.60. The series $\sum e^j$ is wuc in c_{00} with the sup-norm but is not subseries convergent. However, c_{00} being of countable algebraic dimension does not contain a subspace isomorphic to c_0.

We next derive a result of Pelczynski on unconditionally converging operators. A continuous linear operator T from a Banach space X into a Banach space Y is said to be *unconditionally converging* if T carries wuc series into subseries convergent series ([Pl]). A weakly compact operator is unconditionally converging [recall an operator is weakly compact if it carries bounded sets into relatively weakly compact sets; apply Theorem 2.18]. The identity on l^1 gives a example of an unconditionally converging operator which is not weakly compact [recall that a sequence in l^1 is weakly convergent iff the sequence is norm convergent].

Theorem 4.61. *Let X, Y be Banach spaces and $T : X \to Y$ a continuous linear operator which is not unconditionally converging. Then there exist topological isomorphisms $I_1 : c_0 \to X$ and $I_2 : c_0 \to Y$ such that $TI_1 = I_2$ [i.e., T has a bounded inverse on a subspace isomorphic to c_0].*

Proof. By hypothesis there exists a wuc series $\sum_j x_j$ in X such that $\sum_j Tx_j$ is not subseries convergent. Since $\sum_j Tx_j$ contains a subseries which is not convergent, we may as well assume that the series $\sum_j Tx_j$ diverges. Thus, there exist $\delta > 0$ and a subsequence $\{n_j\}$ such that $\|z_j\| \geq \delta$, where $z_j = Tu_j$ and $u_j = \sum_{i=n_j+1}^{n_{j+1}} x_i$. By Proposition 4.51, the series $\sum_j u_j$ and $\sum_j Tu_j$ are both wuc. Since $\|x\| \geq \|Tx\| / \|T\|$ for $x \in X$, $\|u_j\| \geq \delta / \|T\|$. Applying Lemma 4.57 to the series $\sum_j u_j$ and $\sum_j Tu_j$, there is a subsequence $\{m_j\}$ such that

$$I_1\{t_j\} = \sum_{j=1}^{\infty} t_j u_{m_j} \text{ and } I_2 = \sum_{j=1}^{\infty} t_j Tu_{m_j}$$

define isomorphisms from c_0 into X and Y, respectively. Obviously, $TI_1 = I_2$. $\square$

Remark 4.62. The converse of Theorem 4.61 holds and gives an interesting characterization of unconditionally converging operators (see [Ho]).

We next consider wuc series in the strong dual of a LCTVS.

Theorem 4.63. *Let X be a barrelled LCTVS. The following are equivalent:*

(i) $(X', \beta(X', X))$ *contains no subspace isomorphic to c_0,*

(ii) *every wuc series $\sum_j x'_j$ in X' is $\beta(X', X)$ subseries convergent,*

(iii) *every series $\sum_j x'_j$ in X' which satisfies $\sum_{j=1}^{\infty} |\langle x'_j, x \rangle| < \infty$ for every $x \in X$ is $\beta(X', X)$ subseries convergent,*

(iv) *every continuous linear operator $T : X \to l^1$ is compact.*

Proof. Conditions (i) and (ii) are equivalent by Theorem 4.58 since $\beta(X', X)$ is sequentially complete by the barrelledness of X ([Wi2] 6.1.16 and 9.3.8).

Assume that (ii) holds. Let $\sum_j x'_j$ be such that $\sum_{j=1}^{\infty} |\langle x'_j, x \rangle| < \infty$ for every $x \in X$. Then $\{\sum_{j \in \sigma} x'_j : \sigma \text{ finite}\}$ is weak* bounded and, therefore, $\beta(X', X)$ bounded since X is barrelled. Therefore, $\sum_j x'_j$ is wuc in $(X', \beta(X', X))$ by Proposition 4.51. Hence, $\sum_j x'_j$ is $\beta(X', X)$ subseries convergent by (ii) and (iii) holds.

Assume that (iii) holds. Let $T : X \to l^1$ be linear and continuous. Set $x'_j = T'e^j$. Now T' is $\beta(l^\infty, l^1) - \beta(X', X)$ continuous so $\{x'_j\}$ is $\beta(X', X)$ bounded. For $x \in X$, $Tx \in l^1$ we have

$$\sum_{j=1}^{\infty} |\langle x'_j, x \rangle| = \sum_{j=1}^{\infty} |\langle T'e^j, x \rangle| = \sum_{j=1}^{\infty} |\langle e^j, Tx \rangle| < \infty.$$

By (iii), $\sum_j x'_j$ is $\beta(X', X)$ subseries convergent and, therefore, l^∞ multiplier convergent since $\beta(X', X)$ is sequentially complete as noted above (Theorem 3.2). Therefore, if $B \subset X$ is bounded, then

$$\limsup_n \sum_{x \in B}^{\infty} \sum_{j=n} |\langle x'_j, x \rangle| = \limsup_n \sum_{x \in B}^{\infty} \sum_{j=n} |\langle e^j, Tx \rangle| = 0.$$

Hence, TB is relatively compact in l^1 ([Sw2] 10.15) and (iv) holds.

Assume that (iv) holds. Let $\sum_j x'_j$ be wuc in $(X', \beta(X', X))$. Define $T : X \to l^1$ by $Tx = \{\langle x'_j, x \rangle\}$. T is obviously linear and is $\sigma(X, X') - \sigma(l^1, l^\infty)$ continuous since if $t \in l^\infty, x \in X$,

$$t \cdot Tx = \sum_{j=1}^{\infty} t_j \langle x'_j, x \rangle = \left\langle \sum_{j=1}^{\infty} t_j x'_j, x \right\rangle$$

[the series $\sum_j t_j x'_j$ is $\sigma(X', X)$ Cauchy and, therefore, $\sigma(X', X)$ convergent since X is barrelled ([Wi] 9.3.8)]. Thus, T is $\beta(X, X') - \beta(l^1, l^\infty)$ continuous. By (iv), T is compact. If $B \subset X$ is bounded, TB is relatively compact in l^1 so

$$\limsup_n \sum_{x \in B}^{\infty} \sum_{j=n} |\langle e^j, Tx \rangle| = \limsup_n \sum_{x \in B}^{\infty} \sum_{j=n} |\langle x'_j, x \rangle| = 0$$

([Sw2] 10.15) and $\sum_j x'_j$ is $\beta(X', X)$ convergent. The same argument can be applied to every subseries of $\sum_j x'_j$ so (ii) holds. $\qquad \square$

Remark 4.64. If X is barrelled, then X' is weak* sequentially complete so condition (iii) is equivalent to the statement that every series $\sum_j x'_j$ in X' which is $\sigma(X', X)$ subseries convergent is $\beta(X', X)$ subseries convergent.

Without the barrelledness assumption, the conclusion of Theorem 4.63 may fail.

Example 4.65. Let $X = c_{00}$ with the sup-norm. The series $\sum_j e^j$ in $l^1 = X'$ satisfies the condition (iii) in Theorem 4.63 but is not strongly subseries convergent in l^1 and l^1 contains no subspace isomorphic to c_0.

We can obtain a version of Kalton's Theorem for subseries convergent compact operators.

Theorem 4.66. *Let X be a barrelled LCTVS. The following are equivalent:*

(i) *Every series $\sum_j T_j$ which is subseries convergent in the weak operator topology of $K(X,Y)$ is subseries convergent in $K_b(X,Y)$.*

(ii) *Every continuous linear operator $S : X \to (l^1, \sigma(l^1, m_0))$ is compact.*

(iii) *Every continuous linear operator $S : X \to (l^1, \|\cdot\|_1)$ is compact.*

(iv) *$(X', \beta(X', X))$ contains no subspace isomorphic to c_0.*

(v) *X has the DF property.*

Proof. (i), (ii) and (iii) are equivalent by Theorem 30. Since X is barrelled (iii), (iv) and (v) are equivalent by the theorem above.   □

Remark 4.67. If X and Y are Banach spaces, the equivalence of (i) and (v) is Kalton's result except that Kalton uses the hypothesis that X' contains no subspace isomorphic to l^∞ which is equivalent to the DF property by the Diestel/Faires result ([DF]). For Banach spaces the equivalence of (i) and (iv) was established by Bu and Wu ([BW]).

We next give a characterization of Banach–Mackey spaces in terms of multiplier convergent series. Recall that a LCTVS X is a *Banach–Mackey space* if every $\sigma(X, X')$ bounded subset of X is $\beta(X, X')$ bounded; i.e., if $B \subset X$ is pointwise bounded on X', then B is uniformly bounded on $\sigma(X', X)$ bounded subsets of X' ([Wi2] 10.4.3). The Banach–Mackey Theorem states that any sequentially complete LCTVS is a Banach–Mackey space ([Wi2] 10.4.8).

Let X be a LCTVS. Let X^b (X^s) be the space of all bounded (sequentially continuous) linear functionals on X. Since $X' \subset X^s \subset X^b$, (X, X^s) and (X, X^b) both form dual pairs. We will consider these spaces in more detail in Chapter 5 and give examples showing the containments can be proper. We now give a characterization of Banach–Mackey spaces in terms of l^1 multiplier convergent series and the spaces X^s and X^b.

Theorem 4.68. *Let X be a LCTVS. The following are equivalent:*

(i) *X is a Banach–Mackey space.*
(ii) *If $\{x'_j\}$ is $\sigma(X', X)$ bounded and $\{t_j\} \in l^1$, then $\sum_{j=1}^{\infty} t_j x'_j \in X^s$.*
(iii) *If $\{x'_j\}$ is $\sigma(X', X)$ bounded and $\{t_j\} \in l^1$, then $\sum_{j=1}^{\infty} t_j x'_j \in X^b$.*
(iv) *If $\{x'_j\}$ is $\sigma(X', X)$ Cauchy and $\langle x', x \rangle = \lim \langle x'_j, x \rangle$ for $x \in X$, then $x' \in X^b$.*

Proof. Suppose that (i) holds. Let $x_j \to 0$ in X. Then $\{x_j\}$ is bounded in X and, therefore, $\beta(X, X')$ bounded by (i). Hence,

$$M = \sup\{|\langle x'_i, x_j \rangle| : i, j \in \mathbb{N}\} < \infty$$

and

$$\left| \sum_{j=n}^{\infty} t_j \langle x'_j, x_i \rangle \right| \leq M \sum_{j=n}^{\infty} |t_j|$$

for $\{t_j\} \in l^1$. Therefore, the series $\sum_{j=1}^{\infty} t_j \langle x'_j, x_i \rangle$ converge uniformly for $i \in \mathbb{N}$. Hence,

$$\lim_i \sum_{j=1}^{\infty} t_j \langle x'_j, x_i \rangle = \sum_{j=1}^{\infty} t_j \lim_i \langle x'_j, x_i \rangle = 0$$

so $\sum_{j=1}^{\infty} t_j x'_j \in X^s$ and (ii) holds.

That (ii) implies (iii) is immediate.

Assume that (iii) holds. We show that (i) holds. Let $A \subset X$ be $\sigma(X, X')$ bounded and $B \subset X'$ be $\sigma(X', X)$ bounded. We show that

$$\sup\{|\langle x', x \rangle| : x' \in B, x \in A\} < \infty.$$

If this fails to hold, there exist $\{x'_j\} \subset B$ and $\{x_j\} \subset A$ such that

$$(\#) \quad |\langle x'_i, x_i \rangle| > i^2 \text{ for every } i.$$

Consider the matrix

$$M = [m_{ij}] = [(1/j) \langle x'_j, (1/i)x_i \rangle].$$

We claim that M is a $\mathcal{K}$-matrix. First, the columns of M converge to 0 since $\{x_i\}$ is $\sigma(X, X')$ bounded. Given any subsequence $\{m_j\}$ pick a further subsequence $\{n_j\}$ such that $\sum_{j=1}^{\infty} 1/n_j < \infty$. By (iii)

$$\left\langle \sum_{j=1}^{\infty} (1/n_j) x'_{n_j}, (1/i)x_i \right\rangle = \sum_{j=1}^{\infty} (1/n_j) \left\langle x'_{n_j}, (1/i)x_i \right\rangle \to 0.$$

Hence, M is a $\mathcal{K}$-matrix so by the Antosik–Mikusinski Matrix Theorem (Appendix E) the diagonal of M converges to 0. But, this contradicts (#).

We next show that (i) implies (iv). Let $A \subset X$ be bounded. Then $\{x'_j\}$ is $\beta(X', X)$ bounded by (i). Since $\{x'_j\}$ is $\beta(X', X)$ bounded,

$$\{\langle x'_j, x \rangle : x \in A, j \in \mathbb{N}\}$$

is bounded. Therefore, $\{\langle x', x \rangle : x \in A\}$ is bounded. Therefore, $x' \in X^b$ and (iv) holds.

Suppose that (iv) holds. We show that (iii) holds and this will complete the proof. If $x \in X$ and $\{t_j\} \in l^1$, then

$$\lim_n \sum_{j=1}^n t_j \langle x'_j, x \rangle = \sum_{j=1}^\infty t_j \langle x'_j, x \rangle .$$

By (iv), $\sum_{j=1}^\infty t_j x'_j \in X^b$ and (iii) holds. $\square$

Theorem 4.68 is contained in [LS], Theorem 7, where other characterizations of Banach–Mackey spaces are given.

Chapter 5

The Uniform Boundedness Principle

The Uniform Boundedness Principle (UBP) was one of the early abstract results in the history of functional analysis and has found applications in many areas of analysis (see [Di], [Sw8],[Sw9] for the history). The classic version of the theorem asserts that if X is a Banach space, Y is a normed space and Γ is a subset of the space $L(X,Y)$ of continuous linear operators from X into Y which is pointwise bounded on X, then Γ satisfies

$$(\#) \quad \sup\{\|Tx\| : T \in \Gamma, \|x\| \leq 1\} = \sup\{\|T\| : T \in \Gamma\} < \infty.$$

There are two interpretations of the conclusion $(\#)$. First, the family Γ is uniformly bounded on bounded subsets of X and the other is that the family Γ is equicontinuous. In this chapter we will address the first interpretation; the second interpretation will be addressed in the next chapter.

It should be noted that without the completeness assumption on the domain space X the conclusion $(\#)$ of the UBP may fail.

Example 5.1. Let c_{00} be the space of scalar sequences which are eventually 0 with the sup norm. Let e^j be the sequence with 1 in the jth coordinate and 0 in the other coordinates. The sequence $\{je^j\}$ in l^1, the dual of c_{00}, is pointwise bounded on c_{00}, but is not norm bounded, $\|je^j\|_1 = j$.

In order to obtain versions of the UBP which are valid when the domain space is not complete we seek a family of subsets, $\mathcal{F}$, of the domain space X with the property that any family of continuous linear operators Γ from X into a topological vector space Y which is pointwise bounded on X is uniformly bounded on the members of $\mathcal{F}$. Moreover, the family $\mathcal{F}$ should have the property that it coincides with the family of bounded subsets of X when X is complete. It was shown in [AS1],[Sw1] that the family of $\mathcal{K}$ bounded sets has this property (see the definition of $\mathcal{K}$ bounded sets given below). We will give the analogue of the definition of $\mathcal{K}$ bounded sets for abstract triples below and then consider versions of the UBP for triples. Applications to various versions of the UBP for topological vector spaces will be given.

We begin by recalling the definition of a $\mathcal{K}$ convergent sequence.

Definition 5.2. Let (G, τ) be an Abelian topological group. A sequence $\{x_j\}$ in G is $\tau - \mathcal{K}$ convergent if every subsequence of $\{x_j\}$ has a further subsequence $\{x_{n_j}\}$ such that the subseries $\sum_{j=1}^{\infty} x_{n_j}$ is τ convergent to an element of G.

Note that a $\tau - \mathcal{K}$ convergent sequence converges to 0 since any subsequence has a further subsequence which converges to 0. The converse does not hold.

Example 5.3. In c_{00} the sequence $\{e^j/j\}$ converges to 0 but is not $\mathcal{K}$ convergent since any subseries of $\sum e^j/j$ has infinitely many non-zero coordinates.

However, for complete, quasi-normed groups the converse does hold.

Proposition 5.4. *Let $(G, |\cdot|)$ be a complete, quasi-normed group and $x_j \rightarrow 0$ in G. Then $\{x_j\}$ is $\mathcal{K}$ convergent.*

Proof. Given any subsequence, pick a further subsequence $\{x_{n_j}\}$ with

$$|x_{n_j}| < 1/2^j.$$

Then the subseries $\sum_{j=1}^{\infty} x_{n_j}$ is absolutely convergent, and, therefore, convergent by the completeness. $\qquad \square$

We give an example of a $\mathcal{K}$ convergent sequence in a non-normed space.

Example 5.5. Let $E = l^{\infty}, F = l^1$ with the usual duality. Then $\{e^j\}$ is $\sigma(l^{\infty}, l^1) - \mathcal{K}$ convergent but is not $\beta(l^{\infty}, l^1) - \mathcal{K}$ convergent since $\beta(l^{\infty}, l^1) = \|\cdot\|_{\infty}$.

Note that in this example the sequence is $\mathcal{K}$ convergent in the weak topology and although it is not $\mathcal{K}$ convergent in the strong topology the sequence is strongly bounded. This is always the case.

Proposition 5.6. *Let E, F be a pair of vector spaces in duality. If $\{x_k\}$ is $\sigma(E, F) - \mathcal{K}$ convergent, then $\{x_k\}$ is $\beta(E, F)$ bounded.*

Proof. It suffices to show that $\{x'_k(x_k)\}$ is bounded for every $\sigma(F, E)$ bounded sequence $\{x'_k\}$. Let $t_k \to 0$. Consider the matrix

$$M = [t_i x'_i(x_j)].$$

The columns of M converge to 0 since $\{x'_k\}$ is weak bounded. Given any subsequence there is a further subsequence $\{n_j\}$ such that the series $\sum_{j=1}^{\infty} x_{n_j}$ is $\sigma(E, F)$ convergent to some $x \in E$. Then

$$t_i \sum_{j=1}^{\infty} x'_i(x_{n_j}) = t_i x'_i(x) \to 0$$

so M is a $\mathcal{K}$ matrix whose diagonal $t_i x'_i(x_i) \to 0$ by the Antosik–Mikusinski Theorem. Thus, $\{x'_k(x_k)\}$ is bounded. $\square$

We give another example of a $\mathcal{K}$ convergent sequence which will be used later.

Example 5.7. Let G be a quasi-normed group, Σ a σ-algebra of subsets of a set S and let $\{m_k\}$ be a sequence of finitely additive, G valued set functions on Σ. Consider the triple

$$(\Sigma, \{m_k\} : G)$$

under the map $(A, m_k) \to m_k(A)$. If $\{A_j\}$ is a pairwise sequence from Σ, then $\{A_k\}$ is $w(\Sigma, \{m_k\}) - \mathcal{K}$ convergent by Drewnowski's Lemma (Appendix D). For any subsequence has a further subsequence $\{n_j\}$ such that each m_k is countably additive on the σ-algebra generated by $\{A_{n_j}\}$ so

$$\sum_{j=1}^{\infty} m_k(A_{n_j}) = m_k(\cup_{j=1}^{\infty} A_{n_j}).$$

Definition 5.8. A group in which null convergent sequences are $\mathcal{K}$ convergent is called a $\mathcal{K}$ space.

There are $\mathcal{K}$ spaces which are not complete. Klis has given an example of a normed space which is a $\mathcal{K}$ space but is not complete ([Kl]) and Burzyk, Klis and Lipecki have shown that any infinite dimensional F-space contains a subspace which is a Baire space but is not a $\mathcal{K}$ space ([BKL]).

The notion of a $\mathcal{K}$ space was originally introduced in an equivalent form by Mazur and Orlicz ([MO]) where it was observed that the classical UBP holds for such spaces. Alexiewicz also studied this notion ([Al]). The definition of $\mathcal{K}$ convergence was rediscovered in the seminar of J. Mikusinski in Katowice, Poland (hence, the appellation $\mathcal{K}$).

From Proposition 5.4 complete quasi-normed spaces are $\mathcal{K}$ spaces. We give an example of a non-normed $\mathcal{K}$ space.

Example 5.9. Since weakly convergent sequences are norm convergent in l^1,

$$(l^1, \sigma(l^1, l^\infty))$$

is a $\mathcal{K}$–space. Similarly, $(l^1, \sigma(l^1, m_0))$ is a $\mathcal{K}$–space.

We use the analogue of $\mathcal{K}$ convergence to define UB sequences in triples.

Definition 5.10. Let G be an Abelian topological group and $(E, F : G)$ an abstract triple. A sequence $\{x_j\}$ in E is a UB sequence in the triple $(E, F : G)$ if every subsequence of $\{x_j\}$ has a further subsequence $\{x_{n_j}\} = \sigma$ such that the series $\sum_{j=1}^\infty x_{n_j}$ is $w(E, F)$ convergent in E, i.e., there exists $x_\sigma \in E$ such that

$$\sum_{j=1}^\infty x_{n_j} \cdot y = x_\sigma \cdot y$$

for all $y \in F$.

If E, F is a pair of vector spaces in duality, then a sequence $\{x_j\}$ in E is a UB sequence in the triple $(E, F : \mathbb{R})$ iff $\{x_j\}$ is $\sigma(E, F)$-$\mathcal{K}$ convergent. If (E, τ) is a TVS with dual F, then a sequence $\{x_j\}$ which is $\tau - \mathcal{K}$ convergent is $\sigma(E, F) - \mathcal{K}$ convergent and, therefore, a UB sequence in the triple $(E, F : \mathbb{R})$.

We now establish our first UBP for triples. Let G be a TVS. A subset $B \subset F$ is *pointwise bounded* on E in the triple $(E, F : G)$ if

$$\{x \cdot y : y \in B\} = x \cdot B$$

is bounded in G for every $x \in E$. If $A \subset E$, then B is *uniformly bounded* on A if

$$\{x \cdot y : y \in B, x \in A\} = A \cdot B$$

is bounded in G.

Theorem 5.11. (UBP1) *If $B \subset F$ is pointwise bounded on E and $\{x_j\} = A$ is a UB sequence, then $A \cdot B$ is bounded.*

Proof. Since any subsequence of a UB sequence is a UB sequence, it suffices to show $t_i(x_i \cdot y_i) \to 0$ when $t_i \to 0$, $\{y_i\} \subset B$. For this consider the matrix

$$M = [t_i(x_j \cdot y_i)].$$

We show M is a $\mathcal{K}$ matrix. First, the columns of M converge to 0 by the pointwise boundedness assumption. Next, given any subsequence there is a further subsequence $\{x_{n_j}\}$ such that the subseries $\sum_{j=1}^{\infty} x_{n_j}$ is $w(E, F)$ convergent to some $x \in E$. Then

$$\sum_{j=1}^{\infty} t_i(x_{n_j} \cdot y_i) = t_i(x \cdot y_i) \to 0$$

by the pointwise bounded assumption. Hence, M is a $\mathcal{K}$ matrix and by the Antosik–Mikusinski Theorem the diagonal of M, $t_i(x_i \cdot y_i)$, converges to 0 as desired. $\qquad\square$

We indicate an application of Theorem 5.11 to the Nikodym Boundedness Theorem.

Theorem 5.12. (Nikodym) *Assume that G is a semi-convex, quasi-normed TVS. Let Σ be a σ-algebra of subsets of a set S and $\mathcal{M}$ a family of G valued, bounded, finitely additive set functions defined on Σ. If $\mathcal{M}$ is pointwise bounded on Σ, then $\mathcal{M}$ is uniformly bounded on Σ.*

Proof. It suffices to show $\{m_i(A_i)\}$ is bounded for every $\{m_i\} \subset \mathcal{M}$ and pairwise disjoint $\{A_i\}$ from Σ (Appendix C). Consider the triple

$$(\Sigma, \{m_i\} : G)$$

under the map $(A, m_i) \to m_i(A)$. By the computation in Example 5.7, $\{A_j\}$ is a UB sequence so $\{m_i(A_i)\}$ is bounded by Theorem 5.11. $\qquad\square$

We use Theorem 5.11 to derive uniform boundedness principles for linear operators. Let $(E, \tau), G$ be TVS and $LS(E, G)$ the space of all sequentially continuous linear operators from E into G. Let $\Gamma \subset LS(E, G)$. Consider the triple

$$(E, \Gamma : G)$$

under the map $(x, T) \to Tx$. From Theorem 5.11 we obtain the UBP of [AS1], [Sw1] for $\mathcal{K}$ convergent sequences.

Theorem 5.13. *If Γ is pointwise bounded on E and $\{x_k\}$ is a $w(E, \Gamma)$-$\mathcal{K}$ convergent sequence, then*

$$\Gamma \cdot \{x_k\} = \{Tx_k : T \in \Gamma, k \in \mathbb{N}\}$$

is bounded in G.

Note the topology depends on the set Γ which is weaker than the topology $w(E, LS(E, G))$ in the triple

$$(E, LS(E, G) : G)$$

and this topology is weaker than τ, the topology of E. Hence, we have

Corollary 5.14. *If Γ is pointwise bounded on E and $\{x_k\}$ is a $w(E, LS(E, G))$ (or τ) -$\mathcal{K}$ convergent sequence, then*

$$\Gamma \cdot \{x_k\} = \{Tx_k : T \in \Gamma, k \in \mathbb{N}\}$$

is bounded in G.

We will now use the corollary to derive classic versions of the UBP. For this we need an observation.

Proposition 5.15. *Let G be a LCTVS. Γ is uniformly bounded on bounded subsets of E iff Γ is bounded on null sequences in E.*

Proof. Suppose there exists a bounded subset B of E such that ΓB is not bounded. Then there exists a continuous semi-norm p on E such that

$$\sup\{p(Tx) : T \in \Gamma, x \in B\} = \infty.$$

There exist $\{T_k\} \subset \Gamma, \{x_k\} \subset B$ with

$$p(T_k x_k) > k^2.$$

Then $x_k/k \to 0$ but $p(T_k(x_k/k)) > k$ so Γ is not bounded on the null sequence $\{x_k/k\}$. The other implication is obvious. $\square$

From the corollary and the proposition, we have the following UBP.

Theorem 5.16. *Let E be a $\mathcal{K}$ space and G a LCTVS. If Γ is pointwise bounded on E, then Γ is uniformly bounded on bounded subsets of E.*

In particular, if E is a Banach space and G is a normed space the conclusion of the theorem holds; this is the classic form of the UBP for normed spaces (see (#)).

A subset B of a TVS E is *bounded* iff $t_i x_i \to 0$ whenever $\{x_i\} \subset B$ and $t_i \to 0$. We can use the analogue of this statement and $\mathcal{K}$ convergence to strengthen the notion of boundedness.

Definition 5.17. *A subset B of a TVS (E, τ) is $\mathcal{K}$ bounded with respect to τ or is $\tau - \mathcal{K}$ bounded if the sequence $\{t_i x_i\}$ is $\tau - \mathcal{K}$ convergent whenever $\{x_i\} \subset B$ and $t_i \to 0$.*

A $\mathcal{K}$ bounded set is bounded but not conversely.

Example 5.18. The sequence $\{e^k\}$ is bounded in the sup-norm of c_{00} but is not norm $\mathcal{K}$ bounded.

From Proposition 5.4 it follows that in a complete metric linear space a subset is bounded iff it is $\mathcal{K}$ bounded.

Definition 5.19. A TVS in which bounded sets are $\mathcal{K}$ bounded is called an $\mathcal{A}$ space.

From the remarks above, a complete metric linear space is an $\mathcal{A}$ space.
    A $\mathcal{K}$ space is obviously an $\mathcal{A}$ space but there are $\mathcal{A}$ spaces which are not $\mathcal{K}$ spaces; we will give examples of such spaces below.
    A null sequence in a TVS is always bounded, but, unfortunately, a $\mathcal{K}$ convergent sequence is not always $\mathcal{K}$ bounded.

Example 5.20. Let m_0 be the subspace of l^∞ of sequences with finite range. Pick $\{x_k\} \subset l^1$ with $x_k \neq 0$ for all k. Define a norm (induced by $\{x_k\}$) on m_0 by

$$\|\{t_k\}\| = \sum_{k=1}^{\infty} |x_k t_k|.$$

Consider the sequence $\{e^k\}$ in $(m_0, \|\cdot\|)$. The series $\sum e^k$ is subseries convergent with respect to $\|\cdot\|$ since

$$\left\| \sum_{j=1}^{n} e^{k_j} - \chi_{\{k_j : j \in \mathbb{N}\}} \right\| \leq \sum_{j=n}^{\infty} |x_j| \to 0$$

for any subsequence $\{k_j\}$. Hence, $\{e^k\}$ is norm $\mathcal{K}$ convergent. However, $\{e^k\}$ is not norm $\mathcal{K}$ bounded since no subseries of the series $\{e^k/k\}$ converges in m_0 with respect to the norm.

We use the analogue of $\mathcal{K}$ boundedness to define UB sets for triples.
    Let G be a TVS.

Definition 5.21. A subset $B \subset E$ is a UB set for the triple $(E, F : G)$ if whenever $\{x_k\} \subset B$ and $t_k \to 0$ every subsequence has a further subsequence $\{n_k\} = \sigma$ and an element $x_\sigma \in E$ such that

$$\sum_{k=1}^{\infty} t_{n_k} (x_{n_k} \cdot y) = x_\sigma \cdot y$$

for every $y \in F$.

Remark 5.22. If E is a vector space so scalar products are defined, then B is a UB set if the sequence $\{t_k x_k\}$ in the definition is a $w(E, F) - \mathcal{K}$ convergent sequence; this is the analogue of $\mathcal{K}$ boundedness.

We establish a UBP for UB sets.

Theorem 5.23. *Let $(E, F : G)$ be a triple. If $A \subset F$ is pointwise bounded on E and $B \subset E$ is a UB set for the triple, then*

$$B \cdot A = \{x \cdot y : x \in B, y \in A\}$$

is bounded in G.

Proof. Let $\{y_k\} \subset A, \{x_k\} \subset B$ and $t_k \to 0$ with $t_k > 0$. Consider the matrix

$$M = [\sqrt{t_i} \sqrt{t_j} (x_j \cdot y_i)].$$

The columns of M converge to 0 by the pointwise bounded assumption. Next, since B is a UB set, given any subsequence there is a further subsequence $\{n_j\} = \sigma$ and $x_\sigma \in E$ such that

$$\sum_{j=1}^{\infty} \sqrt{t_{n_j}} (x_{n_j} \cdot y_i) = x_\sigma \cdot y_i$$

for every i. Then

$$\sqrt{t_{n_i}} \sum_{j=1}^{\infty} \sqrt{t_{n_j}} (x_{n_j} \cdot y_i) = \sqrt{t_{n_i}} (x_\sigma \cdot y_i) \to 0$$

by pointwise boundedness. Thus, M is a $\mathcal{K}-$ matrix and by the Antosik–Mikusinski Theorem the diagonal of M, $\{t_i(x_i \cdot y_i)\}$, converges to 0 so $B \cdot A$ is bounded. $\qquad\qquad\square$

We can now obtain UBP's for linear operators.

Theorem 5.24. *Let E, G be TVS and $\Gamma \subset LS(E, G)$. If Γ is pointwise bounded on E and $B \subset E$ is $w(E, \Gamma) - \mathcal{K}$ bounded, then $\Gamma(B)$ is bounded in G.*

This follows from Theorem 5.23 applied to the triple $(E, \Gamma : G)$ under the map $(x, T) \to Tx$. See [AS1], [Sw1] for this version of the UBP.

Corollary 5.25. *Let $(E, \tau), G$ be TVS and $\Gamma \subset LS(E, G)$. If Γ is pointwise bounded on E and $B \subset E$ is $w(E, LS(E, G)) - \mathcal{K}$ bounded (or $\tau - \mathcal{K}$ bounded), then $\Gamma(B)$ is bounded in G.*

Corollary 5.26. *If E is an $\mathcal{A}$ space and Γ is pointwise bounded on E, then Γ is uniformly bounded on bounded subsets of E.*

We give examples of $\mathcal{K}$ bounded sets.

Proposition 5.27. *Let E be a TVS. If $B \subset E$ is bounded, absolutely convex and sequentially complete, then B is $\mathcal{K}$ bounded.*

Proof. Let $\{x_k\} \subset B$ and $t_k \to 0$. Choose a subsequence $\{n_k\}$ such that $\sum_{k=1}^{\infty} |t_{n_k}| \leq 1$. The partial sums $s_p = \sum_{j=1}^{p} t_{n_j} x_{n_j}$ of the series $\sum_{k=1}^{\infty} t_{n_k} x_{n_k}$ are Cauchy. To see this, let U be a neighborhood of 0 in E. There exists $\delta > 0$ such that $tB \subset U$ when $|t| \leq \delta$. Pick k such that $\sum_{j=k}^{\infty} |t_{n_j}| \leq \delta$. If $p > q \geq k$ and $t = \sum_{j=q}^{p} t_{n_j}$, then since B is absolutely convex

$$s_p - s_q = t \sum_{j=q+1}^{p} (t_{n_j}/t) x_{n_j} \in tB \subset U$$

justifying the claim. The series $\sum_{k=1}^{\infty} t_{n_k} x_{n_k}$ converges by the completeness assumption and B is $\mathcal{K}$ bounded. □

From Theorem 5.23 and the proposition above, we have a generalization of the Banach–Mackey Theorem

Corollary 5.28. (Banach–Mackey Theorem for Operators) *Let $\Gamma \subset LS(E, G)$ be pointwise bounded on E and $A \subset E$ be absolutely convex, $w(E, \Gamma)$ bounded and $w(E, \Gamma)$ sequentially complete. Then $\Gamma(A)$ is bounded.*

Since a compact set is sequentially complete ([Wi2] 6.1.18), we have

Corollary 5.29. *Let E be a TVS. If $B \subset E$ is absolutely convex and compact, then B is $\mathcal{K}$ bounded.*

In a LCTVS the convex hull of a compact set is compact so we also have

Corollary 5.30. *If E is a sequentially complete LCTVS and $B \subset E$ is compact, then B is $\mathcal{K}$ bounded. Thus, a sequentially complete LCTVS is an $\mathcal{A}$ space.*

The convexity assumptions in the results above are important.

Example 5.31. Let $E = c_{00}$ with the weak topology $\sigma(c_{00}, l^1)$. Then

$$\{e^j : j \in \mathbb{N}\} \cup \{0\}$$

is compact but is not $\mathcal{K}$ bounded.

We have the analogue of Proposition 5.6 for $\mathcal{K}$ bounded sets.

Proposition 5.32. *Let E, F be a pair of vector spaces in duality. If B is $\sigma(E, F) - \mathcal{K}$ bounded, then B is $\beta(E, F)$ bounded.*

Proof. Let $\{x_k\} \subset B$ and $t_k \to 0$ with $t_k > 0$. Then $\{\sqrt{t_k}x_k\}$ is $\sigma(E, F) - \mathcal{K}$ convergent and is $\beta(E, F)$ bounded by Proposition 5.6. Therefore,

$$\sqrt{t_k}\sqrt{t_k}x_k = t_k x_k \to 0$$

in $\beta(E, F)$ and the result follows. $\square$

We thus have

Corollary 5.33. *If E is a sequentially complete LCTVS, then bounded sets of E are $\beta(E, E')$ bounded.*

A LCTVS in which bounded sets are strongly bounded is called a *Banach–Mackey space* ([Wi2] 10.4). Thus, a sequentially complete LCTVS is a Banach–Mackey space; this result is often referred to as the Banach–Mackey Theorem ([Wi2]). More generally, we have

Corollary 5.34. *A locally convex $\mathcal{A}$ space is a Banach–Mackey space.*

Again, since compact sets are sequentially complete, we have

Corollary 5.35. (Mackey Theorem for Operators) *Let $\Gamma \subset LS(E, G)$ be pointwise bounded on E and $A \subset E$ be absolutely convex, $w(E, \Gamma)$ bounded and $w(E, \Gamma)$ compact, then $\Gamma(A)$ is bounded.*

Remark 5.36. Let E, F be a pair of vector spaces in duality. Let $\mathcal{D}$ be the set of all absolutely convex, $\sigma(F, E)$ bounded, $\sigma(F, E)$ sequentially complete subsets of F and let $\mathcal{T}$ be the set of all absolutely convex, $\sigma(F, E)$ bounded, $\sigma(F, E)$ compact subsets of F. Let $\delta(E, F)$ $[\tau(E, F)]$ be the topology of uniform convergence on the elements of $\mathcal{D}$ $[\mathcal{T}]$ so $\tau(E, F)$ is just the Mackey topology and $\delta(E, F) \supset \tau(E, F)$. Mackey's Theorem asserts that every $\sigma(E, F)$ bounded set is $\tau(E, F)$ bounded while the Banach–Mackey Theorem for operators asserts that every $\sigma(E, F)$ bounded set is $\delta(E, F)$ bounded. The topology $\delta(E, F)$ can be strictly stronger than the Mackey topology $\tau(E, F)$. For example, by the Hahn–Schur Theorem, $\sigma(l^1, l^\infty)$ is sequentially complete and

$$\tau(l^\infty, l^1) \subsetneqq \delta(l^\infty, l^1) = \|\cdot\|_\infty = \beta(l^\infty, l^1).$$

Hence, the Banach–Mackey Theorem for operators gives an improvement of Mackey's Theorem.

We can use the results above to give examples of $\mathcal{A}$ spaces which are not $\mathcal{K}$ spaces.

Proposition 5.37. *Let E be a LCTVS.*

(i) *If E is semi-reflexive, then $(E, \sigma(E, E'))$ is an $\mathcal{A}$ space.*

(ii) *If E is barrelled, then $(E', \sigma(E', E))$ is an $\mathcal{A}$ space.*

Proof. Both topologies in (i) and (ii) are sequentially complete ([Köl] 23.3(2), [Kö2] 39.5). $\qquad\square$

Example 5.38. $(l^p, \sigma(l^p, l^q)), 1 < p < \infty, \frac{1}{p} + \frac{1}{q} = 1$, are $\mathcal{A}$ spaces but are not $\mathcal{K}$ spaces (consider $\{e^j\}$).

The examples above are non-metrizable. This is always the case.

Proposition 5.39. *A metrizable $\mathcal{A}$ space E is a $\mathcal{K}$ space.*

Proof. Let $x_k \to 0$ in E. There is a scalar sequence $t_k \uparrow \infty$ such that $t_k x_k \to 0$ (Appendix A). Then $\{t_k x_k\}$ is bounded and, therefore, $\mathcal{K}$ bounded so $\{\frac{1}{t_k}(t_k x_k)\} = \{x_k\}$ is $\mathcal{K}$ convergent. $\qquad\square$

Remark 5.40. Actually the proof above shows that any braked $\mathcal{A}$ space E is a $\mathcal{K}$ space (Appendix A).

We now use the results above to study the spaces of sequentially continuous and bounded linear operators on a TVS. If E is a TVS, the space of all sequentially continuous [bounded] linear functionals on E is denoted by E^s [E^b]. Thus, we have $E' \subset E^s \subset E^b$ and the containments can be proper as the following example shows.

Example 5.41. The identity map from $(c_0, \sigma(c_0, l^1)) \to (c_0, \|\cdot\|_\infty)$ is not sequentially continuous but is bounded. The identity map from $(l^1, \sigma(l^1, l^\infty)) \to (l^1, \|\cdot\|_1)$ is sequentially continuous by the Hahn–Schur Theorem but is not continuous.

For $\mathcal{K}$ spaces the spaces of sequentially continuous linear functionals and the space of bounded linear functionals coincide.

Proposition 5.42. ([LSC]) *If E is a $\mathcal{K}$ space, then $E^s = E^b$.*

Proof. Suppose there exists $x' \in E^b \setminus E^s$. Then there exist $\delta > 0$ and $x_k \to 0$ with $x'(x_k) \geq \delta$ for all k. There exists a subsequence $\{x_{n_k}\}$ of $\{x_k\}$

such that $\sum x_{n_k}$ converges to some $x \in E$. Now $\{\sum_{k=1}^{m} x_{n_k}\}$ is bounded but

$$\left| x'\left(\sum_{k=1}^{m} x_{n_k}\right) \right| = \left| \sum_{k=1}^{m} x'(x_{n_k}) \right| \geq m\delta$$

so x' is not bounded, a contradiction.   □

We can extend this result to operators.

Theorem 5.43. *Let* (E, τ) *be a* $\mathcal{K}$ *space. If* p *is a bounded semi-norm on* E, *then* p *is sequentially continuous.*

Proof. Let $x_k \to 0$ in E. To show $p(x_k) \to 0$ it suffices to show there is a subsequence $\{x_{n_k}\}$ such that $p(x_{n_k}) \to 0$. From the proposition above, we have $(E, p)' \subset (E, \tau)^b = (E, \tau)^s$. We may assume that (E, p) is separable by replacing E by the closed, linear span of $\{x_k\}$ in (E, p), if necessary. For each k there exists $x'_k \in (E, p)'$ with $\|x'_k\| \leq 1$ and $x'_k(x_k) = p(x_k)$. By the Banach–Alaoglu Theorem and the separability of (E, p), there is a subsequence $\{x'_{n_k}\}$ and $x' \in (E, p)'$ such that $x'_{n_k}(x) \to x'(x)$ for all $x \in E$. Consider the matrix

$$M = [x'_{n_i}(x_{n_j})].$$

The columns of M converge by the weak* convergence of $\{x'_{n_k}\}$. Next, given any subsequence of $\{x_{n_j}\}$ there is a further subsequence $\{x_{p_j}\}$ such that the series $\sum_{j=1}^{\infty} x_{p_j}$ is τ convergent to some $x \in E$. As noted above each $x'_i \in E^s$ so

$$\sum_{j=1}^{\infty} x'_{n_i}(x_{p_j}) = x'_{n_i}(x) \to x'(x).$$

Thus, M is a $\mathcal{K}$ matrix and the diagonal of M,

$$\{x'_{n_i}(x_{n_i})\} = p(x_{n_i}) \to 0$$

by the Antosik–Mikusinski Theorem.   □

Corollary 5.44. *Let* E *be a* $\mathcal{K}$ *space and* G *a LCTVS. If* $T : E \to G$ *is linear and bounded, then* T *is sequentially continuous.*

Proof. Let p be a continuous semi-norm on G. Then pT is a bounded semi-norm on E so by the theorem pT is sequentially continuous and the result follows.   □

We also have a sequential completeness result for sequentially continuous linear operators on $\mathcal{K}$ spaces.

Theorem 5.45. [LSC] *Let E be a $\mathcal{K}$ space and G a Hausdorff TVS. If $\{T_k\}$ is sequence of linear, sequentially continuous operators from E into G such that*

$$\lim_k T_k x = T x$$

exists for every $x \in E$, then T is linear and sequentially continuous.

Proof. Let $x_j \to 0$ in E. Consider the matrix

$$M = [T_i x_j].$$

The columns of M converge by hypothesis. Next, given any subsequence of $\{x_j\}$ there is a further subsequence $\{x_{n_j}\}$ such that the series $\sum_{j=1}^{\infty} x_{n_j}$ converges to some $x \in E$. Then

$$T_i \left(\sum_{j=1}^{\infty} x_{n_j} \right) = \sum_{j=1}^{\infty} T_i x_{n_j} = T_i x \to T x.$$

So M is a $\mathcal{K}$ matrix. By the Antosik–Mikusinski Theorem $\lim_i T_i x_j = 0$ uniformly for $j \in \mathbb{N}$. By the Iterated Limit Theorem,

$$\lim_j \lim_i T_i x_j = \lim_i \lim_j T_i x_j = \lim_j T x_j = 0$$

so T is sequentially continuous. $\qquad\square$

Corollary 5.46. ([LSC]) *Let E be a $\mathcal{K}$ space. Then $(E^s, \sigma(E^s, E))$ is sequentially complete.*

We next consider some results for bounded linear functionals. A subset of a TVS is a *bornivore* if it is absolutely convex and absorbs bounded sets. An absolutely convex neighborhood of 0 is a bornivore and a LCTVS in which bornivores are neighborhoods of 0 is called a *bornological space*. We have the following characterization of bornological spaces.

Theorem 5.47. *Let E be a LCTVS. The following are equivalent.*

(i) *E is bornological,*
(ii) *every bounded semi-norm q on E is continuous,*
(iii) *every bounded linear operator from E into a LCTVS G is continuous.*

Proof. (i)$\Longrightarrow$(ii): Let $\epsilon > 0$. It suffices to show that q is continuous at 0. If

$$A = \{x : q(x) \leq \epsilon\},$$

then A is absolutely convex and if $B \subset E$ is bounded, there exists $t > 0$ such that $q(x) \leq t$ for all $x \in B$ so $B \subset (t/\epsilon)A$. Thus, A is a bornivore and a neighborhood of 0 by hypothesis.

(ii)$\Longrightarrow$(iii): Suppose $T : E \to G$ is linear and bounded. Let q be a continuous semi-norm on E. Then qT is a bounded semi-norm on E and is, therefore, continuous by (ii). Hence, T is continuous.

(iii)$\Longrightarrow$(i): Suppose V is a bornivore in E. Then V is absolutely convex and absorbing so its Minkowski functional $p_V = p$ is a semi-norm on E. Consider the identity map $I : E \to (E, p)$. If $B \subset E$ is bounded, there exists $t > 0$ with $B \subset tV$. Since

$$V \subset \{x : p(x) \leq 1\},$$

$p(x) \leq t$ for all $x \in B$. Therefore, I is bounded and continuous by (iii). Since $\{x : p(x) < 1\} \subset V$, V is a neighborhood of 0 and E is bornological. $\qquad\square$

Since any bounded linear operator on a metric linear space is continuous ([Sw2] 5.4), we have

Corollary 5.48. *Any metrizable LCTVS is bornological.*

For bornological spaces we have the following criterion for continuity.

Corollary 5.49. *Let E be bornological, G a LCTVS and $T : E \to G$ linear. The following are equivalent.*

(i) *T is continuous,*
(ii) *T is bounded,*
(iii) *T is sequentially continuous.*

If E, F are a pair of vector spaces in duality, the topology $\beta^*(E, F)$ is the polar topology on E of uniform convergence on the $\beta(F, E)$ bounded subsets of F. Since any $\beta(F, E)$ subset is $\sigma(F, E)$ bounded, $\beta(E, F)$ is stronger than $\beta^*(E, F)$ and can be strictly stronger (see [Sw2]). The topology $\beta^*(E, F)$ is stronger than $\tau(E, F)$ and has the interesting property that it has the same bounded sets as $\tau(E, F)$ ([Sw2] 20.3,20.6). From these observations, we have

Theorem 5.50. *If (E, τ) is bornological, then E has $\beta^*(E, F)$. Therefore,*

$$\tau = \tau(E, F) = \beta^*(E, F).$$

Proof. The identity map $I : (E, \tau) \to (E, \beta^*(E, F))$ is bounded and, therefore, continuous. Since $\beta^*(E, F) \supset \tau(E, F)$, the result follows. $\qquad\square$

Let (E, τ) be a LCTVS. The family of all absolutely convex subsets of E which absorb all the bounded sets, i.e., the *bornivores*, form a subbase for a locally convex topology on E called the *bornological topology* and is denoted by τ^b. The space (E, τ^b) is called the *bornological space* associated with (E, τ).

The topology τ^b has the following properties.

Theorem 5.51.

(i) τ^b *is the strongest locally convex topology on E which has the same bounded sets as (E, τ).*

(ii) (E, τ^b) *is bornological.*

(iii) (E, τ) *is bornological iff $\tau^b = \tau$.*

Proof. (i): Any locally convex topology on E which has the same bounded sets as (E, τ) must have a neighborhood base which consists of absolutely convex bornivores.

(ii): Let G be a LCTVS and $T : E \to G$ be a bounded, linear operator. If V is an absolutely convex neighborhood of 0 in G, then $T^{-1}V = U$ is absolutely convex, and we claim that U is a bornivore so it must be a τ^b neighborhood of 0. Let $B \subset E$ be bounded. Then TB is bounded so there exists $t > 0$ such that $TB \subset tV$. Thus, $B \subset tU$ and U is a bornivore. Since T is continuous with respect to τ^b, (E, τ^b) is bornological by (iii) of Theorem 5.47.

(iii) If $\tau^b = \tau$, then (E, τ) is bornological by (ii). On the other hand, if (E, τ) is bornological, the identity $I : (E, \tau) \to (E, \tau^b)$ is bounded and so continuous. Therefore, $\tau^b \subset \tau$. But, always $\tau^b \supset \tau$ so $\tau^b = \tau$. $\qquad\square$

We have the dual of τ^b.

Theorem 5.52. $(E, \tau^b)' = E^b$.

Proof. First, suppose $f \in (E, \tau^b)'$. Let $B \subset E$ be bounded. Then

$$V = \{x : |f(x)| \le 1\}$$

is a τ^b neighborhood of 0 and, therefore, absorbs B. So there exists $t > 0$ with $tB \subset V$. Hence, $|f(z)| \leq 1/t$ for all $z \in B$ and $f(B)$ is bounded and f is bounded or $f \in E^b$. On the other hand, if $f \in E^b$, then for $\epsilon > 0$ set

$$V = \{t \in E : |f(t)| \leq \epsilon\}.$$

Then V is absolutely convex. If B is bounded, $f(B)$ is bounded so

$$\sup\{x \in B : |f(x)|\} = M < \infty.$$

Then $B \subset (M/\epsilon)V$ so V is a bornivore and a τ^b neighborhood of 0 so $f \in (E, \tau^b)'$. $\qquad\square$

From the Banach–Mackey Theorem for operators, we have the following

Proposition 5.53. *If (E, τ^b) is sequentially complete, then every bounded subset of (E, τ) is $\beta(E, E^b)$ bounded.*

Remark 5.54. Since (E, τ) sequentially complete implies (E, τ^b) is sequentially complete, this result gives an improvement of Corollary 33. Moreover, (E, τ^b) can be sequentially complete without (E, τ) being sequentially complete as the following example shows.

Example 5.55. $(E, \tau) = (c_0, \sigma(c_0, l^1))$ is not sequentially complete whereas

$$(E, \tau^b) = (c_0, \|\cdot\|_\infty)$$

is sequentially complete.

Every locally convex topology is a polar topology of uniform convergence on some family of weak* bounded sets. We now give a description of τ^b as a polar topology. Let $\mathcal{F}$ be the family of all $\sigma(E^b, E) - \mathcal{K}$ bounded subsets of E^b and let

$$\mathcal{KB}(E, E^b) = \tau_{\mathcal{F}}$$

be the polar topology of uniform convergence on the members of $\mathcal{F}$.

Theorem 5.56. $\tau^b = \mathcal{KB}(E, E^b)$.

Proof. Let $\mathcal{G}$ be the family of $\sigma(E^b, E)$ bounded sets with the property that τ^b is the polar topology, $\tau_{\mathcal{G}}$, of uniform convergence on the members of $\mathcal{G}$. Thus, if A is bounded in (E, τ) and $B \in \mathcal{G}$, then

$$B(A) = \{x'(x) : x' \in B, x \in A\}$$

is bounded.

Now let $B \in \mathcal{G}$. We claim that B is $\sigma(E^b, E) - \mathcal{K}$ bounded. Let $\{x'_k\} \subset B$ and $t_k \to 0$ with $t_k > 0$. Given any subsequence pick a further subsequence $\{n_k\}$ such that $\sum_{k=1}^{\infty} t_{n_k} < \infty$. For every $x \in E$ the series

$$\sum_{k=1}^{\infty} t_{n_k} x'_{n_k}(x) = f(x)$$

converges and defines a linear functional f on E. Since B is uniformly bounded on the bounded subsets A of (E, τ), the set

$$f(A) = \{f(x) : x \in A\}$$

is bounded so $f \in E^b$. This justifies the claim. Hence,

$$\tau_{\mathcal{G}} = \tau^b \subset \tau_{\mathcal{F}} = \mathcal{KB}(E, E^b).$$

On the other hand, since a subset $A \subset E$ is τ bounded iff it is τ^b bounded, Theorem 5.23 implies that the topology of uniform convergence on the $\sigma(E^b, E) - \mathcal{K}$ subsets of E^b has the same bounded sets as τ (or $\sigma(E, E')$) so this topology is weaker than τ^b. $\qquad \square$

Proposition 5.57. $\mathcal{KB}(E, E')$ *is stronger than* $\tau(E, E')$ *and* $\mathcal{KB}(E, E')$ *has the same bounded sets as* τ.

Proof. The first statement follows from Corollary 5.29. The second statement follows from the proof of the theorem above. $\qquad \square$

This gives another proof of Mackey's theorem. It is also the case that $\mathcal{KB}(E, E')$ can be strictly stronger than $\tau(E, E')$.

Example 5.58. Consider the dual pair (l^{∞}, l^1). The $\sigma(l^{\infty}, l^1) - \mathcal{K}$ bounded sets are exactly the norm bounded sets of l^{∞}. Thus, $\mathcal{KB}(l^{\infty}, l^1)$ is just the norm topology of l^{∞}. Since the dual of l^{∞} is ba, $\mathcal{KB}(l^{\infty}, l^1)$ is strictly stronger than $\tau(l^{\infty}, l^1)$.

We also have that $\mathcal{KB}(E, E')$ is weaker than the strong topology $\beta(E, E')$ and can be strictly weaker.

Example 5.59. Consider the dual pair c_{00}, c_{00} under the pairing

$$(s, t) \to \sum_{j=1}^{\infty} s_j t_j.$$

A subset $A \subset c_{00}$ is $\sigma(c_{00}, c_{00})$ or $\mathcal{KB}(c_{00}, c_{00})$ bounded iff the elements of A are coordinate bounded and A is strongly bounded iff there exists n such that $t_i = 0$ for $i \geq n$, $t \in A$ ([Köl] 21.11). Thus, the strong topology is strictly stronger than $\mathcal{KB}(c_{00}, c_{00})$.

We can give another characterization of the topology τ^b.

Theorem 5.60. $\tau^b = \tau(E, E^b)$.

Proof. Let $B \subset E^b$ be absolutely convex and $\sigma(E^b, E)$ compact so B_0, the polar of B in E, is a basic $\tau(E, E^b)$ neighborhood of 0. By Corollary 5.29 and Proposition 5.32, B is $\beta(E^b, E)$ bounded and, therefore, τ^b equicontinuous. Hence, B_0 is a τ^b neighborhood of 0 and $\tau(E, E^b) \subset \tau^b$.

On the other hand, $E^b = (E, \tau^b)'$ and $\tau(E, E^b)$ is the strongest locally convex topology on E with dual E^b so $\tau^b \subset \tau(E, E^b)$. $\qquad\square$

From [Köl] 21.4 and the fact that $\tau^b = \mathcal{KB}(E, E^b)$, we have

Corollary 5.61. *If $B \subset E^b$ is $\sigma(E^b, E) - \mathcal{K}$ bounded, then B is contained in an absolutely convex, $\sigma(E^b, E)$ compact subset of E^b.*

We can use E^b to give another example of an $\mathcal{A}$ space. Let E be a LCTVS and let

$$b(E^b, E)$$

be the vector topology on E^b of uniform convergence on the bounded subsets of E.

Proposition 5.62. $(E^b, b(E^b, E))$ *is an $\mathcal{A}$ space.*

Proof. Let $B \subset E^b$ be $b(E^b, E)$ bounded. Let $\{x'_k\} \subset B$ and $t_k \to 0$. Given a subsequence, pick a further subsequence $\{t_{n_k}\}$ such that $\sum_{k=1}^{\infty} |t_{n_k}| \leq 1$. Then

$$f(x) = \sum_{k=1}^{\infty} t_{n_k} x'_{n_k}(x)$$

defines a linear functional f on E which is bounded since

$$\sup\{|x'(x)| : x' \in B, x \in A\} < \infty$$

for every bounded set $A \subset E$. Hence, $f \in E^b$ and the series $\sum_{k=1}^{\infty} t_{n_k} x'_{n_k}$ is $b(E^b, E)$ convergent to f. Therefore, $\{t_k x'_k\}$ is $b(E^b, E) - \mathcal{K}$ convergent and B is $b(E^b, E) - \mathcal{K}$ bounded. $\qquad\square$

We give a description of the strongest admissible locally convex topology on a dual pair E, F which has the same bounded sets as $\sigma(E, F)$ ([WCC]). Recall an *admissible topology* for a dual pair is a locally convex topology which lies between the weak and strong topologies. For this the following lemma is useful. Let

$$(E, F : G)$$

be an abstract triple with G a TVS.

Lemma 5.63. *Let $A \subset E$ and $B \subset F$. The following are equivalent.*

(1) *$A \cdot B$ is bounded,*

(2) *for every $\{x_i\} \subset A, \{y_i\} \subset B$, $\lim_i \frac{1}{i} x_i \cdot y_i = 0$,*

(3) *for every $\{x_i\} \subset A, \{y_i\} \subset B$, $\lim_i \frac{1}{i} x_j \cdot y_i = 0$ uniformly for $j \in \mathbb{N}$,*

(4) *for every $\{x_i\} \subset A, \{y_i\} \subset B$, there exists $n_i \uparrow \infty$*

such that $\lim_i \frac{1}{n_i} x_{n_j} \cdot y_{n_i} = 0$ uniformly for $j \in \mathbb{N}$,

(5) *for every $\{x_i\} \subset A, \{y_i\} \subset B$, there exists $n_i \uparrow \infty$*

such that $\lim_i \frac{1}{n_i} x_{n_i} \cdot y_{n_i} = 0$.

Proof. (1) and (2) are equivalent by the characterization of boundedness. (1)$\Longrightarrow$(3): Let U be a balanced neighborhood of 0 in G. There exists i_0 such that $\frac{1}{i_0} A \cdot B \subset U$. Since U is balanced, $\frac{1}{i} A \cdot B \subset U$ for $i \geq i_0$. Therefore, $\frac{1}{i} x_j \cdot y_i \in U$ for $i \geq i_0, j \in \mathbb{N}$. That (3) implies (2) is clear as is (3)$\Longrightarrow$(4)$\Longrightarrow$(5). (5) and (1) are equivalent by the characterization of null sequences. $\square$

Definition 5.64. ([WCC]) Let E, F be a dual pair. A $\sigma(F, E)$ bounded subset of $B \subset F$ is a $\sigma(F, E) - U$ set if any of the conditions (1)-(5) of the lemma hold for any $\sigma(E, F)$ bounded subset $A \subset E$.

Let $\mathcal{U}$ be the family of $\sigma(F, E) - U$ sets and

$$U(E, F)$$

be the polar topology of uniform convergence on the members of $\mathcal{U}$.

Remark 5.65. From Theorem 5.23 any $\sigma(F, E) - \mathcal{K}$ bounded set is a $\sigma(F, E) - U$ set so

$$\mathcal{K}\mathcal{B}(E, F) \subset U(E, F) \subset \beta(E, F).$$

These containments can be proper ([WCC]).

Example 5.66. Let $E = ba$ and $F = m_0$ with the pairing

$$(m, f) \rightarrow \int_{\mathbb{N}} f \, dm.$$

Let $\{A_j\}$ be pairwise disjoint from $\mathbb{N}$. Then $\{\chi_{A_j}\}$ is not $\sigma(F, E) - \mathcal{K}$ bounded but is a $\sigma(F, E) - U$ set. Thus, $\mathcal{KB}(E, F) \subsetneqq U(E, F)$.

Example 5.67. Let $E = F = c_{00}$ with the usual pairing. Set $A = \{ie^j\}$. Then A is $\sigma(E, F)$ bounded and is $U(E, F)$ bounded by the theorem below but not $\beta(E, F)$ bounded. Thus, $U(E, F) \subsetneqq \beta(E, F)$.

Theorem 5.68. ([WCC]) $U(E, F)$ *is the strongest admissible topology with the same bounded sets as* $\sigma(E, F)$.

Proof. Let τ be an admissible topology with the same bounded sets as $\sigma(E, F)$. Let $\tau = \tau_{\mathcal{B}}$ be the polar topology of uniform convergence on the members of $\mathcal{B}$. Let $B \in \mathcal{B}$, $\{y_i\} \subset B$, and $\{x_i\} \subset E$ be $\sigma(E, F)$ bounded. Then $\{x_j \cdot y_i\}$ is bounded so $\lim_i \frac{1}{i} x_i \cdot y_i = 0$. Therefore, condition (2) of the lemma is satisfied and B is a $\sigma(F, E) - U$ set. Hence, $\tau \subset U(E, F)$. $\square$

Corollary 5.69. (E, τ) *is a Banach–Mackey space iff each* $\sigma(E', E)$ *bounded set is a* $\sigma(E', E) - U$ *set.*

Proof. Each $\sigma(E', E)$ bounded set $B \subset E'$ is a $\sigma(E', E) - U$ set iff B is uniformly bounded on each $\sigma(E, E')$ bounded set $A \subset E$ iff A is $\beta(E, E')$ bounded iff (E, τ) is a Banach–Mackey space. $\square$

We will give several continuity results for the topology $\mathcal{KB}(E, F)$.

Hellinger–Toeplitz

Hellinger and Toeplitz proved one of the early automatic continuity results for matrix transformations acting on l^2 ([HT]). We will show how the UBP can be used to establish their result. Let $A = [a_{ij}]$ be an infinite matrix which satisfies the condition

$$(\&) \sum_{i=1}^{\infty} t_i \sum_{j=1}^{\infty} a_{ij} s_j$$

converges for every $s = \{s_j\}, t = \{t_j\} \in l^2$. Hellinger and Toeplitz showed that if this condition is satisfied there exists a constant $M > 0$ such that

$$(\&\&) \qquad \left| \sum_{i=1}^{\infty} t_i \sum_{j=1}^{\infty} a_{ij} s_j \right| \le M \text{ for } \|s\|_2 \le 1, \|t\|_2 \le 1.$$

If $(\&)$ holds, then

$$As = \left\{ \sum_{j=1}^{\infty} a_{ij} s_j \right\}_i \in l^2$$

for $s \in l^2$ so the matrix A maps l^2 into itself. Condition $(\&\&)$ then asserts that $\|As\|_2 \le M$ for $\|s\|_2 \le 1$, i.e., A is a continuous linear operator. We show the Uniform Boundedness Principle (UBP) can be used to establish this conclusion.

Theorem 5.70. *If the matrix A maps l^2 into itself, then A is continuous.*

Proof. Let R_i be the ith row of the matrix $A = [a_{ij}]$. Since $A : l^2 \to l^2$, $R_i \in l^2$. Define $A_n : l^2 \to l^2$ by

$$A_n s = \sum_{i=1}^{n} (R_i \cdot s) e^i.$$

Then A_n is linear, continuous and

$$\lim_n t \cdot A_n s = t \cdot As$$

for $s, t \in l^2$. By UBP, $\{A_n s : n\}$ is norm bounded in l^2. By UBP again, $\{\|A_n\| : n\}$ is bounded, say, by M. Since $|t \cdot A_n s| \le M$ for $\|s\|_2 \le 1, \|t\|_2 \le 1$ and $t \cdot A_n s \to t \cdot As$, $(\&\&)$ follows. $\qquad \square$

A result such as this is referred to as an *automatic continuity* result since an algebraic condition implies a continuity condition. A more general automatic continuity result for matrix mappings can be established by employing the Closed Graph Theorem (see [Wi1] 11.3.5).

This classic result of Hellinger and Toeplitz has been given a more abstract form and it asserts that a symmetric, linear operator on a Hilbert space is continuous ([Sw2] 35.10). This is another automatic continuity result in the sense that an algebraic condition, symmetry, implies continuity. We will establish a result for the topology $\mathcal{KB}(E, F)$ which has the Hellinger–Toeplitz type result as a corollary.

Let G be a LCTVS.

Theorem 5.71. *Let E be a metric linear space whose topology is generated by the quasi-norm $|\cdot|$ and let $T : E \to G$ be linear. If T is $|\cdot| - \sigma(G, G')$ continuous, then T is $|\cdot| - \mathcal{KB}(G, G')$ continuous.*

Proof. Let $|x_k| \to 0$. Pick $n_k \uparrow \infty$ such that $|n_k x_k| \to 0$ (Appendix A). Let $B \subset G'$ be $\sigma(G', G) - \mathcal{K}$ bounded. We need to show that

$$\sup\{|y'(Tx_k)| : y' \in B\} \to 0.$$

For this it suffices to show

$$y'_k(Tx_k) \to 0$$

for every $\{y'_k\} \subset B$. Consider the matrix

$$M = [(1/n_j)y'_j(T(n_i x_i))].$$

The columns of M converge to 0 by the continuity assumption. Given any subsequence there is a further subsequence $\{q_j\}$ such that the series $\sum_{j=1}^{\infty}(1/n_{q_j})y'_{q_j}$ is $\sigma(G', G)$ convergent to some $y' \in G'$. Then

$$\sum_{j=1}^{\infty}(1/n_{q_j})y'_{q_j}(T(n_i x_i)) = y'(T(n_i x_i)) \to 0$$

by the continuity assumption. Hence M is a $\mathcal{K}$ matrix and by the Antosik–Mikusinski Theorem the diagonal of M, $\{y'_j(Tx_j)\}$, converges to 0 as desired. $\square$

Since $\tau(G, G') \subset \mathcal{KB}(G, G')$, if the hypothesis of the theorem is satisfied, then T is continuous with respect to the original topology of G. This will give the Hellinger–Toeplitz result.

Corollary 5.72. (Hellinger–Toeplitz) *Let H be a Hilbert space and $A : H \to H$ be a symmetric linear operator. Then A is norm continuous.*

Proof. Let $x, y \in H$. Since $y \cdot Ax = Ay \cdot x$, A is $\|\cdot\| - \sigma(H, H)$ continuous. By the observation above, A is norm continuous. $\square$

If E, G are LCTVS and $T : E \to G$ is linear, the domain of the *adjoint operator* T' is defined to be

$$\mathcal{D}(T') = \{y' \in G' : y'T \in E'\}$$

and the adjoint operator $T' : \mathcal{D}(T') \to G'$ is defined to be $T'y' = y'T$. If $A : H \to H$ is a symmetric linear operator, then the domain of A is $H' = H$. We use the theorem to establish an automatic continuity result for an operator whose adjoint has domain equal to the dual.

Corollary 5.73. *Let (E, τ) be a metrizable LCTVS. If $T : E \to G$ is linear and $\mathcal{D}(T') = G'$, then T is $\tau - \mathcal{KB}(G, G')$ continuous.*

Proof. If $x_k \to 0$ in $\tau = \tau(E, E')$ and $\mathcal{D}(T') = G'$, then for $y' \in G'$,

$$y'(Tx_k) = T'y'(x_k) \to 0.$$

Thus, T is $\tau - \sigma(G, G')$ continuous and the result follows from the theorem.

$\square$

The condition $\mathcal{D}(T') = G'$ holds iff T is weakly continuous so this is the case when the assumptions in the corollary are satisfied.

The proof of the theorem can also be used to establish a boundedness result for $\mathcal{K}B(G, G')$.

Theorem 5.74. *Let E, G be LCTVS. If $T : E \to G$ is linear and weakly sequentially continuous, then T carries bounded subsets of E into $\mathcal{K}B(G, G')$ bounded subsets of G.*

Proof. Let $A \subset E$ be bounded and $B \subset G'$ be $\sigma(G', G) - \mathcal{K}$ bounded. It suffices to show $\{y'_k(Tx_k)\}$ is bounded when $\{x_k\} \subset A, \{y'_k\} \subset B$ or

$$t_k y'_k(Tx_k) \to 0$$

with $t_k \to 0, t_k > 0$. Define a matrix M by

$$M = [\sqrt{t_j} y'_j(T(\sqrt{t_i} x_i))].$$

As in the proof of Theorem 71, M is a $\mathcal{K}$ matrix whose diagonal $t_k y'_k(x_k) \to 0$. $\square$

We also use the UBP to establish a result from summability theory, the Silvermann–Toeplitz Theorem on regular matrix transformations. A matrix A which maps c into c, i.e., maps convergent sequences to convergent sequences is said to be *conservative* and if A preserves limits it is said to be *regular*. The Silvermann–Toeplitz Theorem characterizes regular matrices.

Theorem 5.75. *A matrix $A = [a_{ij}]$ maps c into c (i.e., is conservative) iff*

(i) $\sup_i \sum_{j=1}^{\infty} |a_{ij}| < \infty$,

(ii) $\lim_i a_{ij} = a_j$ *exists for every j,*

(iii) $\lim_i \sum_{j=1}^{\infty} a_{ij}$ *exists.*

Proof. We prove only the necessity; the sufficiency is classic and can be found, for example, in [Bo]. For (i) each row R_i of A induces a continuous linear functional on c by $R_i s = \sum_{j=1}^{\infty} a_{ij} s_j$ with $\|R_i\| = \sum_{j=1}^{\infty} |a_{ij}|$. Since A maps c into c, $\lim_i R_i s$ exists for every $s \in c$. By UBP,

$$\sup_i \|R_i\| = \sup_i \sum_{j=1}^{\infty} |a_{ij}| < \infty$$

or (i) holds. If A is conservative, (ii) and (iii) follow by setting $s = e^j$ and $s = (1, 1, ...)$. $\qquad\square$

A matrix A is regular iff (i) and

(ii') $\lim_i a_{ij} = 0$ for every j and

(iii') $\lim_i \sum_{j=1}^{\infty} a_{ij} = 0$.

Again these conditions can be seen to be necessary by setting $s = e^j$ and $s = (1, 1, ...)$. For sufficiency see [Bo].

It is interesting that the proof of (i) by the UBP can be found in Banach's book ([Ba]) and is one of the earliest applications of the UBP. It is also of historic interest to note that the Hellinger–Toeplitz Theorem was established by Hahn with a use of an abstract Uniform Boundedness Principle ([Ha]). Hahn established an abstract set-up which he used to prove several uniform boundedness results which were previously known; his methods were more cumbersome to use than those of Banach and have fallen by the wayside.

Chapter 6

Banach–Steinhaus

As noted earlier there are two possible interpretations for the conclusion of the classical Uniform Boundedness Principle (UBP). If E is a Banach space, G is a normed space and Γ is a subset of the continuous linear operators from E into G which is pointwise bounded on E, then

$$(\#) \quad \sup\{\|Tx\| : \|x\| \leq 1, T \in \Gamma\} = \sup\{\|T\| : T \in \Gamma\} < \infty.$$

The first interpretation of the conclusion in $(\#)$ is that Γ is uniformly bounded on bounded subsets of E. The generalization of this interpretation for abstract triples was addressed in the previous chapter. The second interpretation of the conclusion in $(\#)$ is that the family Γ is equicontinuous. In this chapter we will consider generalizations of this interpretation for abstract triples.

In what follows

$$(E_i, F_i : G)$$

will denote abstract triples for $i = 1, 2$, where E_i, F_i are LCTVS and G a TVS. Let $\mathcal{F}_i$ be a family of subsets of F_i and let $\tau_{\mathcal{F}_i}(E_i) = \tau_i$ be the topology on E_i of uniform convergence on the members of $\mathcal{F}_i$ so a net $\{x_\alpha\}$ in E_i converges to $x \in E_i$ iff

$$x_\alpha \cdot y \to x \cdot y$$

uniformly for y belonging to a member of $\mathcal{F}_i$. Let Γ be a family of mappings $T : E_1 \to E_2$. We consider conditions which guarantee that Γ is $\tau_1 - \tau_2$ equicontinuous. We then establish several versions of the Banach–Steinhaus Theorem for abstract triples and give applications to continuous linear operators between locally convex spaces. We also give applications to the Nikodym Convergence Theorem and summability results of Hahn and Schur.

To motivate a condition which guarantees that Γ is $\tau_1 - \tau_2$ equicontinuous, we consider the case of continuous linear operators between locally convex spaces. Let $(E_1, F_1), (E_2, F_2)$ be dual pairs and let τ_i be the polar topology of uniform convergence on the members of $\mathcal{F}_i$ and let Γ be a family of weakly continuous linear operators $T : E_1 \to E_2$ with adjoint operator T'. Suppose

$(*)$ for every $B \in \mathcal{F}_2$ there exists $A \in \mathcal{F}_1$ such that
$$T'B = BT \subset A \text{ for every } T \in \Gamma,$$

or, taking polars in E_1, E_2,

$$(**) \ (T'B)_0 = T^{-1}B_0 \supset A_0 \ for \ T \in \Gamma.$$

Since B_0 is a basic τ_2 neighborhood of 0 and A_0 is a basic τ_1 neighborhood of 0, condition $(**)$ implies that Γ is $\tau_1 - \tau_2$ equicontinuous.

We consider abstracting condition $(*)$ to abstract triples. For this, regard the elements y of F_i as functions from $E_i \to G$ defined by $y(x) = x \cdot y$ for $x \in E_i$. We say that the pair $(\mathcal{F}_1, \mathcal{F}_2)$ satisfies the equicontinuity condition (E) if

(E) for every $B \in \mathcal{F}_2$ there exists $A \in \mathcal{F}_1$ such that
$$B\Gamma = \{y \circ T : y \in B, T \in \Gamma\} \subset A$$

[note if $x \in E_1$, $(y \circ T)(x) = y(Tx) = y \cdot Tx$].

Theorem 6.1. *If* $(\mathcal{F}_1, \mathcal{F}_2)$ *satisfies condition* (E), *then* Γ *is* $\tau_1 - \tau_2$ *equicontinuous.*

Proof. Suppose the net $\{x_\delta\}$ in E_1 converges to $x \in E_1$ with respect to τ_1 so

$$x_\delta \cdot y \to x \cdot y$$

uniformly when y belongs to a member of $\mathcal{F}_1$. Let $B \in \mathcal{F}_2, z \in B$ and let A be as in condition (E). Then $z \circ T \in A$ for every $T \in \Gamma, z \in B$ so

$$z \cdot Tx_\delta \to z \cdot Tx$$

uniformly for $T \in \Gamma, z \in B$ by the definition of convergence in τ_1. Therefore, $Tx_\delta \to Tx$ in τ_2 uniformly for $T \in \Gamma$. □

The case of a single operator satisfying condition (E) is of interest.

Corollary 6.2. *Suppose* $T : E_1 \to E_2$ *is such that for every* $B \in \mathcal{F}_2$ *there exists* $A \in \mathcal{F}_1$ *such that* $BT \subset A$. *Then* T *is* $\tau_1 - \tau_2$ *continuous.*

Corollary 6.3. (Banach–Steinhaus) *Suppose $\{T_\alpha\}$ is a net of maps from E_1 to E_2 such that*

$$\tau_2 - \lim_\alpha T_\alpha x = Tx$$

exists for every $x \in E_1$. If $\Gamma = \{T_\alpha\}$ satisfies condition (E), then T is $\tau_1 - \tau_2$ continuous.

Proof. Suppose the net $\{x_\delta\}$ is τ_1 convergent to $x \in E_1$. Then by hypothesis

$$\tau_2 - \lim_\alpha T_\alpha x_\delta = Tx_\delta$$

for each δ. Also, by Theorem 1,

$$\tau_2 - \lim_\delta T_\alpha x_\delta = T_\alpha x$$

uniformly with respect to α. Therefore,

$$\lim_\delta Tx_\delta = \lim_\delta \lim_\alpha T_\alpha x_\delta = \lim_\alpha \lim_\delta T_\alpha x_\delta = \lim_\alpha T_\alpha x$$

by the Iterated Limit Theorem (Appendix A) and T is $\tau_1 - \tau_2$ continuous. $\square$

We next consider conditions for which (E) holds and establish versions of the Banach–Steinhaus Theorem for topological vector spaces.

In what follows G will be a Hausdorff topological vector space.

We first give a motivation for the conditions which appear in a version of the Banach–Steinhaus Theorem for abstract triples.

Suppose (E, F) is a dual pair and $\tau_{\mathcal{F}}$ is a polar topology on E of uniform convergence on the members of $\mathcal{F}$. Recall a subset $C \subset E$ is $\tau_{\mathcal{F}}$ bounded iff

$$BC = \{\langle y, x \rangle : y \in B, x \in C\}$$

is bounded for every $B \in \mathcal{F}$. We abstract this condition to abstract triples.

Definition 6.4. A subset $C \subset E_2$ is $\mathcal{F}_2$ bounded if

$$C \cdot B = \{x \cdot y : y \in B, x \in C\}$$

is bounded in G for every $B \in \mathcal{F}_2$.

We give an equicontinuity version of the Banach–Steinhaus Theorem for abstract triples. Let $(E, F : G)$ be a triple. If $\mathcal{B}$ is any family of subsets of F, $\tau_{\mathcal{B}}$ is the topology of uniform convergence on the elements of $\mathcal{B}$.

Theorem 6.5. *Suppose Γ is pointwise $\mathcal{F}_2$ bounded on E_1 (i.e., for every $x \in E_1$ the set Γx is $\mathcal{F}_2$ bounded in E_2). Let $\mathcal{B}$ be the family of subsets of F_1 which are pointwise bounded on E_1. Then the pair $(\mathcal{B}, \mathcal{F}_2)$ satisfies condition (E). Hence, Γ is $\tau_{\mathcal{B}} - \tau_2$ equicontinuous.*

Proof. Let $B \in \mathcal{F}_2$. We claim $B\Gamma \in \mathcal{B}$. Let $x \in E_1$. Since Γx is $\mathcal{F}_2$ bounded,

$$B(\Gamma x) = \{y \cdot Tx : T \in \Gamma, y \in B\}$$

is bounded in G so $B\Gamma \in \mathcal{B}$. Therefore, $(\mathcal{B}, \mathcal{F}_2)$ satisfies condition (E) and the result follows from Theorem 1.   □

From Corollary 3 we have another version of the Banach–Steinhaus Theorem.

Corollary 6.6. (Banach–Steinhaus) *Let $\{T_\alpha\}$ be a net of maps from $E_1 \to E_2$ which is pointwise $\mathcal{F}_2$ bounded on E_1. If*

$$\tau_2 - \lim_\alpha T_\alpha x = Tx$$

exists for every $x \in E_1$, then T is $\tau_\mathcal{B} - \tau_2$ continuous.

We also have the more familiar form of the Banach–Steinhaus Theorem for sequences.

Corollary 6.7. (Banach–Steinhaus) *Let $T_k : E_1 \to E_2$ and suppose*

$$\tau_2 - \lim_k T_k x = Tx$$

exists for each $x \in E_1$. Then $\{T_k\}$ is $\tau_\mathcal{B} - \tau_2$ equicontinuous and T is $\tau_\mathcal{B} - \tau_2$ continuous.

Proof. For each $x \in E_1$, $\{T_k x\}$ is $\mathcal{F}_2$ bounded so the corollary above applies.   □

In the case when (E_i, F_i) are vector spaces in duality the set $\mathcal{B}$ in Theorem 5 is the family of $\sigma(F_1, E_1)$ bounded sets so $\tau_\mathcal{B}$ is just the strong topology $\beta(E_1, F_1)$ on E_1. If τ_2 is the polar topology of uniform convergence on $\mathcal{F}_2$, by the theorem we have

Corollary 6.8. *Let Γ be a family of $\tau_1 - \tau_2$ continuous linear operators which is pointwise bounded on E_1, then Γ is $\beta(E_1, F_1) - \tau_2$ equicontinuous.*

Note that if E_1 is a LCTVS with dual F_1, the equicontinuity is with respect to $\beta(E_1, F_1)$ not the original topology of E_1 unless E_1 is barrelled. If E_1 is barrelled, this is a usual form of the UBP for barrelled spaces. Thus, Corollary 8 is a form of the Uniform Boundedness Principle without assumptions on the domain space.

Corollary 6.9. *If $\Gamma = \{T_k\}$ is a sequence of continuous linear operators such that $\lim_k T_k x = Tx$ exists for all $x \in E_1$, then Γ is $\beta(E_1, F_1) - \tau_2$ equicontinuous and the limit operator T is $\beta(E_1, F_1) - \tau_2$ continuous.*

In the corollary above, the limit operator is $\beta(E_1, F_1) - \tau_2$ continuous so if E_1 is barrelled this gives the usual form of the Banach–Steinhaus Theorem for barrelled spaces. Thus, Corollary 9 is a form of the Banach–Steinhaus Theorem which holds without assumptions on the domain space.

There are LCTVS which carry the strong topology but are not barrelled. See Wilansky ([Wi2] 15.4.6, 15.4.18).

We can also give a generalization of Theorem 1. Let $\mathcal{A}_1$ be a family of subsets of E_1 and let $\mathcal{B}_1$ be a family of subsets of F_1 which is uniformly bounded on members of $\mathcal{A}_1$ (i.e., $\mathcal{B}_1$ is $\mathcal{A}_1$ bounded).

Theorem 6.10. *Suppose ΓA is $\mathcal{F}_2$ bounded for every $A \in \mathcal{A}_1$. Then Γ is $\tau_{\mathcal{B}_1} - \tau_2$ equicontinuous.*

Proof. As in the proof of Theorem 5 the pair $(\mathcal{B}_1, \mathcal{F}_2)$ satisfies condition (E). □

In the case of Theorem 5, the family $\mathcal{A}_1$ consists of singletons.

We can use this theorem along with Theorem 5.23 to obtain an interesting corollary.

Corollary 6.11. *Let $\mathcal{A}_1$ be the family of UB sets in the triple $(E_1, F_1 : G)$ and let Γ be a family of maps from E_1 into E_2 such that*

$$(\&) \quad B\Gamma \subset F_1 \text{ for every } B \in \mathcal{F}_2.$$

If Γ is pointwise bounded on E_1, then Γ is uniformly $\mathcal{F}_2$ bounded on $\mathcal{A}_1$, and, hence, $\tau_{\mathcal{B}_1} - \tau_2$ equicontinuous.

Proof. Let $A \in \mathcal{A}_1$. Then A is a UB set in the triple $(E_1, B\Gamma : G)$ by condition $(\&)$. Then $B\Gamma$ is pointwise bounded and by Theorem 5.23, $B\Gamma$ is uniformly bounded on A. That is, ΓA is $\mathcal{F}_2$ bounded so Theorem 10 above applies. □

If E_i are LCTVS with duals F_i, then condition $(\&)$ is just the condition that

$$\{T'f : T' \in \Gamma, f \in B\} \subset F_1$$

for $B \in \mathcal{F}_2$, i.e., the transpose T' is defined on F_2 and has values in F_1. This is equivalent to T being weakly continuous. Thus, we obtain

Corollary 6.12. *Let E_i be LCTVS with duals F_i and Γ a family of weakly continuous linear operators from E_1 into E_2 which is pointwise bounded on E_1. If $A \subset E_1$ is $\sigma(E_1, F_1) - \mathcal{K}$ bounded, then Γ is uniformly bounded on A.*

Proof. See Definitions 2 and 10 in Chapter 5 for the fact that A is a UB set in the triple $(E_1, F_1 : \mathbb{R})$ so the result follows from the corollary above. $\square$

If A is $\mathcal{K}$ bounded in the original topology of E_1, then A is $\sigma(E_1, F_1) - \mathcal{K}$ bounded so the corollary applies.

Remark 6.13. Let $\mathcal{A}_1$ be the family of $\sigma(E_1, F_1) - \mathcal{K}$ bounded subsets of E_1 and $\mathcal{B}_1$ the family of all subsets of F_1 which are uniformly bounded on the members of $\mathcal{A}_1$. It follows from Theorem 10 and the corollary above that if Γ is a pointwise bounded family of continuous linear operators from E_1 into E_2, then Γ is $\tau_{\mathcal{B}_1} - \tau_2$ equicontinuous. In particular, if $\Gamma = \{T_k\}$ is a sequence of continuous linear operators such that $\lim_k T_k x = Tx$ exists for every $x \in E_1$, then $\{T_k\}$ is $\tau_{\mathcal{B}_1} - \tau_2$ equicontinuous and the limit operator T is $\tau_{\mathcal{B}_1} - \tau_2$ continuous.

Recall an $\mathcal{A}$ space is a space in which bounded sets are $\mathcal{K}$ bounded. We make some observations on the remarks above for $\mathcal{A}$ spaces.

Remark 6.14. If E_1 is an $\mathcal{A}$ space, then $(E_1, \sigma(E_1, F_1))$ is an $\mathcal{A}$ space so the family $\mathcal{A}_1$ above consists of all the bounded subsets of E_1 and $\mathcal{B}_1$ is the family of subsets of F_1 which are strongly, $\beta(F_1, E_1)$, bounded. Then $\tau_{\mathcal{B}_1}$ is the locally convex topology $\beta^*(E_1, F_1)$ (Appendix A); this topology is denoted by $b(E)$ by Wilansky ([Wi2])). If Γ is as in the remark above, then Γ is $\beta^*(E_1, F_1) - \tau_2$ equicontinuous when Γ is pointwise bounded. If $\Gamma = \{T_k\}$ is such that $\lim_k T_k x = Tx$ exists for every $x \in E_1$, then $\{T_k\}$ is $\beta^*(E_1, F_1) - \tau_2$ equicontinuous and the limit operator is $\beta^*(E_1, F_1) - \tau_2$ continuous.

Since any sequentially complete LCTVS is an $\mathcal{A}$ space, these remarks apply and give a Banach–Steinhaus Theorem for sequentially complete LCTVS.

More generally, we can consider these remarks for Banach–Mackey spaces.

We can obtain an improvement of Corollary 8 for Banach–Mackey spaces. Recall a locally convex space E_1 is a *Banach–Mackey space* if the bounded subsets of E_1 are strongly bounded ([Wi2]10.4.3). For example, any sequentially complete locally convex space is a Banach–Mackey space ([Wi2] 10.4.8).

Theorem 6.15. *Suppose E_1 is a Banach–Mackey space. If $\Gamma \subset L(E_1, E_2)$ is pointwise bounded on E_1, then Γ is $\beta^*(E_1, F_1) - \tau_2$ equicontinuous.*

Proof. By the Banach–Mackey property the family $\mathcal{B}$ of Theorem 5 is the family of all $\beta(F_1, E_1)$ bounded subsets of F_1 so $\tau_{\mathcal{B}} = \beta^*(E_1, F_1)$ and the result follows from Theorem 5. □

Note $\beta^*(E_1, F_1) \subset \beta(E_1, F_1)$ so Theorem 15 improves the conclusion of Corollary 8 for Banach–Mackey spaces. We can also obtain an improvement of Corollary 9 for Banach–Mackey spaces.

Theorem 6.16. *Suppose E_1 is a Banach–Mackey space. Let $T_k : E_1 \to E_2$ be a sequence of $\tau_1 - \tau_2$ continuous linear operators such that $\lim_k T_k x = Tx$ exists for every $x \in E_1$. Then T is $\beta^*(E_1, F_1) - \tau_2$ continuous and $\{T_k\}$ is $\beta^*(E_1, F_1) - \tau_2$ equicontinuous.*

We can also obtain a corollary of Theorem 10.

Corollary 6.17. *Let $\mathcal{A}_1$ be the family of all $\sigma(E_1, F_1)$ bounded subsets of E_1 and $\mathcal{B}_1$ be the family of all $\beta(F_1, E_1)$ bounded subsets of F_1. If $\Gamma \subset L(E_1, E_2)$ is uniformly bounded on members of $\mathcal{A}_1$, then Γ is $\beta^*(E_1, F_1) - \tau_2$ equicontinuous.*

Proof. $\tau_{\mathcal{B}_1} = \beta^*(E_1, F_1)$ so the result follows from Theorem 10. □

Corollary 2 about a single mapping also has an interesting application to bounded linear operators.

Corollary 6.18. *Suppose $T : E_1 \to E_2$ is a bounded linear operator. Then T is $\beta^*(E_1, F_1) - \tau_2$ continuous.*

Note that T may not be continuous with respect to the original topology of E_1. Consider the identity operator on an infinite dimensional normed space when the domain has the weak topology and the range the norm topology.

The result in Corollary 2 also has an application to a Hellinger–Toeplitz result for linear operators. Let X, Y be locally convex spaces with duals X', Y'. A property $\mathcal{P}$ of subsets B of a dual space Y' is said to be *linearly invariant* if for every continuous linear operator $T : X \to Y$ there exists $A \subset X'$ with property $\mathcal{P}$ such that $BT = T'B \subset A$. For example, the family of subsets with finite cardinal, the weak* compact sets, the weak* convex compact sets, the weak* bounded sets, etc.

If $\mathcal{P}$ is a linearly invariant property, let

$$P(X, X')$$

be the locally convex topology of uniform convergence on the members of X' with property $\mathcal{P}$. From Corollary 2 we have a Hellinger–Toeplitz result in the spirit of Wilansky ([Wi2]11.2.6).

Corollary 6.19. *If $T : X \to Y$ is a continuous linear operator, then T is $P(X, X') - P(Y, Y')$ continuous.*

In particular, T is continuous with respect to the Mackey topologies and strong topologies ([Wi2]11.2.6).

Finally we indicate an application concerning automatic continuity of matrix transformations between sequence spaces. If λ, μ are two vector spaces of scalar sequences and $A = [a_{ij}]$ is an infinite matrix, we say that A maps λ into μ, if the series $\sum_{j=1}^{\infty} a_{ij}s_j$ converges for every $s = \{s_j\} \in \lambda$ and $\{\sum_{j=1}^{\infty} a_{ij}s_j\}_i \in \mu$ for every $s \in \lambda$. We denote the space of all matrices which map λ into μ by

$$M(\lambda, \mu).$$

Let λ_1, λ_2 be scalar sequence spaces containing c_{00}, the space of sequences with finite range, and if $a = \{a_j\} \in \lambda_1^{\beta}$, the β-dual of λ_1, $t = \{t_j\} \in \lambda_1$, we write

$$a \cdot t = \sum_{j=1}^{\infty} a_j t_j.$$

Assume that λ_i has a locally convex polar topology τ_i from the duality pair $\lambda_i, \lambda_i^{\beta}$ and that $A = [a_{ij}]$ is an infinite matrix which maps λ_1 into λ_2. Under assumptions on the sequence spaces, we use Corollary 8 to show that A is continuous with respect to appropriate topologies. First, we assume that the β-dual of λ_1 is contained in the topological dual λ_1' and then we assume that λ_2 is an AK-space under its topology (i.e., the canonical unit vectors $\{e^i\}$ form a Schauder basis for λ_2 (Appendix B, [Sw2] 4.2.13, [Sw4] B.2). Now let a^i be the ith row of the matrix A so $a^i \in \lambda_1^{\beta} \subset \lambda_1'$ and define $A_k : \lambda_1 \to \lambda_2$ by $A_k t = \sum_{i=1}^{k}(a^i \cdot t)e^i$. Then A_k is $\tau_1 - \tau_2$ continuous and

$$\tau_2 - \lim_k A_k t = \sum_{i=1}^{\infty}(a^i \cdot t)e^i = At$$

by the AK assumption. By the Banach–Steinhaus result in Corollary 8, $\{A_k\}$ is $\beta(\lambda_1, \lambda_1^{\beta}) - \tau_2$ equicontinuous and A is $\beta(\lambda_1, \lambda_1^{\beta}) - \tau_2$ continuous, an automatic continuity result. In particular, if $\lambda_1 = \lambda_2 = l^2$, then this result implies that any matrix mapping l^2 into itself is continuous; this is the classic theorem of Hellinger and Toeplitz ([Kö2] 34.7 and Chapter 5).

Further automatic continuity theorems for matrix mappings can be found in [Kö2] 34.7 and [Sw1] 12.6.

There is often another conclusion included in the statements of the Banach–Steinhaus Theorem. Namely, if X is a Banach space, Y is a normed space and $\{T_k\}$ is a sequence of continuous linear operators from X into Y such that $\lim_k T_k x = Tx$ exists for every $x \in X$, then $\{T_k\}$ is equicontinuous, T is continuous and, moreover,

$$\lim_k T_k x = Tx$$

uniformly for x belonging to compact subsets of X.

We first observe that equicontinuity guarantees this uniform convergence.

Theorem 6.20. *Let E, G be TVS and $\{T_k\} \subset L(E, G)$ be such that*

$$\lim_k T_k x = Tx$$

exists for every $x \in E$. If $\{T_k\}$ is equicontinuous and $K \subset E$ is compact, then $\lim_k T_k x = Tx$ uniformly for $x \in K$.

Proof. Let U be a closed neighborhood of 0 in G and pick a closed, symmetric neighborhood of 0, W, such that $W + W + W \subset U$. There exists a closed neighborhood, V, of 0 in E with $T_k V \subset W$ for all k so $TV \subset W$. There exists finite $A \subset E$ such that

$$K \subset \cup_{x \in A}(x + V).$$

There exists N such that $k \geq N$ implies $T_k x - Tx \in W$ for $x \in A$. Suppose $z \in K$. Then there exists $x \in A$ with $z \in x + V$. If $k \geq N$, then

$$T_k z - Tz = T_k(z - x) + (T_k x - Tx) + T(x - z) \in W + W + W \subset U.$$

Hence, the result. $\qquad\square$

It should be noted that the equicontinuity and the compactness must be with respect to the same topology. The topologies in the equicontinuity results above are often not with respect to the original topologies of the domain spaces.

We can use the notion of $\mathcal{K}$ convergence to obtain a version of the Banach–Steinhaus Theorem without assumptions on the domain space.

Theorem 6.21. (General Banach–Steinhaus Theorem) *Let E, G be TVS and*

$$\{T_k\} \subset LS(E, G)$$

with
$$\lim_k T_k x = T x$$

existing for every $x \in E$. *If* $\{x_j\} \subset E$ *is* $w(E, \{T_k\}) - \mathcal{K}$ *convergent in the triple* $(E, \{T_k\} : G)$ *under the map* $(x, T_k) \to T_k x$, *then*

(i) $\lim_j T x_j = 0$,

(ii) $\lim_j T_i x_j = 0$ *uniformly for* $i \in \mathbb{N}$, *and*

(iii) $\lim_i T_i x_j = T x_j$ *uniformly for* $j \in \mathbb{N}$.

Proof. Consider the matrix
$$M = [T_i x_j].$$
The columns of M converge to $T x_j$ for each i. Given a subsequence, there is a further subsequence $\{n_j\}$ with the series $\sum_{j=1}^{\infty} x_{n_j}$ being $w(E, \{T_k\})$ convergent to some $x \in E$. Then
$$\sum_{j=1}^{\infty} T_i x_{n_j} = T_i x$$
and
$$\lim_i T x = T x$$
so M is a $\mathcal{K}$ matrix and (ii) and (iii) follow from the Antosik–Mikusinski Theorem. (i) follows from the Iterated Limit Theorem,
$$\lim_i \lim_j T_i x_j = 0 = \lim_j \lim_i T_i x_j = \lim_j T x_j.$$

$\square$

Since any sequence which converges in the original topology of E or with respect to $w(E, LS(E, G))$ converges with respect to $w(E, \{T_k\})$, conclusions (i), (ii) and (iii) hold for any sequence which is $\mathcal{K}$ convergent in the original topology of E. This observation gives a Banach–Steinhaus Theorem for metric linear $\mathcal{K}$ spaces.

Theorem 6.22. *Let* E *be a metric linear* $\mathcal{K}$ *space*, G *a Hausdorf TVS and* $\{T_k\} \subset L(E, G)$. *Assume*
$$\lim_k T_k x = T x$$
exists for every $x \in E$. *Then*

(i) T *is continuous*,

(ii) $\lim_k T_k x = T x$ *uniformly for* x *belonging to compact subsets of* E,

(iii) *if* $x_j \to 0$ *in* E, *then* $\lim_j T_i x_j = 0$ *uniformly for* $i \in \mathbb{N}$.

Proof. (i) is immediate from the General Banach–Steinhaus Theorem. For (ii) it suffices to consider the case when the compact set is a sequence $\{x_j\}$ which converges to some $x \in E$. The sequence $\{x_j - x\}$ is $\mathcal{K}$ convergent so by the General Banach–Steinhaus Theorem

$$\lim_i T_i(x_j - x) = T(x_j - x)$$

uniformly in j. Since $\lim_i T_i x = Tx$, $\lim_i T_i x_j = Tx_j$ uniformly in j. (iii) follows from (ii). $\qquad\square$

We will give applications of the General Banach–Steinhaus Theorem to the Nikodym Convergence Theorem and summability results of Hahn and Schur.

First, we consider the Nikodym Convergence Theorem. Let Σ be a σ algebra of subsets of a set S and $\{m_k\}$ a sequence of strongly additive set functions from Σ into G. See Chapter 2 for the definitions.

Theorem 6.23. (Nikodym) *Let G be a quasi-normed LCTVS. If*

$$\lim_k m_k(A) = m(A)$$

exists for every $A \in \Sigma$, then

(i) *m is strongly additive,*

(ii) *$\{m_k\}$ is uniformly strongly additive.*

Proof. Let $\mathcal{S}(\Sigma)$ be the space of all Σ simple functions with the sup-norm. Define a linear operator $T_k : \mathcal{S}(\Sigma) \to G$ by

$$T_k f = \int_S f \, dm_k;$$

we are only integrating simple functions so no elaborate integration theory is involved. Note each T_k is continuous (this can be seen by using the fact that strongly additive set functions have bounded range; Appendix C) and

$$\lim_k \int_S f \, dm_k = \lim_k T_k f$$

exists for every $f \in \mathcal{S}(\Sigma)$. Let $\{A_j\}$ be pairwise disjoint from Σ. Now $\{\chi_{A_j}\}$ is $w(\mathcal{S}(\Sigma), \{T_k\}) - \mathcal{K}$ convergent in the triple

$$(\mathcal{S}(\Sigma), \{T_k\} : G)$$

under the map $(f, T_k) \to T_k f$. For by Drewnowski's Lemma (Appendix D), there is a subsequence $\{A_{n_j}\}$ such that each m_k is countably additive on the σ-algebra generated by $\{A_{n_j}\}$ so

$$\sum_{j=1}^{\infty} T_k(\chi_{A_{n_j}}) = \sum_{j=1}^{\infty} m_k(A_{n_j}) = m_k(\cup_{j=1}^{\infty} A_{n_j}) = T_k(\chi_{\cup_{j=1}^{\infty} A_{n_j}})$$

which implies that $\{\chi_{A_j}\}$ is $w(\mathcal{S}(\Sigma), \{T_k\}) - \mathcal{K}$ convergent. By the General Banach–Steinhaus Theorem,

$$\lim_k T_k(\chi_{A_j}) = \lim_k m_k(A_j) = 0$$

uniformly in j. By the Iterated Limit Theorem, $\lim_j m_k(A_j) = 0$ uniformly in k so $\{m_k\}$ is uniformly strongly additive. (i) follows from (ii). $\qquad\square$

We treated the Nikodym Convergence Theorem in Chapter 2 by different methods.

We next use the General Banach–Steinhaus Theorem to establish a summability result of Hahn and Schur. If λ, μ are two vector spaces of scalar sequences and $A = [a_{ij}]$ is an infinite matrix, we say that A maps λ into μ, if the series $\sum_{j=1}^{\infty} a_{ij} s_j$ converges for every $s = \{s_j\} \in \lambda$ and $\{\sum_{j=1}^{\infty} a_{ij} s_j\}_i \in \mu$ for every $s \in \lambda$. We denote the space of all matrices which map λ into μ by

$$M(\lambda, \mu).$$

One of the problems of summability theory is to characterize matrix mappings from one concrete sequence space into another such space. The result of Hahn and Schur give characterizations of the spaces $M(m_0, c)$ and $M(l^{\infty}, c)$. We first establish a lemma.

Lemma 6.24. *Assume $\sum_{j=1}^{\infty} |a_{ij}| < \infty$ for every i and*

$$\lim_i a_{ij} = a_j$$

exists for every j. Let $\mathcal{F} = \{A \subset \mathbb{N} : A \text{ finite}\}$. If

$$\lim_i \sum_{j \in A} a_{ij} = \sum_{j \in A} a_j$$

is not uniform for $A \in \mathcal{F}$, then there exist $\epsilon > 0$, an increasing sequence $\{i_j\}$ of positive integers and a pairwise disjoint sequence of finite sets $\{B_j\}$ with $\max B_j < \min B_{j+1}$ and

$$\left| \sum_{k \in B_j} (a_{i_j k} - a_k) \right| \geq \epsilon$$

for all j.

Proof. Suppose the limit is not uniform for $A \in \mathcal{F}$. Then there exists $\epsilon > 0$ such that for every i there exist $k_i > i$ and a finite set A_i such that

$$\left| \sum_{k \in A_i} (a_{k_i k} - a_k) \right| \geq 2\epsilon.$$

Put $i = 1$ and let finite A_1, i_1 be such that

$$\left| \sum_{k \in A_1} (a_{i_1 k} - a_k) \right| \geq 2\epsilon.$$

Set $B_1 = A_1$ and $M_1 = \max A_1$. There exists n_1 such that

$$\sum_{j=1}^{M_1} |a_{ij} - a_j| < \epsilon$$

for $i \geq n_1$. There exist $i_2 > \max\{i_1, n_1\}$ and finite A_2 with

$$\left| \sum_{k \in A_2} (a_{i_2 k} - a_k) \right| \geq 2\epsilon.$$

Set $B_2 = A_2 \setminus A_1$ so

$$\left| \sum_{k \in B_2} (a_{i_2 k} - a_k) \right| \geq \left| \sum_{k \in A_2} (a_{i_2 k} - a_k) \right| - \sum_{k \in A_1} |a_{i_2 k} - a_k| \geq 2\epsilon - \epsilon = \epsilon.$$

Continue the construction. $\qquad\qquad\qquad\qquad\qquad\qquad\qquad\qquad\qquad$ $\square$

Theorem 6.25. *For a matrix $A = [a_{ij}]$ the following conditions are equivalent:*

(i) $A \in M(l^\infty, c)$,

(ii) $A \in M(m_0, c)$,

(iii) (a) $\lim\limits_i a_{ij} = a_j$ *exists for every* j,

 (b) $\{a_{ij}\}_j$ *and* $\{a_j\}$ *belong to* l^1 *for every* i,

 (c) $\lim\limits_i \sum\limits_{j=1}^\infty |a_{ij} - a_j| = 0$,

(iv) (a) *and*

 (d) $\{a_{ij}\}_j \in l^1$ *for every* i *and* $\sum\limits_j |a_{ij}|$ *converge uniformly for* $i \in \mathbb{N}$.

Proof. Clearly (i) implies (ii). Suppose (ii) holds. Then (a) follows by setting $s = e^j$. Each row R_i of M induces a linear functional on m_0 by

$$R_i(s) = \sum_{j=1}^{\infty} a_{ij} s_j$$

which is continuous since m_0 is barrelled (2.53). Moreover, $\|R_i\| = \sum_{j=1}^{\infty} |a_{ij}|$. We claim that

$$\lim_i \sum_{j \in A} a_{ij} = \sum_{j \in A} a_j$$

uniformly for finite subsets A of $\mathbb{N}$. For assume this is not the case and let the notation be as in the lemma. The sequence $\{\chi_{B_j}\}$ is $\sigma(l^1, m_0) - \mathcal{K}$ convergent in the triple

$$(m_0, l^1 : \mathbb{R})$$

under the map $(\{s_j\}, \{t_j\}) \to \sum_{j=1}^{\infty} s_j t_j$ so it is $w(\{R_i\}, m_0) - \mathcal{K}$ convergent in the triple $(\{R_i\}, m_0 : \mathbb{R})$. Since $\lim_i R_i(s) = R(s)$ exists for every $s \in m_0$, the General Banach–Steinhaus Theorem implies that

$$\lim_i R_i(\chi_{B_j}) = \lim_i \sum_{k \in B_j} a_{ik} = R(\chi_{B_j}) = \sum_{k \in B_j} a_k$$

uniformly for j. But, this contradicts the conclusion of the lemma and the claim is established.

Let $\epsilon > 0$. Then $\left| \sum_{j \in A} (a_{ij} - a_j) \right| < \epsilon$ for large i and all finite sets A. Hence,

$$\sum_{j=1}^{\infty} |a_{ij} - a_j| \leq 2\epsilon$$

for large i (Lemma 2.50) and (b), (c) follow and (iii) is established.

Assume (iii). Let $\epsilon > 0$. There exists N such that $\sum_{j=1}^{\infty} |a_{ij} - a_j| < \epsilon/2$ for $i \geq N$. There exists $M > 0$ such that

$$\sum_{j=M}^{\infty} |a_{ij}| < \epsilon$$

for $1 \leq i \leq N - 1$ and

$$\sum_{j=M}^{\infty} |a_j| < \epsilon/2.$$

If $i \geq N$,

$$\sum_{j=M}^{\infty} |a_{ij}| \leq \sum_{j=M}^{\infty} |a_{ij} - a_j| + \sum_{j=M}^{\infty} |a_j| < \epsilon$$

so (iv) holds.

Assume (iv) holds. Let $\epsilon > 0$. There exists M such that $\sum_{j=M}^{\infty} |a_{ij}| < \epsilon/4$ for all i. Let $t = \{t_j\} \in l^{\infty}$ and assume $\|t\|_{\infty} \leq 1$. The series $\sum_{j=1}^{\infty} a_{ij} t_j$ converges for every i since

$$\left| \sum_{j=M}^{M+p} a_{ij} t_j \right| \leq \sum_{j=M}^{\infty} |a_{ij}| < \epsilon/4$$

for all $p > 0$. There exists N such that $i, k \geq N$ implies $|a_{ij} - a_{kj}| < \epsilon/2M$ for $j = 1, ..., M - 1$. If $i, k \geq N$, then

$$\left| \sum_{j=1}^{\infty} a_{ij} t_j - \sum_{j=1}^{\infty} a_{kj} t_j \right| \leq \sum_{j=1}^{M-1} |a_{ij} - a_{ik}| + \sum_{j=M}^{\infty} |a_{ij}| + \sum_{j=M}^{\infty} |a_{kj}|$$

$$\leq M\epsilon/2M + \epsilon/4 + \epsilon/4 = \epsilon$$

so $\lim_i \sum_{j=1}^{\infty} a_{ij} t_j$ exists. □

The equivalence of (i) and (iv) is due to Schur ([Sch]). Schur's result was generalized to the equivalence of (i) and (ii) by Hahn ([Ha]). It is interesting that Hahn obtained his result from an early form of an abstract UBP which he established. The abstract result which Hahn employed was more cumbersome than the development of normed spaces by Banach and lost favor. Hahn's paper is very interesting and contains several more applications of his UBP to Lebesgue spaces and matrix transformations.

- As an aside, this theorem can be used to establish a summability result due to Steinhaus. Steihaus showed that a regular matrix cannot sum every bounded sequence. Suppose $A = [a_{ij}]$ is a regular matrix and that $A : m_0 \to c$. By the theorem, $\lim_i a_{ij} = a_j$ exists for every j, $\{a_j\} \in l^1$ and

$$\lim_i \sum_{j=1}^{\infty} a_{ij} = \sum_{j=1}^{\infty} a_j.$$

For $n \in \mathbb{N}$ let t^n be the sequence with 0 in the first $n - 1$ coordinates and 1 in the other coordinates. Since A is regular,

$$\lim_i \sum_{j=1}^{\infty} a_{ij} t_j^n = \lim_i \sum_{j=n}^{\infty} a_{ij} = \sum_{j=n}^{\infty} a_j = 1$$

for each n. But

$$\lim_n \sum_{j=n}^{\infty} a_j = 0$$

since $\{a_j\} \in l^1$. This contradiction means A cannot map m_0 into c and, thus, cannot map l^{∞} into c.

Chapter 7

Biadditive and Bilinear Operators

In this chapter we consider continuity results for biadditive and bilinear operators. These operators fit nicely into the framework of abstract triples. Let E, G be Hausdorff Abelian topological groups and F a set. Let $b : E \times F \to G$ be such that $b(\cdot, y) : E \to G$ is additive and continuous for every $y \in F$. We consider the triple

$$(E, F : G)$$

under the map b.

We first consider hypocontinuity results for this triple. Bourbaki introduced the concept of hypocontinuity for bilinear operators. For bilinear operators the property of hypocontinuity falls between separate continuity and joint continuity. After giving our definition of hypocontinuity for triples we will give examples showing the relationship between these concepts of continuity for bilinear operators between LCTVS.

Let $\mathcal{F}$ be a family of subsets of F. Then b is $\mathcal{F}$-hypocontinuous (sequentially $\mathcal{F}$-hypocontinuous) if whenever $\{x_\delta\}$ is a net ($\{x_i\}$ is a sequence) in E which converges to 0 and $B \in \mathcal{F}$,

$$\lim_\delta b(x_\delta, y) = 0 \quad (\lim_i b(x_i, y) = 0)$$

uniformly for $y \in B$.

The situation considered by Bourbaki is for separately continuous bilinear operators from the product of LCTVS into a LCTVS and they take $\mathcal{F}$ as the family of bounded subsets of F. In this case it is clear that (joint) continuity implies hypocontinuity [if W is a neighborhood of 0 in G, there exist neighborhoods of 0, U, V, such that $b(U, V) \subset W$ and if $B \subset F$ is bounded there exists $\epsilon > 0$ with $\epsilon B \subset V$ so $b(U, \epsilon B) = b(\epsilon U, B) \subset W$]. We give examples below showing that hypocontinuity does not imply continuity and separate continuity does not imply hypocontinuity.

Example 7.1. Define $b : c_{00} \times c_{00} \to \mathbb{R}$ by

$$b(s,t) = \sum_{j=1}^{\infty} s_j t_j.$$

If c_{00} has the sup-norm, b is separately continuous but not sequentially hypocontinuous with respect to the family of bounded sets. For $x_i = \sum_{j=1}^{i} e^j / i \to 0$, $\{y_i\} = \{\sum_{j=1}^{i} e^j\}$ is bounded but $b(x_i, y_i) = 1$ for every i.

Example 7.2. Let E be a LCTVS with dual E' and let E'_b be E' with the strong topology $\beta(E', E)$. Denote the bilinear pairing by $\langle \cdot, \cdot \rangle$. This bilinear form is $\mathcal{B}(E)$ hypocontinuous since if $A \subset E$ is bounded, A^0, the polar of A, is a strong neighborhood of 0. Therefore, if $\epsilon > 0$, then

$$\sup\{|\langle \epsilon x', x \rangle| : x' \in A^0, x \in A\} \le \epsilon.$$

Example 7.3. The bilinear form above is (jointly) continuous iff the space E is normed. If E is normed, then $\langle \cdot, \cdot \rangle$ is clearly continuous. Suppose the form is continuous. Then there exist a convex neighborhood of 0, U, in E and a closed, bounded, absolutely convex set $B \subset E$ such that

$$\sup\{|\langle x', x \rangle| : x' \in B^0, x \in U\} \le 1.$$

By the Bipolar Theorem, $U \subset B^{00} = B$. Therefore, U is a bounded, convex neighborhood of 0 and E is normed by Kolmogorov's Theorem ([Sw2] 13.1.1).

We now consider hypocontinuity results for the triple $(E, F : G)$. Let $\mathcal{UB}_s$ be all UB sequences in F with respect to $(E, F : G)$ (see Definition 5.10).

Theorem 7.4. *b is sequentially $\mathcal{UB}_s$-hypocontinuous.*

Proof. Let $x_i \to 0$ in E and $\{y_j\} \in \mathcal{UB}_s$. Define the matrix M by

$$M = [b(x_i, y_j)].$$

We claim M is a $\mathcal{K}$-matrix. First, the columns converge to 0 since $b(\cdot, y_j)$ is continuous. Next, given any sequence there is a further subsequence $\{n_j\}$ and $y \in F$ such that

$$\sum_{j=1}^{\infty} b(x_i, y_{n_j}) = b(x_i, y)$$

for every i and $b(x_i, y) \to 0$. Hence, M is a $\mathcal{K}$ matrix and $\lim_i b(x_i, y_j) = 0$ uniformly for $j \in \mathbb{N}$ by the Antosik–Mikusinski Matrix Theorem. $\qquad \square$

Let F be a Hausdorff, Abelian topological group. Let $\mathcal{K}_s(F, E)$ be all $w(F, E)$ $\mathcal{K}$-convergent sequences in F and let $\mathcal{K}_s(F)$ be all $\mathcal{K}$-convergent sequences in F. From Theorem 4, we have

Corollary 7.5. *b is sequentially $\mathcal{K}_s(F, E)$-hypocontinuous and $\mathcal{K}_s(F)$-hypocontinuous.*

We can use the corollary to give a generalization of the Mazur–Orlicz theorem on separately continuous bilinear operators.

Corollary 7.6. *Let E, F be quasi-normed groups with F a $\mathcal{K}$ space. If b is separately continuous, the b is (jointly) continuous.*

Proof. Let $x_i \to 0$ in E and $y_i \to 0$ in F. Then $\{y_i\}$ is $\mathcal{K}$-convergent so $b(x_i, y_i) \to 0$ by the corollary. $\qquad\square$

The Mazur–Orlicz result is for bilinear operators from the product of two metric linear spaces one of which is complete. The corollary above clearly gives a generalization of this result since a complete metric linear space is a $\mathcal{K}$ space. The result above applies to topological groups.(See also Corollary 2.71.)

The Mazur–Orlicz result has some interesting applications to early treatments of metric linear spaces. First, consider the definition of metric linear spaces due to Banach ([Ba],[Köl] 15.13). Let X be a vector space with a translation invariant metric d. Banach gives the following axioms for a metric linear space.

(a) $t_n x \to 0$ for each $x \in X$ and $t_n \to 0$,

(b) $t x_n \to 0$ for each scalar t and $x_n \to 0$,

(c) X is complete with respect to d.

It is shown in [Köl], 15.13, that E is a TVS under the metric d when the conditions (a),(b) and (c) are satisfied. This follows directly for the Mazur–Orlicz Theorem when X is a $\mathcal{K}$ space, not necessarily complete. Köthe uses the Baire Category Theorem in $\mathbb{R} \times X$ which is not available in the $\mathcal{K}$ space case.

A similar treatment concerning quasi-normed spaces is given in Yosida ([Y] I.2). If X is vector space, Yosida defines a quasi-norm to be a function $\|\cdot\| : X \to \mathbb{R}$ satisfying

(i) $\|x\| \geq 0$ and $\|x\| = 0$ iff $x = 0$,

(ii) $\|x + y\| \leq \|x\| + \|y\|$,

(iii) $\|-x\| = \|x\|$,

and conditions (a),(b). Yosida shows that X is a TVS under the metric $d(x, y) = \|x - y\|$. This follows directly from the Mazur–Orlicz result. Yosida uses results in measure theory to establish his result.

Let E, F be TVS and $\mathcal{UB}_b$ be all UB subsets of F in the triple $(E, F : G)$ (see Definition 5.21). We assume that b is bilinear, F is a TVS and b is separately continuous.

Theorem 7.7. *If E is braked, then b is sequentially $\mathcal{UB}_b$-hypocontinuous.*

Proof. Let $x_i \to 0$ in E and $B \in \mathcal{UB}_b$. It suffices to show $\lim_i b(x_i, y_j) = 0$ uniformly for $\{y_j\} \subset B$. There exists $t_i \to \infty$ such that $t_i x_i \to 0$ in E. Then $\{(1/t_j)y_j\} \in \mathcal{UB}_s$. By Theorem 7.4,

$$\lim_i b(t_i x_i, (1/t_j)y_j) = \lim_i b(x_i, y_j) = 0$$

uniformly for $j \in \mathbb{N}$.   □

The proof above also gives another hypocontinuous result. Let $\mathcal{B}(F)$ be all bounded subsets of F.

Theorem 7.8. *Assume E is braked. If b is sequentially jointly continuous, then b is sequentially $\mathcal{B}(F)$-hypocontinuous.*

Proof. Let $x_j \to 0$ in E and $B \subset F$ be bounded. It suffices to show $\lim_j b(x_j, y_j) = 0$ for $\{y_j\} \subset B$. Let $t_j \to \infty$ with $t_j x_j \to 0$. Then $(1/t_j)y_j \to 0$ so

$$\lim_j b(t_j x_j, (1/t_j)y_j) = \lim_j b(x_j, y_j) = 0.$$

                                                                                  □

Let $\mathcal{K}_b(F, E)$ be all $w(F, E)$ $\mathcal{K}$ bounded sets and $\mathcal{K}_b(F)$ all $\mathcal{K}$ bounded sets in F. From Theorem 7.7, we have

Corollary 7.9. *If E is braked, then b is sequentially $\mathcal{K}_b(F, E)$ and $\mathcal{K}_b(F)$ hypocontinuous.*

Let $\mathcal{B}(F)$ be all bounded subsets of F.

Corollary 7.10. *If E is braked and F is an $\mathcal{A}$ space, then b is sequentially $\mathcal{B}(F)$ hypocontinuous.*

Example 1 shows the $\mathcal{A}$ space assumption in the corollary cannot be dropped.

We next consider boundedness for bilinear operators. The separately continuous bilinear operator b is *bounded* if $b(A, B)$ is bounded for every bounded set $A \subset E, B \subset F$. First, we observe that a separately continuous bilinear operator may not be bounded.

Example 7.11. Consider the bilinear operator in Example 1. If

$$A = B = \left\{ \sum_{j=1}^{i} e^j : i \right\},$$

then $b(A, B) = \mathbb{N}$.

We show that hypocontinuity results imply boundedness results.

Proposition 7.12. *If b is sequentially $\mathcal{F}$-hypocontinuous, $A \subset E$ is bounded and $B \in \mathcal{F}$, then $b(A, B)$ is bounded. In particular, if b is sequentially $\mathcal{B}(F)$-hypocontinuous, then b is bounded.*

Proof. Let $t_j \to 0$, $\{x_j\} \subset A$ and $\{y_j\} \subset B$. Then

$$\lim_j t_j b(x_j, y_j) = \lim_j b(t_j x_j, y_j) = 0.$$

$\square$

From this result and Corollary 10, we have

Theorem 7.13. *Assume E is braked and F is an $\mathcal{A}$ space. If b is jointly sequentially continuous, then b is bounded.*

Theorem 7.14. *If $A \subset E$ is bounded and B belongs to $\mathcal{UB}_s$ or $\mathcal{UB}_b$, then $b(A, B)$ is bounded.*

Proof. The set $\{b(\cdot, y) : y \in B\} \subset LS(E, G)$ is pointwise bounded by the separate continuity. The result follows from Theorems 5.11 and 5.23. $\square$

Corollary 7.15. *If $A \subset E$ is bounded and B belongs to $\mathcal{K}_s(F, E)$ or $\mathcal{K}_b(F, E)$, then $b(A, B)$ is bounded.*

Corollary 7.16. *If F is an $\mathcal{A}$ space, then b is bounded.*

Example 1 above shows the $\mathcal{A}$ space assumption cannot be dropped. In general, boundedness does not imply $\mathcal{B}(F)$ hypocontinuity.

Example 7.17. Define $b : c_0 \times l^1 \to \mathbb{R}$ by

$$b(s,t) = \sum_{j=1}^{\infty} s_j t_j.$$

Equip c_0 with $\sigma(c_0, l^1)$ and l^1 with $\sigma(l^1, c_0)$. Then b is bounded but not $\mathcal{B}(l^1)$ sequentially hypocontinuous. Consider $e^i \to 0$ in $\sigma(c_0, l^1)$ and $\{e^i : i\}$ $\sigma(l^1, c_0)$ bounded but $b(e^i, e^i) = 1$ for all i.

However, for braked (metric linear) spaces, we do have

Theorem 7.18. *If E is a braked space and b is bounded, then b is jointly sequentially continuous.*

Proof. Let $x_i \to 0$ in E and $y_i \to 0$ in F. There exists $t_i \to \infty$ such that $t_i x_i \to 0$. Then $\{b(t_i x_i, y_i) : i\}$ is bounded so

$$b(x_i, y_i) = (1/t_i)b(t_i x_i, y_i) \to 0.$$

$\square$

Corollary 7.19. *If E, F are metric linear spaces and b is bounded, the b is continuous.*

Families of Bilinear Operators

We next consider families of bilinear operators. Let E, F, G be TVS and let $b_i : E \times F \to G$ be separately continuous, bilinear operators for $i \in I$. Set $\mathcal{B} = \{b_i : i \in I\}$. First, we establish a Uniform Boundedness Principle for bilinear operators.

Theorem 7.20. *Suppose $\mathcal{B}$ is pointwise bounded on $E \times F$. Then $\mathcal{B}$ is bounded on $A \times B$ when A (B) belongs to either $\mathcal{K}_s(E, F)$ ($\mathcal{K}_s(F, E)$) or $\mathcal{K}_b(E, F)$ ($\mathcal{K}_b(F, E)$).*

Proof. Consider first the case when $A \in \mathcal{K}_s(E, F)$, $B \in \mathcal{K}_b(F, E)$. Fix $y \in F$. Then

$$\mathcal{B}_y = \{b_i(\cdot, y) : i \in I\} \subset LS(E, G)$$

is pointwise bounded on E. By Theorem 5.11, $\mathcal{B}_y$ is uniformly bounded on A, i.e. $\{b_i(x, y) : x \in A, i \in I\}$ is bounded. Thus, the family

$$\{b(x, \cdot) : i \in I, x \in A\} \subset LS(F, G)$$

is pointwise bounded on F. By Theorem 5.23 the family is uniformly bounded on B. That is, $\{b_i(x, y) : x \in A, y \in B, i \in I\}$ is bounded.

The other cases are treated similarly using the general UBP's of Chapter 5. $\qquad\square$

Corollary 7.21. *Assume that E, F are $\mathcal{A}$ spaces. If $\mathcal{B}$ is pointwise bounded on $E \times F$, then $\mathcal{B}$ is uniformly bounded on sets of the form $A \times B$ when A and B are bounded.*

We give an example showing the $\mathcal{A}$ space assumption cannot be dropped.

Example 7.22. Define $b_i : l^\infty \times c_{00} \to \mathbb{R}$ by

$$b_i(s, t) = \sum_{j=1}^{i} s_j t_j$$

for $i \in \mathbb{N}$. Then each b_i is separately continuous when both spaces have the sup-norm and $\{b_i : i \in \mathbb{N}\}$ is pointwise bounded on $l^\infty \times c_{00}$. However, $\{b_i : i \in \mathbb{N}\}$ is not uniformly bounded on the set $\{(e, \sum_{j=1}^{i} e^j) : i \in \mathbb{N}\}$, where e is the constant sequence with 1 in each coordinate $(b_i(e, \sum_{j=1}^{i} e^j) = i)$.

Theorem 7.20 can be used to derive sequential equicontinuity results as in Chapter 6.

Theorem 7.23. *Let $b_i : E \times F \to G$ be bilinear and separately continuous for $i \in \mathbb{N}$. Let E be a braked $\mathcal{K}$ space. Assume $\{b_i : i\}$ is pointwise bounded on $E \times F$. If $x_j \to 0$ in E and $\{y_j\}$ is $\mathcal{K}$ convergent in F, then*

$$\lim_j b_i(x_j, y_j) = 0$$

uniformly for $i \in \mathbb{N}$.

Proof. If the conclusion fails, there exists a neighborhood of 0, W, in G such that for every i there exist $m_i, n_i > i$ with

$$b_{m_i}(x_{n_i}, y_{n_i}) \notin W.$$

Put $i_1 = 1$ and apply the condition above to obtain m_1, n_1 with

$$b_{m_1}(x_{n_1}, y_{n_1}) \notin W.$$

By Corollary 5 there exists $i_2 > n_1$ such that $j \geq i_2$ implies $b_i(x_j, y_j) \in W$ for $1 \leq i \leq m_1$. Applying the observation above there exist $m_2, n_2 > \max\{n_1, m_1\}$ with

$$b_{m_2}(x_{n_2}, y_{n_2}) \notin W.$$

Note $m_2 > m_1, n_2 > n_1$. Continuing this construction produces increasing sequences $\{m_i\}, \{n_i\}$ with

$$b_{m_i}(x_{n_i}, y_{n_i}) \notin W.$$

Pick a sequence of scalars $t_i \to \infty$ with $t_i x_{n_i} \to 0$. Since E is a $\mathcal{K}$ space, $\{t_i x_{n_i}\}$ is $\mathcal{K}$ convergent so by Theorem 7.20 $\{b_{m_i}(t_i x_{n_i}, y_{n_i})\}_i$ is bounded. Thus,

$$\lim_i (1/t_i) b_{m_i}(t_i x_{n_i}, y_{n_i}) = \lim_i b_{m_i}(x_{n_i}, y_{n_i}) = 0$$

contradicting the construction above. $\qquad\square$

Corollary 7.24. *Let E, F be metric linear $\mathcal{K}$ spaces and $b_i : E \times F \to G$ be bilinear and separately continuous for $i \in \mathbb{N}$. If $\{b_i\}_i$ is pointwise bounded on $E \times F$, then $\{b_i\}_i$ is equicontinuous.*

The $\mathcal{K}$ space assumption in this result cannot be dropped.

Example 7.25. Let b_i be as in the Example 22 above. Then $y^i = (1/\sqrt{i}) \sum_{j=1}^i e^j \to 0$ in the sup-norm, but $b_i(y^i, y^i) = 1$ for every i. Thus, $\{b_i\}_i$ is not equicontinuous.

We can also obtain a Banach–Steinhaus type theorem for bilinear operators.

Corollary 7.26. *Let E, F be metric linear $\mathcal{K}$ spaces and $b_i : E \times F \to G$ be bilinear and separately continuous for $i \in \mathbb{N}$. If*

$$\lim_i b_i(x, y) = b(x, y)$$

exists for every $x \in E, y \in F$, then $\{b_i\}_i$ is equicontinuous and b is bilinear and continuous.

Proof. That b is bilinear is clear. The equicontinuity follows from the previous corollary. If $x_j \to 0$ in E and $y_j \to 0$ in F, then $\lim_j b_i(x_j, y_j) = 0$ uniformly for $i \in \mathbb{N}$ by Theorem 7.23. Then

$$\lim_j b(x_j, y_j) = \lim_j \lim_i b_i(x_j, y_j) = \lim_i \lim_j b_i(x_j, y_j) = 0$$

by the Iterated Limit Theorem. $\qquad\square$

The $\mathcal{K}$ space assumption in the corollary cannot be dropped.

Example 7.27. Let b_i be as in Example 22. Then

$$\lim_i b_i(s,t) = b(s,t) = \sum_{j=1}^{\infty} s_j t_j.$$

However, b is not even separately continuous. If $s^i = \sum_{j=1}^{i} e^j/i$, then $s^i \to 0$ but $b(e, s^i) = 1$ for every i.

We next consider separate equicontinuity for bilinear operators. If $\mathcal{B}$ is a family of separately continuous bilinear operators from $E \times F$ into G, then $\mathcal{B}$ is *left (sequentially) equicontinuous* if the family of linear operators

$$\mathcal{B}_y = \{b(\cdot, y) : b \in \mathcal{B}\}$$

is (sequentially) equicontinuous. Right equicontinuity is defined similarly. $\mathcal{B}$ is *separately (sequentially) equicontinuous* if $\mathcal{B}$ is both left and right (sequentially) equicontinuous.

We first observe that a family can be equicontinuous in one variable and not so in the other variable.

Example 7.28. Let $\{b_i\}$ be as in Example 22 above. Fix $t \in c_{00}$ and assume that $t_i = 0$ for $i > n$. Then for $i > n$ and $s \in l^{\infty}$, $b_i(s,t) = \sum_{j=1}^{n} s_j t_j$ and $\|b_i(\cdot, t)\| \leq \sum_{j=1}^{n} |t_j|$. Therefore, $\{b_i(\cdot, t) : i\} \subset L(l^{\infty}, \mathbb{R})$ is equicontinuous. However, $\|b_i(e, \cdot)\| = i$ for every i so $\{b_i(e, \cdot) : i\} \subset L(c_{00}, \mathbb{R})$ is not equicontinuous. That is $\{b_i : i\}$ is left equicontinuous but not right equicontinuous.

We show below that a left equicontinuous family is separately equicontinuous if both E and F are metric linear $\mathcal{K}$ spaces.

We consider boundedness for left sequentially equicontinuous families.

Theorem 7.29. *Let $\mathcal{B}$ be a family of left sequentially equicontinuous, separately continuous bilinear operators. If A is bounded and B is either $\mathcal{K}$ bounded or the range of a $\mathcal{K}$ convergent sequence, then*

$$\{b(A, B) : b \in \mathcal{B}\} = \{b(x, y) : x \in A, y \in B, b \in \mathcal{B}\}$$

is bounded. In particular, $\mathcal{B}$ is pointwise bounded on $E \times F$.

Proof. Let $\{x_i\} \subset A$, $t_i \to 0$ and $y \in F$. Then $\lim_i b(t_i x_i, y) = 0$ uniformly for $b \in \mathcal{B}$. Thus, if $\{b_i\} \subset \mathcal{B}$, $t_i b_i(x_i, y) \to 0$ so $\mathcal{B}(A, y)$ is bounded. That is, the family $\{b(A, \cdot) : b \in \mathcal{B}\} \subset LS(F, G)$ is pointwise bounded on F. The result follows from Theorems 5.11 and 5.23. $\qquad \square$

Corollary 7.30. *Let E, F be metric linear $\mathcal{K}$ spaces. If $\mathcal{B}$ is left sequentially equicontinuous, then $\mathcal{B}$ equicontinuous.*

Proof. This follows from the theorem above and Corollary 24. □

Köthe establishes a similar result for separately equicontinuous bilinear operators when E and F are metric linear, barrelled spaces. Note the result above only uses equicontinuity in one variable and is valid for $\mathcal{K}$ spaces.

The left equicontinuity in the result above cannot be dropped.

Example 7.31. For $i \in \mathbb{N}$, define $b_i : l^\infty \times l^\infty \to \mathbb{R}$ by $b_i(s, t) = \sum_{j=1}^{i} s_j t_j$. Each b_i is continuous but $\{b_i : i\}$ is not even pointwise bounded on $l^\infty \times l^\infty$ (consider $b_i(e, e)$).

The sequence in Example 22 shows that a pointwise bounded sequence of separately continuous bilinear operators needn't be right (left) sequentially equicontinuous. However, for metric linear $\mathcal{K}$ spaces, we have a partial converse.

Theorem 7.32. *Let $\mathcal{B}$ be a family of separately continuous bilinear operators and E be a metric linear $\mathcal{K}$ space. If $\mathcal{B}$ is pointwise bounded on $E \times F$, then $\mathcal{B}$ is left equicontinuous.*

Proof. If $y \in F$, then $\{b(\cdot, y) : b \in \mathcal{B}\} \subset L(E, G)$ is pointwise bounded on E. The result follows from Theorem 5.23. □

We have boundedness results for left equicontinuous bilinear operators.

Proposition 7.33. *Let $\mathcal{B}$ be left sequentially equicontinuous and F an $\mathcal{A}$ space. If $A \subset E$ and $B \subset F$ are bounded, then*

$$\mathcal{B}(A, B) = \{b(A, B) : b \in \mathcal{B}\}$$

is bounded.

Proof. This follows from Theorem 29. □

Example 22 shows that the conclusion in the proposition above does not hold without the $\mathcal{A}$ space assumption. However, for equicontinuous bilinear operators, we do have

Proposition 7.34. *If $\mathcal{B}$ is a sequentially equicontinuous family of separately continuous bilinear operators, then $\mathcal{B}(A, B)$ is bounded when $A \subset E, B \subset F$ are bounded.*

Proof. Let $b_i \in \mathcal{B}, \{x_i\} \subset A, \{y_i\} \subset B$ and $t_i \to 0$ with $t_i \geq 0$. Then $\sqrt{t_i} x_i \to 0, \sqrt{t_i} y_i \to 0$ so

$$b_i(\sqrt{t_i} x_i, \sqrt{t_i} y_i) = t_i b(x_i, y_i) \to 0$$

so the result follows. $\qquad\qquad\square$

We give an example which shows that equicontinuity cannot be replaced with separate equicontinuity in the result above.

Example 7.35. For each i define $b_i : c_{00} \times c_{oo} \to \mathbb{R}$ by $b_i(s, t) = \sum_{j=1}^{i} s_j t_j$. Then $\{b_i\}$ is separately sequentially equicontinuous but not equicontinuous (consider $s^i = \sum_{j=1}^{i} e^j$ and $b_i(s^i/\sqrt{i}, s^i/\sqrt{i}) = (1/i)b(s^i, s^i)$). Also, $\{b_i(s^i, s^i)\}$ is not bounded.

We now define equihypocontinuity for families of bilinear operators. Let $\mathcal{B}$ be a family of separately continuous bilinear operators and $\mathcal{R}$ a family of bounded subsets of F. Then $\mathcal{B}$ is $\mathcal{R}$ *equihypocontinuous* if for every neighborhood, W, in G and $B \in \mathcal{R}$, there exists a neighborhood of 0, U, in E with $\mathcal{B}(U, B) \subset W$.

Theorem 7.36. *Let E be metrizable and $\mathcal{B}$ left sequential equicontinuous. Then $\mathcal{B}$ is*

(i) $\mathcal{K}_s(F)$ *equihypocontinuous and*
(ii) $\mathcal{K}_b(F)$ *equihypocontinuous.*

Proof. For (i) it suffices to show $\lim_i b_i(x_i, y_j) = 0$ uniformly for $j \in \mathbb{N}$ whenever $\{b_i\} \subset \mathcal{B}, x_i \to 0$ in E and $\{y_j\}$ is $\mathcal{K}$ convergent in F. For this consider the matrix

$$M = [b_i(x_i, y_j)].$$

The columns of M converge to 0 by left sequential equicontinuity. Given any subsequence there is a further subsequence $\{n_j\}$ such that the subseries $\sum_{j=1}^{\infty} y_{n_j}$ converges to some $y \in F$. By separate continuity

$$\sum_{j=1}^{\infty} b_i(x_i, y_{n_j}) = b_i(x_i, y)$$

and $\lim_i b_i(x_i, y) = 0$ by left sequential equicontinuity. Hence, M is a $\mathcal{K}$ matrix and by the Antosik–Mikusinski Theorem,

$$\lim_i b_i(x_i, y_j) = 0 \text{ uniformly for } j \in \mathbb{N}.$$

For (ii) it suffices to show $\lim_i b_i(x_i, y_j) = 0$ uniformly for $j \in \mathbb{N}$ whenever $\{b_i\} \subset \mathcal{B}, x_i \to 0$ in E and $\{y_j\}$ is $\mathcal{K}$ bounded in F. If this condition fails to hold, then there exists W, a neighborhood of 0 in G, such that for every i there exist $k_i > i, j_i$ with

$$b_{k_i}(x_{k_i}, y_{j_i}) \notin W.$$

Applying this condition to $i = 1$, there exist $k_1 > 1, j_1$ with

$$b_{k_1}(x_{k_1}, y_{j_1}) \notin W.$$

By left sequential equicontinuity, there exists $n_1 > k_1$ such that $b_i(x_i, y_j) \in W$ for $i \geq n_1, 1 \leq j \leq j_1$. There exist $k_2 > \max\{n_1, k_1\}, j_2$ such that

$$b_{k_2}(x_{k_2}, y_{j_2}) \notin W.$$

Note $j_2 > j_1, k_2 > k_1$. Continuing this construction produces increasing sequences $\{k_i\}, \{j_i\}$ such that

$$(\#) \ b_{k_i}(x_{k_i}, y_{j_i}) \notin W$$

for all i. Pick a sequence of scalars $\{t_i\}$ such that $t_i x_i \to 0$ with $t_i \to \infty$. Then $\{(1/t_i)x_i\}$ is $\mathcal{K}$ convergent in E so by (i),

$$\lim_i b_{k_i}(t_i x_{k_i}, (1/t_i)y_{j_i}) = \lim_i b_{k_i}(x_{k_i}, y_{j_i}) = 0$$

contradicting (#). $\qquad\square$

Corollary 7.37. *Let E be metrizable and F an $\mathcal{A}$ space. If $\mathcal{B}$ is left sequentially equicontinuous, then $\mathcal{B}$ is $\mathcal{B}(F)$ equihypocontinuous.*

The sequence in Example 22 shows the $\mathcal{A}$ space assumption cannot be dropped.

Corollary 37 is similar to a result in 40.2.3(b) of [Kö2]; Köthe's result is for separately equicontinuous bilinear operators and F barrelled. The result above only uses the equicontinuity in one variable, replaces the barrelledness assumption with an $\mathcal{A}$ space assumption but requires metrizability.

Chapter 8

Triples with Projections

In the paper [NS], D. Noll and W. Stadler introduced an abstract notion of sectional operators modeled on the natural sectional projections of sequence spaces. They established a number of results for such sectional operators and gave applications to scalar sequence and function spaces. In [ZCL] Zheng, Cui and Li gave a generalization of sectional operators to abstract triples and established several uniform convergence results along with applications.

In this chapter we present an abstraction of the coordinate projections for scalar and vector valued sequence spaces much in the spirit of the abstraction of sectional operators given by Zheng, Cui and Li ([ZCL]). We establish a uniform boundedness principle for abstract duality pairs which generalizes a scalar uniform boundedness result of Wu, Luo and Cui ([WLC]). We also establish several uniform convergence results analogous to those of Zheng, Cui and Li ([ZCL]). In some sense our presentation requires fewer assumptions and offers simpler notation than that of sectional operators and the conclusions are stated in terms of series. Of course, it is possible to transfer back and forth between projection and sectional operators. We indicate several applications which give general uniform convergence results for vector valued sequence spaces, for vector valued integrable functions and vector valued measures.

Let E, G be Hausdorff TVS and let F be a vector space. Let $b : E \times F \to G$ be a bilinear operator with $b(\cdot, y) : E \to G$ continuous for every $y \in F$. We consider the triple

$$(E, F : G)$$

under the map b.

We assume that there exist a sequence of projection operators $\{P_j\}$,

$$P_j : E \to E,$$

which are continuous with respect to the topology of E. We often write

$$x \cdot y = b(x, y) = \langle x, y \rangle$$

for the map from $E \times F \to G$. We refer to the triple

$$(E, F : G)$$

and the projections $\{P_j\}$ as an abstract triple or abstract duality pair with projections. One can construct sectional operators s_k as in [NS], [WLC], [ZCL] by setting

$$s_k = \sum_{j=1}^{k} P_j;$$

if one assumes $P_i P_j = 0$ when $i \neq j$, then the sectional operators will satisfy $s_k s_l = s_l$ if $l \leq k$ as required in [NS], [WLC], [ZCL]. It should be noted that these authors require that the space F is also equipped with sectional operators, an assumption we do not make.

We give examples of abstract duality pairs with projections.

Example 8.1. Let λ be a scalar valued sequence space which contains c_{00}, the space of all scalar sequences which are eventually 0. If $t \in \lambda$, we write $t = \{t_j\}$ so t_j is the jth coordinate of t. The β-dual, λ^β, of λ is

$$\lambda^\beta = \left\{ \{s_j\} : \sum_{j=1}^{\infty} s_j t_j \text{ converges for every } t \in \lambda \right\}.$$

Then λ, λ^β form a dual pair with respect to the pairing $t \cdot s = \sum_{j=1}^{\infty} s_j t_j$. Let e^j be the sequence with 1 in the jth coordinate and 0 in the other coordinates. Then for every $j \in \mathbb{N}$,

$$P_j(t) = t_j e^j$$

defines the coordinate projection $P_j : \lambda \to \lambda$ from t onto its jth coordinate. Each P_j is obviously $\sigma(\lambda, \lambda^\beta)$ continuous and in many cases λ will carry a locally convex topology with respect to which each P_j will be continuous.

More generally, we have

Example 8.2. Let X, Y be topological vector spaces and let E be a vector space of X valued sequences which contains $c_{00}(X)$, the space of all X valued sequences which are eventually 0. Let $L(X, Y)$ be the space of all continuous linear operators from X into Y. The β-dual of E with respect to Y, $E^{\beta Y}$, is defined to be

$$E^{\beta Y} = \left\{ \{T_j\} \subset L(X,Y) : \sum_{j=1}^{\infty} T_j x_j \text{ converges for every } x = \{x_j\} \in E \right\}$$

and we have a bilinear operator $(x,T) \to \sum_{j=1}^{\infty} T_j x_j = x \cdot T$ from $E \times E^{\beta Y} \to Y$. If $w(E, E^{\beta Y})$ is the weakest topology on E such that all the linear maps $x \to x \cdot T$ from E into Y are continuous for every $T \in E^{\beta Y}$, then $E, E^{\beta Y}$ is an abstract duality pair with respect to Y or an abstract triple,

$$(E, E^{\beta Y} : Y).$$

If $z \in X$ and $j \in \mathbb{N}$, let $e^j \otimes z$ be the sequence with z in the jth coordinate and 0 in the other coordinates. The space E has the natural coordinate projection operators P_j defined by

$$P_j x = e^j \otimes x_j$$

which are continuous with respect to $w(E, E^{\beta Y})$. Then $(E, E^{\beta Y} : Y)$ is an abstract triple with projections.

We give several non-sequence space examples. Let (S, Σ, μ) be a σ-finite measure space with $\{A_j\}$ a pairwise disjoint sequence from Σ.

Example 8.3. Then $L^1(\mu), L^\infty(\mu)$ form a dual pair under the bilinear map

$$(f,g) \to \int_S fg d\mu$$

which is continuous with respect to the natural topologies and

$$P_j f = \chi_{A_j} f$$

defines a projection operator on $L^1(\mu)$ which is $\|\cdot\|_1$ continuous, where χ_A is the characteristic function of A.

More generally, let X be a Banach space.

Example 8.4. Let $L^1(\mu, X)$ be the space of all X valued, Bochner μ integrable functions with the L^1 norm,

$$\|f\|_1 = \int_S \|f(s)\| d\mu(s)$$

(see Chapter 1). Then

$$\langle f, g \rangle = \int_S g(s) f(s) d\mu(s)$$

defines a continuous bilinear operator

$$\langle \cdot, \cdot \rangle : L^1(\mu, X) \times L^\infty(\mu) \to X$$

when $L^\infty(\mu)$ has its natural topology and

$$P_j f = \chi_{A_j} f$$

defines a continuous projection operator on $L^1(\mu, X)$. Then

$$(L^1(\mu, X), L^\infty(\mu) : X)$$

is an abstract triple with projections. Similarly, if $L^\infty(\mu, X')$ is the space of essentially μ-bounded, X' valued functions with its natural topology,

$$(L^1(\mu, X), L^\infty(\mu, X'))$$

is a dual pair with projections $\{P_j\}$. Dually,

$$(L^\infty(\mu), L^1(\mu, X) : X), \; (L^\infty(\mu, X), L^1(\mu, X')) \text{ and } (L^\infty(\mu), L^1(\mu))$$

are abstract triples with the projections defined as above.

Similarly,

Example 8.5. Let $D(\mu, X)$ $[P(\mu, X)]$ be the space of X valued Dunford [Pettis] μ integrable functions with the norm

$$\|f\|_D = \sup \left\{ \int_S |x'f| \, d\mu : \|x'\| \le 1 \right\}$$

(see Chapter 1 for these integrals). Then

$$\langle f, g \rangle = \int_S g(s) f(s) d\mu(s)$$

defines a continuous bilinear operator

$$\langle \cdot, \cdot \rangle : D(\mu, X) \times L^\infty(\mu) \to X''$$

$[\langle \cdot, \cdot \rangle : P(\mu, X) \times L^\infty(\mu) \to X \;]$ (if $x' \in X', \|x'\| \le 1$,

$$\left| x' \int_S g(s) f(s) d\mu(s) \right| = \left| \int_S g(s) x' f(s) d\mu(s) \right| \le \|g\|_\infty \int_S |x' f(s)| \, d\mu(s)$$

so

$$\|\langle f, g \rangle\| \le \|f\|_D \|g\|_\infty;$$

for the integrability of gf in the Pettis integrable case, see Chapter 1). Also

$$P_j f = \chi_{A_j} f$$

defines a continuous projection operator on $D(\mu, X)$ $[P(\mu, X)]$. Then

$$(D(\mu, X), L^\infty(\mu) : X'') \; [(P(\mu, X), L^\infty(\mu) : X)]$$

is an abstract triple with respect to X'' $[X]$ with projections. Dually,

$$(L^\infty(\mu), D(\mu, X) : X'') \; [(L^\infty(\mu), P(\mu, X) : X)]$$

is an abstract triple with the projections defined similarly.

Note in all of the examples above the projections satisfy $P_i P_j = 0$ when $i \neq j$ so we can define sectional operators $s_k = \sum_{j=1}^{k} P_j$ which satisfy $s_k s_l = s_k$ when $k \leq l$ as in [NS], [WLC], [ZCL].

We will give further examples of abstract triples with projections defined on spaces of vector valued measures later.

We now define the β-dual with respect to an abstract duality pair with projections.

Definition 8.6. The β-dual of E with respect to the abstract triple $(E, F : G)$ and projections $\{P_j\}$ is defined to be

$$E^\beta = \left\{ y \in F : \sum_{j=1}^{\infty} \langle P_j x, y \rangle \text{ converges in } G \text{ for every } x \in E \right\}.$$

We write

$$x \cdot y = \sum_{j=1}^{\infty} \langle P_j x, y \rangle$$

when $x \in E$ and $y \in E^\beta$ and we set

$$P_I = \sum_{j \in I} P_j$$

when I is a finite subset of $\mathbb{N}$.

A few remarks are in order. First, the β-dual of E is a subset of F and our notation does not reflect the dependence of the β-dual on the abstract triple with projections. We have also used the same notation for the β-duals of sequence spaces previously. Hopefully, the statements will be clear from the context.

It is also the case that the β-dual can be a proper subset of F.

Example 8.7. To see this consider the abstract triple

$$(l^\infty, D(\mu, c_0) : l^\infty)$$

of Example 5, where $S = \mathbb{N}$ and μ is counting measure. Define $f : \mathbb{N} \to c_0$ by $f(j) = e^j$. Then f is Dunford integrable (but not Pettis integrable; see Example 13 of Chapter 1) so $f \in D(\mu, c_0)$ and if $t = \{t_j\} \in l^\infty$,

$$\int_A tf d\mu = \chi_A t$$

[coordinate product]. Let P_j be the coordinate projections as defined in Example 5 so

$$(l^\infty, D(\mu, c_0) : l^\infty)$$

is an abstract triple with projections. Let $e = \{1\}$ be the constant sequence of 1's. Then

$$\sum_{j=1}^{\infty} \langle P_j e, f \rangle = \sum_{j=1}^{\infty} \int_{\mathbb{N}} P_j e f d\mu = \sum_{j=1}^{\infty} e^j,$$

a series which does not converge in l^{∞} with respect to $\|\cdot\|_{\infty}$. Hence, $f \notin (l^{\infty})^{\beta}$ with respect to this abstract triple.

We say that E (or the triple $(E, F : G)$ with projections $\{P_j\}$) is a *weak AK space* if for each $x \in E$,

$$x = \sum_{j=1}^{\infty} P_j x,$$

where the series is convergent with respect to $w(E, E^{\beta})$. If E is a weak AK space, then we have $E^{\beta} = F$ so this is a sufficient condition for equality between F and the β dual.

Next, we should compare this definition with the previous definitions of β-duals.

If λ is a scalar sequence space as in Example 1, then the definition in Example 1 above coincides with the definition above when we consider the abstract duality pair λ, λ^{β} with respect to the scalar field and the projections defined in Example 1.

To see the dependence of the β-dual on the abstract duality pair consider the space c_c of scalar sequences which are eventually constant. In the duality pair c_c, l^1 the β-dual in this pair is l^1 while the β-dual in the classical setting of sequence spaces is cs, the space of convergent series ([KG]).

If X, Y, E are as in Example 2, then the β-dual as defined in Example 2 coincides with the definition above when we consider the abstract triple $(E, E^{\beta Y} : Y)$ and the projections defined in Example 2.

In Examples 4 and 5 for the Bochner and Pettis integrals, the β-dual is just $L^{\infty}(\mu)$. This follows from the countable additivity of the integrals in each case. In each case we would have

$$\sum_{j=1}^{\infty} \langle P_j f, g \rangle = \sum_{j=1}^{\infty} \int_{A_j} g f d\mu = \int_{\cup_{j=1}^{\infty} A_j} g f d\mu$$

for $g \in L^{\infty}(\mu)$ (Chapter 1).

We now consider one of our main results which depends on a gliding hump assumption. An interval I in $\mathbb{N}$ is a subset of the form

$$I = \{j \in \mathbb{N} : m \le j \le n\}, \ m \le n;$$

a sequence of intervals $\{I_j\}$ is increasing if $\max I_j < \min I_{j+1}$. Let $w(E, E^\beta)$ be the weakest topology on E such that the linear maps

$$x \to x \cdot y = \sum_{j=1}^{\infty} \langle P_j x, y \rangle$$

are continuous from E into G for every $y \in E^\beta$.

Definition 8.8. The space E (or the triple $(E, F : G)$ with projections $\{P_j\}$) has the zero gliding hump property (0-GHP) if whenever $x^k \to 0$ in E and $\{I_k\}$ is an increasing sequence of intervals, there is a subsequence $\{n_k\}$ such that the series

$$\sum_{k=1}^{\infty} P_{I_{n_k}} x^{n_k}$$

is $w(E, E^\beta)$ convergent in E.

In the case when E is a sequence space, in the usual definition of the β-dual the series $\sum_{k=1}^{\infty} P_{I_{n_k}} x^{n_k}$ in the definition above is required to converge pointwise or coordinatewise. Of course, in this abstract setting there is no notion of pointwise convergence; in this case there are 2 natural choices for the convergence of the series

$$\sum_{k=1}^{\infty} P_{I_{n_k}} x^{n_k},$$

namely, the topology $w(E, E^\beta)$ or the original topology of E. We have chosen the topology $w(E, E^\beta)$ because it is often the weaker topology and it seems to be the correct topology for the proof of Theorem 16. Also, if λ is a scalar sequence space and if the series $\sum_{k=1}^{\infty} P_{I_{n_k}} x^{n_k}$ is coordinatewise convergent to an element $x \in \lambda$, then the series is $\sigma(\lambda, \lambda^\beta)$ convergent to x so we have agreement with the definition above. (The β-dual of a K-space is often contained in the topological dual so the weak topology $w(E, E^\beta)$ is weaker than the original topology (see [KG], page 68, for examples).) In the papers [WCL] and [ZCL], the authors have chosen the original topology of E and used sectional operators.

Examples of sequence spaces with 0-GHP can be found in Appendix B and [Sw4], Appendices B and C; further examples of function and measure spaces with 0-GHP will be given later. The concept of 0-GHP was introduced by Lee Peng Yee ([LPY]).

Theorem 8.9. *Let E be a Banach space having projections $\{P_j\}$ which satisfy the condition that $P_i P_j = 0$ when $i \neq j$ and*

$$\sup\{\|P_I\| : I \subset \mathbb{N} \text{ finite}\} = M < \infty.$$

If $(E, F : G)$ is any triple with $b(\cdot, y)$ continuous for each $y \in F$, then

$$(E, F : G)$$

E has 0-GHP.

Proof. For if $x^k \to 0$ in E and $\{I_k\}$ is an increasing sequence of intervals, there is a subsequence $\{n_k\}$ such that $\sum_{k=1}^{\infty} \|x^{n_k}\| < \infty$ and then

$$\sum_{k=1}^{\infty} \left\| P_{I_{n_k}} x^{n_k} \right\| \le M \sum_{k=1}^{\infty} \|x^{n_k}\| < \infty$$

and the series $\sum_{k=1}^{\infty} P_{I_{n_k}} x^{n_k}$ converges to some $x \in E$ by completeness. But then the series is $w(E, F)$ convergent to x. $\square$

More generally, any normed $\mathcal{K}$ space where the projections satisfy the conditions above has 0-GHP (see Chapter 5 for the definition of $\mathcal{K}$ space and examples; there exist non-complete normed $\mathcal{K}$ spaces). In particular, these remarks give the following examples of triples with 0-GHP.

Example 8.10. $(L^1(\mu, X), L^{\infty}(\mu) : X), (L^{\infty}(\mu), L^1(\mu, X) : X)$ and the other triples in Example 4 have 0-GHP. As another example, let $1 < p < \infty$ and $\frac{1}{p} + \frac{1}{q} = 1$ and $L^p(\mu, X)$ be the space of strongly measurable functions which are p^{th} power Bochner integrable with the norm

$$\|f\|_p = \left(\int_S \|f(s)\|^p \, d\mu(s) \right)^{1/p}.$$

Then $(L^p(\mu, X), L^q(\mu, X'))$ form a dual pair under the pairing

$$\langle f, g \rangle = \int_S \langle f(s), g(s) \rangle \, d\mu(s)$$

and if $\{A_j\}$ is a pairwise disjoint sequence from Σ, then $P_j f = \chi_{A_j} f$ defines a sequence of continuous projections satisfying the condition above and

$$(L^p(\mu, X), L^q(\mu, X'))$$

has 0-GHP.

Example 8.11. Let $m : \Sigma \to X$ be countably additive. The triple

$$(B(\Sigma), L^1(m) : X)$$

under the integration map $(g, f) \to \int_S gf \, dm$ of Example 31 of Chapter 1 with the projections $P_j g = \chi_{A_j} g$ satisfies the conditions of the theorem above so this triple has 0-GHP.

We also have a dual result.

Example 8.12. The triple

$$(L^1(m), B(\Sigma) : X)$$

under the integration map $(f, g) \to \int_s fg\,dm$ and the projections $P_j f = \chi_{A_j} f$ has 0-GHP. Let $\{I_k\}$ be an increasing sequence of intervals and $f_k \to 0$ in $L^1(m)$. For convenience assume $\|f_k\|_1 < 1/2^k$ and set $B_k = \cup_{j \in I_k} A_j$. Let f be the pointwise limit of the series

$$\sum_{k=1}^{\infty} P_{I_k} f_k = \sum_{k=1}^{\infty} \chi_{B_k} f_k.$$

We need to show that the series converges to f with respect to $w(B(\Sigma), L^1(m))$. Let $g \in B(\Sigma)$. The series

$$\sum_{k=1}^{\infty} \chi_{B_k} |f_k g|$$

converges pointwise to $|fg|$ so if $x' \in X'$, by the Monotone Convergence Theorem

$$\sum_{k=1}^{\infty} \int_{B_k} |fg|\, d\,|x'm| = \int_S |fg|\, d\,|x'm|.$$

Now

$$\int_{B_k} |fg|\, d\,|x'm| \le \|g\|_\infty \|\chi_{B_k} f_k\|_1 \le \|g\|_\infty \|f_k\|_1 \le \|g\|_\infty 1/2^k$$

so fg is scalarly m integrable. If $A \in \Sigma$,

$$\left\| \int_{A \cap B_k} f_k g\,dm \right\| \le \|f_k\|_1 \|g\|_\infty \le \|g\|_\infty 1/2^k$$

so the series

$$\sum_{k=1}^{\infty} \int_{A \cap B_k} f_k g\,dm$$

is absolutely convergent and converges to some $x_A \in X$. The computation above shows that

$$x_A = \int_A fg\,dm$$

so fg is m integrable and the series $\sum_{k=1}^{\infty} P_{I_k} f_k$ converges to f with respect to $w(L^1(m), B(\Sigma))$ as desired.

Example 8.13. The triple $(L^\infty(\mu), P(\mu, X) : X)$ of Example 5 has 0-GHP by Theorem 9. Similarly, the triple $(L^\infty(\mu), D(\mu, X) : X)$ has 0-GHP.

As a dual result, we have

Example 8.14. The triple

$$(P(\mu, X), L^\infty(\mu) : X)$$

and the projections of Example 5 has 0-GHP. (Recall that $P(\mu, X)$ is not complete so the theorem above does not apply ([BS] Example 5.13, [Pe]).) Suppose $\{f_k\} \subset P(\mu, X)$ converges to 0 and for convenience assume $\|f_k\|_1 < 1/2^k$. Let $\{I_k\}$ be an increasing sequence of intervals and set $B_k = \cup_{j \in I_k} A_j$. Let f be the pointwise limit of the series

$$\sum_{k=1}^{\infty} P_{I_k} f_k = \sum_{k=1}^{\infty} \chi_{B_k} f_k.$$

We claim that $f \in P(\mu, X)$. Obviously f is scalarly measurable and if $x' \in X'$, then

$$x'f = \sum_{k=1}^{\infty} x' \chi_{B_k} f_k$$

and

$$|x'f| = \sum_{k=1}^{\infty} |x' \chi_{B_k} f_k|$$

pointwise. By the Monotone Convergence Theorem,

$$\int_S |x'f| \, d\mu = \sum_{k=1}^{\infty} \int_{B_k} |x'f_k| \, d\mu \leq \|x'\| \sum_{k=1}^{\infty} \|f_k\|_1 \leq \|x'\| \sum_{k=1}^{\infty} 1/2^k$$

so f is scalarly or Dunford integrable. We claim that f is Pettis integrable. Let $A \in \Sigma$. Since

$$\sum_{k=1}^{\infty} \left\| \int_{A \cap B_k} f_k d\mu \right\| \leq \sum_{k=1}^{\infty} \|f_k\|_1 \leq \sum_{k=1}^{\infty} 1/2^k,$$

the series $\sum_{k=1}^{\infty} \int_{A \cap B_k} f_k d\mu$ is absolutely convergent and converges to some $x_A \in X$ since X is complete. Now

$$x'(x_A) = \sum_{k=1}^{\infty} \int_{A \cap B_k} x' f_k d\mu$$

and

$$|x'f| \geq \left| \sum_{k=1}^{n} \chi_{A \cap B_k} x' f_k \right|$$

for $n \in \mathbb{N}$ and $x' \in X'$ so the Dominated Convergence Theorem implies

$$\int_A x' f d\mu = \sum_{k=1}^{\infty} \int_{A \cap B_k} x' f_k d\mu.$$

Hence,

$$x_A = \int_A f d\mu \in X$$

and f is Pettis integrable. Let $g \in L^{\infty}(\mu)$. Then

$$\left\| gf - \sum_{k=1}^{n} g \chi_{B_k} f_k \right\|_1 = \sup \left\{ \int_S \left| x' \sum_{k=n+1}^{\infty} g f_k \right| d\mu : \|x'\| \leq 1 \right\}$$

$$\leq \sup \left\{ \sum_{k=n+1}^{\infty} \int_{B_k} |x' g f_k| \, d\mu : \|x'\| \leq 1 \right\}$$

$$\leq \sum_{k=n+1}^{\infty} \|f_k\|_1 \|g\|_{\infty} \leq \|g\|_{\infty} \sum_{k=n+1}^{\infty} 1/2^k$$

so the series $\sum_{k=1}^{\infty} \chi_{B_k} f_k$ is $w(P(\mu, X), L^{\infty}(\mu))$ convergent to f. Hence, the triple has 0-GHP.

The proof of Example 14 also shows that the triple $(D(\mu, X), L^{\infty}(\mu) : X)$ has 0-GHP.

Example 8.15. We give an example of a non-complete normed space whose projections satisfy the conditions in Theorem 9 above and which has 0-GHP. Consider Example 4 and let Y be a subspace of X and let $L^1(\mu, Y)$ be the subspace of $L^1(\mu, X)$ which consists of those functions with values in Y. Suppose $\{f_k\} \to 0$ in $L^1(\mu, Y)$ and $\{I_k\}$ is an increasing sequence of intervals. Note each P_{I_k} is a projection of norm 1 and $P_i P_j = 0$ for $i \neq j$. There is a subsequence $\{n_k\}$ such that $\sum_{k=1}^{\infty} \|f_{n_k}\|_1 < \infty$. Then the series $\sum_{k=1}^{\infty} P_{I_{n_k}} f_{n_k}$ is absolutely convergent in $L^1(\mu, X)$, the series

$$\sum_{k=1}^{\infty} P_{I_{n_k}} f_{n_k}$$

is pointwise convergent to a function f with values in Y and the series is $w(L^1(\mu, Y), L^{\infty}(\mu))$ convergent to f. Thus, the triple

$$(L^1(\mu, Y), L^{\infty}(\mu) : X)$$

has 0-GHP.

We now establish the first of our main results which is a *uniform boundedness* result. For this and later results we use a result of Antosik and Mikusinski for infinite matrices which has been employed earlier (see Appendix E).

Theorem 8.16. *Assume that E has 0-GHP and for every j, $P_j E$ is barrelled under the topology of E. If $A \subset E$ is bounded and $B \subset E^\beta$ is pointwise bounded on E, then*

$$\left\{ x \cdot y = \sum_{j=1}^{\infty} \langle P_j x, y \rangle : y \in B, x \in A \right\} = A \cdot B$$

is bounded in G.

Proof. If the conclusion fails to hold, there exist a closed, symmetric neighborhood of 0, U, in G, $\{y^k\} \subset B$, $\{x^k\} \subset A$, $0 < s_k \to 0$ such that

$$x^k \cdot s_k y^k \notin U.$$

Pick a closed, symmetric neighborhood of 0, V, such that $V + V \subset U$.

For $k_1 = 1$ pick m_1 such that

$$s_{k_1} \sum_{j=1}^{m_1} \langle P_j x^{k_1}, y^{k_1} \rangle \notin U.$$

By the continuity of the P_j, $P_j A$ is bounded in E with respect to the topology of E, $\langle \cdot, y^k \rangle : P_j E \to G$ is continuous with respect to the topologies of E and G and $\{\langle \cdot, y^k \rangle : k \in \mathbb{N}\}$ is pointwise bounded on $P_j E$. Since $P_j E$ is barrelled, for every j, $\{P_j x^k \cdot y^k : k\}$ is bounded in G. Therefore,

$$\lim_k s_k \langle P_j x^k, y^k \rangle = 0$$

in G for every j. Hence, there exists $k_2 > k_1$ such that

$$s_k \sum_{j=1}^{m_1} \langle P_j x^k, y^k \rangle \in V$$

for $k \geq k_2$. Pick $m_2 > m_1$ such that

$$s_{k_2} \sum_{j=1}^{m_2} \langle P_j x^{k_2}, y^{k_2} \rangle \notin U.$$

Put $I_2 = [m_1 + 1, m_2]$. Then

$$s_{k_2} \sum_{j \in I_2} \langle P_j x^{k_2}, y^{k_2} \rangle = s_{k_2} \sum_{j=1}^{m_2} \langle P_j x^{k_2}, y^{k_2} \rangle - s_{k_2} \sum_{j=1}^{m_1} \langle P_j x^{k_2}, y^{k_2} \rangle \notin V.$$

Continuing this construction produces an increasing sequence $\{k_p\}$ and an increasing sequence of intervals $\{I_p\}$ such that

$$(\#) \quad s_{k_p} \sum_{j \in I_p} \langle P_j x^{k_p}, y^{k_p} \rangle \notin V.$$

Define an infinite matrix

$$M = [m_{pq}] = \left[\sqrt{s_{k_p}} \sum_{l \in I_q} \langle \sqrt{s_{k_q}} P_l x^{k_q}, y^{k_p} \rangle \right] = \left[\sqrt{s_{k_p}} \left\langle \sum_{l \in I_q} \sqrt{s_{k_q}} P_l x^{k_q}, y^{k_p} \right\rangle \right].$$

We claim that M is a $\mathcal{K}$ matrix. First, the columns of M converge to 0 since $\{y^p\}$ is pointwise bounded on E. Next, since $\sqrt{s_{k_q}} x^{k_q} \to 0$ in E, the 0-GHP implies that given any subsequence there is a further subsequence $\{r_q\}$ such that

$$x = \sum_{q=1}^{\infty} \sqrt{s_{k_{r_q}}} \sum_{l \in I_{r_q}} P_l x^{k_{r_q}} \in E,$$

where the series converges in $w(E, E^\beta)$. By the $w(E, E^\beta)$ convergence of the series to x, we have

$$\sum_{q=1}^{\infty} m_{pr_q} = \sum_{q=1}^{\infty} \sqrt{s_{k_p}} \left\langle \sqrt{s_{k_{r_q}}} \sum_{l \in I_{r_q}} P_l x^{k_{r_q}}, y^{k_p} \right\rangle = x \cdot \sqrt{s_{k_p}} y^{k_p}$$

and $x \cdot \sqrt{s_{k_p}} y^{k_p} \to 0$ since $\{y^p\}$ is pointwise bounded on E. Therefore, M is a $\mathcal{K}$ matrix and by the Antosik–Mikusinski Matrix Theorem the diagonal of M converges to 0. But, this contradicts $(\#)$. $\qquad \square$

Remark 8.17. The proof of Theorem 16 shows that the assumption in Theorem 16 that the spaces $P_j E$ are barrelled can be replaced by the assumption that these are $\mathcal{A}$ spaces; $\mathcal{A}$ spaces have the property that pointwise bounded families of continuous linear operators on these spaces are uniformly bounded on bounded subsets (see Chapter 5 for the definition and properties).

The scalar version of Theorem 16 when E, F is a dual pair of vector spaces gives a generalization of one part of Theorem 1 of [WLC] where it is assumed that E is a normed AK space. See also Theorem 8.13 of [Sw4] for a scalar sequence space version of the result.

We give an application of Theorem 16 to vector valued sequence spaces as in Example 2.

Example 8.18. Assume the notation as in Example 2 and further assume that E has a locally convex topology, the projections $P_j x = e^j \otimes x_j$ are continuous and E has 0-GHP in the triple

$$(E, E^{\beta Y} : Y).$$

If the space X is barrelled (an $\mathcal{A}$ space) and if the spaces X and $P_j E = e^j \otimes X$ are isomorphic, Theorem 16 (Remark 17) implies that if $A \subset E$ is bounded in E and $B \subset E^{\beta Y}$ is pointwise bounded on E, then

$$\left\{ x \cdot T = \sum_{j=1}^{\infty} T_j x_j : x \in A, T \in B \right\}$$

is bounded in Y.

In particular, Example 18 is applicable to the spaces $c_0(X), l^p(X)$ $(1 \leq p \leq \infty)$, $cs(X), bs(X)$ when X is a normed, barrelled ($\mathcal{A}$ space) space (see Appendix C of [Sw4] for the definitions and topologies of these spaces). It should be noted that there are non-complete, normed, barrelled ($\mathcal{A}$) spaces X so the spaces above may fail to be complete.

For example, let X, Y be normed spaces and consider the triple

$$(l^{\infty}(X), l^1(L(X, Y)) : Y)$$

under the map

$$(x, T) \to x \cdot T = \sum_{j=1}^{\infty} T_j x_j$$

with the projections of Example 2. Then $l^{\infty}(X)$ has 0-GHP by Theorem 9 and if X is a Banach space, Example 18 is applicable. Similar remarks apply to such triples as

$$(l^1(X), l^{\infty}(L(X, Y)) : Y), \quad (c_0(X), l^1(L(X, Y)) : Y),$$

etc.

In the case of scalar sequence spaces as in Example 1, the spaces $P_j E$ are trivially barrelled so if E has 0-GHP, then $\sigma(E^{\beta}, E)$ bounded subsets are uniformly bounded on bounded subsets of E. Therefore, if $E' = E^{\beta}$, then E is a Banach–Mackey space ([Wi2] 10.4) in this case, and if E is also normed, E is barrelled. These statements are similar to those in Theorem 1 of [NS] and Corollaries 1 and 2 of [WCL] where different assumptions are made.

Drewnowski, Florencio and Paul have shown that if μ is a finite measure, the space, $P(\mu, X)$, of Pettis integrable functions is barrelled ([DFP]). We can use this and Theorem 16 to obtain a uniform boundedness result for $P(\mu, X)$.

Corollary 8.19. *Suppose μ is σ-finite with $S = \cup_{j=1}^{\infty} A_j, \mu(A_j) < \infty$ and $\{A_j\}$ pairwise disjoint. Let $A \subset L^{\infty}(\mu)$ be bounded and $B \subset P(\mu, X)$ be $w(P(\mu, X), L^{\infty}(\mu))$ bounded in the triple $(L^{\infty}(\mu), P(\mu, X) : X)$ of Example 5. Then*

$$H = \left\{ \int_S gf d\mu : g \in A, f \in B \right\}$$

is bounded.

Proof. If $f \in P(\mu, X), g \in L^{\infty}(\mu)$, then

$$g \cdot f = \sum_{j=1}^{\infty} \langle P_j g, f \rangle = \sum_{j=1}^{\infty} \int_{A_j} gf d\mu = \int_S gf d\mu.$$

$\square$

It follows from Theorem 16 that if $B \subset P(\mu, X)$ is pointwise bounded on $L^{\infty}(\mu)$, then

$$\sup_{f \in B} \sup \left\{ \left\| \int_C f d\mu \right\| : C \in \Sigma \right\} < \infty.$$

The expression

$$\sup \left\{ \left\| \int_C f d\mu \right\| : C \in \Sigma \right\} = \|f\|'$$

defines a norm on $P(\mu, X)$ which is equivalent to the norm previously defined (Chapter 1) so subsets of $P(\mu, X)$ which are pointwise bounded on $L^{\infty}(\mu)$ are norm bounded, a conclusion like that of the classical Uniform Boundedness Principle. Recall the space of Pettis integrable functions is not complete but is barrelled ([DFP]).

We next establish several uniform convergence results for abstract duality pairs with projections.

Theorem 8.20. *Assume that E has 0-GHP. If $y \in E^{\beta}$ and $x^i \to 0$ in E, then the series*

$$\sum_{j=1}^{\infty} \langle P_j x^i, y \rangle$$

converge uniformly for $i \in \mathbb{N}$.

Proof. If the conclusion fails, there exists a symmetric neighborhood of 0, U, in G such that for every k there exist $m_k > k$, p_k such that

$$\sum_{j=m_k}^{\infty} \langle P_j x^{p_k}, y \rangle \notin U.$$

Choose a symmetric neighborhood V such that $V + V \subset U$. For $k = 1$ let m_1 and p_1 satisfy this condition so

$$\sum_{j=m_1}^{\infty} \langle P_j x^{p_1}, y \rangle \notin U.$$

There exists $n_1 > m_1$ such that $\sum_{j=n_1+1}^{\infty} \langle P_j x^{p_1}, y \rangle \in V$. Then

$$\sum_{j=m_1}^{n_1} \langle P_j x^{p_1}, y \rangle \notin V.$$

There exists N_1 such that $\sum_{j=m}^{n} \langle P_j x^i, y \rangle \in V$ for $1 \leq i \leq p_1, n > m \geq N_1$. Let $p_2, m_2 > N_1, n_2 > m_2$ satisfy the condition above for $k = N_1$ so

$$\sum_{j=m_2}^{n_2} \langle P_j x^{p_2}, y \rangle \notin V$$

(this is an abuse of the notation above but avoids multiple subscripts, should cause no difficulty and makes the notation more palatable). Then $p_2 > p_1$ by the choice of N_1. Continuing this construction produces increasing sequences $\{p_k\}, \{m_k\}, \{n_k\}, m_{k+1} > n_k > m_k$ with

$$(*) \quad \sum_{j=m_k}^{n_k} \langle P_j x^{p_k}, y \rangle \notin V.$$

Set $I_k = [m_k, n_k]$ so $\{I_k\}$ is an increasing sequence of intervals. Since $x^k \to 0$, the 0-GHP implies there exists a subsequence $\{q_k\}$ of $\{p_k\}$ such that

$$x = \sum_{k=1}^{\infty} P_{I_{q_k}} x^{q_k} \in E,$$

where the series is $w(E, E^\beta)$ convergent. Condition $(*)$ implies the series

$$\sum_{j=1}^{\infty} \langle P_j x, y \rangle$$

doesn't converge which gives the desired contradiction. $\square$

A similar result for sectional operators was established in Theorem 12 of [ZCL]; see also Theorem 2.22 of [Sw4] for a scalar sequence version.

Without assumptions on the triple the conclusion of Theorem 20 may fail to hold without some assumptions on the space.

Example 8.21. Let s be the vector space of all scalar sequences. Consider the triple $(c_{00}, s : \mathbb{R})$ under the map

$$(x, y) \to \sum_{j=1}^{\infty} x_j y_j,$$

$x = \{x_j\} \in c_{00}, y = \{y_j\} \in s$, and the projections $P_j(x) = x_j e^j$. Let c_{00} have the sup-norm. The sequence $x^i = \sum_{j=1}^{i} e^j / i \to 0$ in c_{00} in the sup-norm. If $y = \{j\}_{j=1}^{\infty}$, then the series

$$\sum_{j=1}^{\infty} P_j x^i \cdot y = \sum_{j=1}^{i} y_j / i = \sum_{j=1}^{i} j / i$$

do not converge uniformly for $i \in \mathbb{N}$.

We now continue to establish other uniform convergence results which require different gliding hump assumptions.

Definition 8.22. The space E (or the triple $(E, F : G)$ with projections $\{P_j\}$) has the signed weak gliding hump property (signed WGHP) if whenever $x \in E$ and $\{I_k\}$ is an increasing sequence of intervals, there exist a subsequence $\{p_k\}$ and a sequence of signs $\{s_k\}$ such that the series

$$\sum_{k=1}^{\infty} s_k P_{I_{p_k}} x$$

is $w(E, E^{\beta})$ convergent in E. If all of the signs can be chosen equal to 1, E is said to have the weak gliding hump property (WGHP).

See Appendix B for the sequence space definitions and examples where as noted earlier the series $\sum_{k=1}^{\infty} s_k P_{I_{p_k}} x$ is required to converge pointwise, an option not available in this absract setting. In [ZCL] there is a similar definition. Note that the signed-WGHP does not depend on the topology of E but on the topology $w(E, E^{\beta})$; the signed-WGHP is an algebraic condition on the abstract triple.

We give a condition which is sufficient for a triple to have WGHP.

Notation 8.23. If $I \subset \mathbb{N}$ is an infinite set whose elements are arranged in a sequence $\{n_j\}$ and $\{x_j\} \subset E$, we write

$$\sum_{j \in I} x_j = \sum_{j=1}^{\infty} x_{n_j}$$

provided the series $\sum_{j=1}^{\infty} x_{n_j}$ is $w(E, E^\beta)$ convergent to an element of E.

Definition 8.24. The space E (or the triple $(E, F : G)$ with projections $\{P_j\}$) is monotone if for every $x \in E$ and $I \subset \mathbb{N}$ the series $\sum_{j \in I} P_j x$ is $w(E, E^\beta)$ convergent to an element in E, denoted by $P_I x$.

Remark 8.25. A scalar (or vector) sequence space λ is monotone if $\chi_I x \in \lambda$ when $x \in \lambda$ and $I \subset \mathbb{N}$, where $\chi_I x$ is the coordinate product of χ_I and x. This means the series $\sum_{j \in I} x_j e^j$ is coordinatewise convergent to $\chi_I x$. If the element $\chi_I x \in \lambda$, then the series $\sum_{j \in I} x_j e^j$ is $\sigma(\lambda, \lambda^\beta)$ convergent to $\chi_I x$ so the definition above agrees with the scalar (vector) definition of monotone.

Examples of monotone sequence spaces are given in Appendix B and Appendix B of [Sw4].

As is in the sequence space case, a monotone space has WGHP.

Proposition 8.26. *If E is monotone, then E has WGHP.*

Proof. Suppose $x \in E$ and $\{I_j\}$ is an increasing sequence of intervals. If $I = \cup_{j=1}^{\infty} I_j$, then

$$P_I x = \sum_{k=1}^{\infty} \sum_{j \in I_k} P_j x = \sum_{k=1}^{\infty} P_{I_k} x$$

is $w(E, E^\beta)$ convergent to an element of E. $\square$

Example 8.27. The triple

$$(L^1(\mu, X), L^\infty(\mu) : X)$$

of Example 4 is monotone and, therefore, has WGHP. Suppose $f \in L^1(\mu, X)$ and $I \subset \mathbb{N}$. Let h be the pointwise limit of the series

$$\sum_{j \in I} P_j f = \sum_{j \in I} \chi_{A_j} f.$$

Note $h \in L^1(\mu, X)$. We claim $P_I f = h$ with convergence in $w(L^1(\mu, X), L^\infty(\mu))$. For this, let $g \in L^\infty(\mu) = L^1(\mu, X)^\beta$. Then, by countable additivity,

$$\sum_{j \in I} P_j f \cdot g = \sum_{j \in I} \int_{A_j} g f d\mu = \int_{\cup_{j \in I} A_j} g f d\mu = \int_S g h d\mu$$

justifying the claim. Similarly, $(L^\infty(\mu), L^1(\mu, X) : X)$ and $(L^p(\mu, X), L^q(\mu, X'))$ are monotone and have WGHP.

Example 8.28. The abstract triple

$$(L^\infty(\mu), P(\mu, X) : X)$$

with the projections $P_j g = \chi_{A_j} g$ in Example 5 is monotone. Note $L^\infty(\mu)^\beta = P(\mu, X)$ by the countable additivity of the Pettis integral (2.11). We show that $L^\infty(\mu)$ is monotone and so has WGHP. Let $I = \{n_j\} \subset \mathbb{N}$ and $g \in L^\infty(\mu)$. Let

$$h = \chi_{\cup_{j=1}^\infty A_{n_j}} g.$$

We claim that the series

$$\sum_{j=1}^\infty P_{n_j} g = \sum_{j=1}^\infty \chi_{A_{n_j}} g$$

is $w(L^\infty(\mu), P(\mu, X))$ convergent to h. By the countable additivity of the Pettis integral, if $f \in P(\mu, X)$, then

$$\sum_{j=1}^\infty P_{n_j} g \cdot f = \sum_{j=1}^\infty \int_{A_{n_j}} g f d\mu = \int_{\cup_{j=1}^\infty A_{n_j}} g f d\mu = \int_S h f d\mu$$

justifying the claim. Thus, $L^\infty(\mu)$ is monotone and has WGHP. The same proof shows that the triple

$$(P(\mu, X), L^\infty(\mu) : X)$$

is monotone.

Example 8.29. The triple

$$(B(\Sigma), L^1(m) : X)$$

with projections $P_j g = \chi_{A_j} g$ in Example 11 is monotone. Note $B(\Sigma)^\beta = L^1(m)$ by the countable additivity of the integral (2.12). Let $g \in B(\Sigma)$ and $I \subset \mathbb{N}$. The series

$$\sum_{j \in I} P_j g = \sum_{j \in I} \chi_{A_j} g$$

converges in $w(B(\Sigma), L^1(m))$ since if $f \in L^1(m)$,

$$\sum_{j \in I} \int_{A_j} fg\,dm = \int_{\cup_{j \in I} A_j} fg\,dm$$

by countable additivity.

Similarly,

Example 8.30. The triple

$$(L^1(m), B(\Sigma) : X)$$

with projections $P_j g = \chi_{A_j} g$ of Example 12 is monotone. For if $f \in L^1(m), I \subset \mathbb{N}$, $\chi_{\cup_{j \in I} A_j} f = h$ is integrable by the norm countable additivity of the integral (2.12) and the series $\sum_{j \in I} P_j f$ converges to h in $w(L^1(m), B(\Sigma))$.

Further examples of monotone spaces other than sequence spaces and spaces of integrable functions will be given later.

We establish a uniform convergence result for triples with signed-WGHP.

Theorem 8.31. *Assume E has signed-WGHP. If $\{y^k\} \subset E^\beta$ is such that*

$$\lim_k \langle x, y^k \rangle$$

exists for each $x \in E$, then for each x the series

$$\sum_{j=1}^{\infty} \langle P_j x, y^k \rangle$$

converge uniformly for $k \in \mathbb{N}$.

Proof. If the conclusion fails, there is a symmetric neighborhood of 0, U, in G such that for every k there exist $p_k, n_k > m_k > k$ such that

$$(*) \qquad \sum_{j=m_k}^{n_k} \langle P_j x, y^{p_k} \rangle \notin U.$$

For $k = 1$ this condition implies there exist $p_1, n_1 > m_1 > 1$ such that $\sum_{j=m_1}^{n_1} \langle P_j x, y^{p_1} \rangle \notin U$. There exists $m' > n_1$ such that

$$\sum_{j=m}^{n} \langle P_j x, y^k \rangle \in U$$

for $1 \leq k \leq p_1, n > m > m'$. The condition $(*)$ for $k = m'$ implies there exist $p_2, n_2 > m_2 > m'$ such that

$$\sum_{j=m_2}^{n_2} \langle P_j x, y^{p_2} \rangle \notin U.$$

Then $p_2 > p_1$. Continuing this construction produces increasing sequences $\{p_k\}, \{m_k\}, \{n_k\}$ with $m_{k+1} > n_k > m_k$ and

$$(**) \qquad \sum_{j=m_k}^{n_k} \langle P_j x, y^{p_k} \rangle \notin U.$$

Set $I_k = [m_k, n_k]$ so $\{I_k\}$ is an increasing sequence of intervals. Define a matrix

$$M = [m_{ij}] = \left[\sum_{l \in I_j} \langle P_l x, y^{p_i} \rangle \right].$$

We claim that M is a signed $\mathcal{K}$-matrix. First, the columns of M converge by hypothesis. Next, given any subsequence there exist a further subsequence $\{r_j\}$ and signs $\{s_j\}$ such that the series

$$\sum_{j=1}^{\infty} s_j \sum_{l \in I_{r_j}} P_l x = z$$

is $w(E, E^{\beta})$ convergent in E. Then

$$\sum_{j=1}^{\infty} s_j m_{ir_j} = \sum_{j=1}^{\infty} s_j \sum_{l \in I_{r_j}} \langle P_l x, y^{p_i} \rangle = \langle z, y^{p_i} \rangle$$

and $\{\langle z, y^{p_i} \rangle\}$ converges in G by hypothesis. Hence, M is a signed $\mathcal{K}$-matrix and the diagonal of M converges to 0 by the signed version of the Antosik–Mikusinski Matrix Theorem (Appendix E). But, this contradicts $(**)$. $\square$

A similar result was established in Theorem 5 of [ZCL]; see also Theorems 2.26 and 11.14 of [Sw4] for the sequence space result. The results in [Sw4] were used to establish the weak sequential completeness of β-duals (see also [St1],[St2]). Again without assumptions on the space E the conclusion of Theorem 33 may fail.

Example 8.32. Consider the triple $(c, l^1 : \mathbb{R})$ under the map $(s, t) \to \sum_{j=1}^{\infty} s_j t_j$ so $c^{\beta} = l^1$. Then $\{e^k\}$ is $w(l^1, c) = \sigma(l^1, c)$ Cauchy but if e is the constant sequence $\{1\}$, the series

$$\sum_{j=1}^{\infty} e_j^k e^j$$

do not converge uniformly for $k \in \mathbb{N}$.

Using the triple

$$(L^1(m), B(\Sigma) : X)$$

of Example 12 and Theorem 33, Example 32, we can obtain a uniform countable additivity result.

Corollary 8.33. *Let $\{g_k\} \subset B(\Sigma)$ be such that*

$$\lim_k \int_S f g_k \, dm$$

exists for every $f \in L^1(m)$. Then for every $f \in L^1(m)$, the sequence of vector measures

$$H = \left\{ \int_. f g_k \, dm : k \in \mathbb{N} \right\}$$

is uniformly countably additive.

This result follows since we can take the pairwise disjoint sequence $\{A_j\} \subset \Sigma$ arbitrarily.

We will give an improvement of this result later in Corollary 38.

The results in Theorems 20 and 33 require different gliding hump assumptions and these gliding hump assumptions are independent of one another; the space c has 0-GHP but not WGHP while the space c_{00} has WGHP but not 0-GHP (see Proposition B.29 of [Sw4] for a relationship).

Using Theorems 20 and 33 we can obtain a more general result for spaces with both 0-GHP and signed-WGHP.

Theorem 8.34. *Assume E has 0-GHP and signed-WGHP. If $\{y^k\} \subset E^\beta$ is such that*

$$\lim_k \langle x, y^k \rangle$$

exists for each $x \in E$ and $x^k \to 0$ in E, then the series

$$\sum_{j=1}^{\infty} \langle P_j x^k, y^l \rangle$$

converge uniformly for $k, l \in \mathbb{N}$.

Proof. If the conclusion fails, as in the proof above, there exists a neighborhood, U, of 0 in G such that for every k there exist $k < m_k < n_k, p_k, q_k$ such that

$$\sum_{j=m_k}^{n_k} \langle P_j x^{p_k}, y^{q_k} \rangle \notin U.$$

By this condition for $k = 1$ there exist $p_1, q_1, n_1 > m_1 > 1$ such that

$$\sum_{j=m_1}^{n_1} \langle P_j x^{p_1}, y^{q_1} \rangle \notin U.$$

Now by Theorems 20 and 33 above there exists $m' > n_1$ such that

$$\sum_{j=p}^{q} \langle P_j x^k, y^l \rangle \in U$$

for $k \in \mathbb{N}, 1 \le l \le q_1$ and $1 \le k \le p_1, l \in \mathbb{N}, q > p > m'$. By the condition above for $k = m'$ there exist $p_2, q_2, n_2 > m_2 > m'$ such that

$$\sum_{j=m_2}^{n_2} \langle P_j x^{p_2}, y^{q_2} \rangle \notin U.$$

By the choice of m' we have $p_2 > p_1, q_2 > q_1$. Continuing this construction produces increasing sequences $\{p_k\}, \{q_k\}, \{m_k\}, \{n_k\}, m_{k+1} > n_k > m_k$ with

$$(\#) \quad \sum_{j=m_k}^{n_k} \langle P_j x^{p_k}, y^{q_k} \rangle \notin U.$$

Set $I_k = [m_k, n_k]$ so $\{I_k\}$ is an increasing sequence of intervals. Define a matrix

$$M = [m_{ij}] = \left[\sum_{l=m_j}^{n_j} \langle P_l x^{p_j}, y^{q_i} \rangle \right] = [\langle P_{I_j} x^{p_j}, y^{q_i} \rangle].$$

We claim that M is a $\mathcal{K}$-matrix. First the columns of M converge by hypothesis. Next, by 0-GHP, given any increasing sequence of integers, there is a subsequence $\{r_k\}$ such that the series

$$x = \sum_{k=1}^{\infty} P_{I_{r_k}} x^{p_{r_k}}$$

is $w(E, E^\beta)$ convergent in E. Then

$$\sum_{k=1}^{\infty} m_{ir_k} = \sum_{k=1}^{\infty} \left\langle P_{I_{r_k}} x^{p_{r_k}}, y^{q_i} \right\rangle = \langle x, y^{q_i} \rangle$$

and $\{\langle x, y^{q_i} \rangle\}$ converges. Hence, M is a $\mathcal{K}$-matrix and by the Antosik–Mikusinski Matrix Theorem the diagonal of M converges to 0. But this contradicts $(\#)$. $\qquad \square$

There is a version of this result for scalar sequences given in Theorem 2.39 of [Sw4] where the hypothesis that λ has signed WGHP is needed.

We give an application of Theorem 36 to weak convergence in $L^\infty(\mu)$.

Corollary 8.35. *Suppose* $\{g_k\} \subset L^\infty(\mu)$ *is such that*

$$\lim \int_S g_k f d\mu$$

exists for every $f \in L^1(\mu, X)$. *Then if* $\{f_k\} \subset L^1(\mu, X)$ *converges to 0 in* $L^1(\mu, X)$, *the family of vector measures*

$$H = \left\{ \int_. g_k f_j d\mu : j, k \in \mathbb{N} \right\}$$

is uniformly countably additive.

Proof. This follows from Theorem 36 and Examples 10 and 29 applied to the triple

$$(L^1(\mu, X), L^\infty(\mu) : X)$$

since we can take $\{A_j\}$ to be an arbitrary pairwise disjoint sequence from Σ.   □

From Examples 10 and 30 and Theorem 36 applied to the triple

$$(L^\infty(\mu), L^1(\mu, X) : X),$$

we also have a dual result for sequences $\{f_k\} \subset L^1(\mu, X)$ such that $\lim \int_S g f_k d\mu$ exists for every $g \in L^\infty(\mu)$ and $g_k \to 0$ in $L^\infty(\mu)$.

A similar result holds for the triple $(L^p(\mu, X), L^q(\mu, X'))$ (Examples 10 and 30, Theorem 36).

From Examples 12 and 32 and Theorem 36 applied to the triple

$$(L^1(m), B(\Sigma) : X),$$

we have

Corollary 8.36. *Suppose* $\{g_k\} \subset B(\Sigma)$ *is such that*

$$\lim \int_S g_k f d\mu$$

exists for every $f \in L^1(m)$. *Then if* $\{f_k\} \subset L^1(m)$ *converges to 0 in* $L^1(m)$, *the family of vector measures*

$$H = \left\{ \int_. g_k f_j d\mu : j, k \in \mathbb{N} \right\}$$

is uniformly countably additive.

A dual result applied to the triple

$$(B(\Sigma), L^1(m) : X)$$

can be obtained from Theorem 36 by using Examples 11 and 31. If $\{f_k\} \subset L^1(m)$ is such that $\lim \int_S g f_k d\mu$ exists for each $g \in B(\Sigma)$ and $\{g_k\}$ converges to 0 in $B(\Sigma)$, then the family of vector measures

$$H = \left\{ \int g_k f_j d\mu : j, k \in \mathbb{N} \right\}$$

is uniformly countably additive.

We give a similar application of Theorem 36 to weak convergence in the space of Pettis integrable functions (Example 5).

Corollary 8.37. *Suppose $\{f_k\} \subset P(\mu, X)$, the space of Pettis integrable functions (Example 5), is such that*

$$\lim_k \int_S g f_k d\mu$$

exists for every $g \in L^\infty(\mu)$. Then if $g_k \to 0$ in $L^\infty(\mu)$, the family of vector measures

$$H = \left\{ \int g_j f_k d\mu : j, k \in \mathbb{N} \right\}$$

is uniformly countably additive.

Proof. This follows from Examples 13 and 30 and Theorem 36 applied to the triple

$$(L^\infty(\mu), P(\mu, X) : X).$$

$\square$

A similar dual result can be established for weak convergence in the triple

$$(P(\mu, X), L^\infty(\mu) : X).$$

By Examples 14 and 30 and Theorem 36, if the sequence $\{g_k\} \subset L^\infty(\mu)$ is such that $\lim \int_S g_k f d\mu$ exists for every $f \in P(\mu, X)$ and $f_k \to 0$ in $P(\mu, X)$, then the family of vector measures

$$H = \left\{ \int g_k f_j d\mu : j, k \in \mathbb{N} \right\}$$

is uniformly countably additive.

The conclusions in Corollaries 37 and 39 and the observation above also imply that the families H in the conclusions are uniformly μ continuous. This follows from Theorem 3.14.1 of [Sw3] since each indefinite Bochner or Pettis integral is μ continuous. See [DS] IV.8.9, IV.8.11 and IV.9.1 for applications of uniform countable additivity and uniform μ-continuity.

We give an example of a space of measures with WGHP and 0-GHP and give an application of Theorem 36. Let (S, Σ, μ) be a measure space with $\{A_j\}$ a pairwise disjoint sequence from Σ.

Example 8.38. Let $B(\Sigma)$ be the space of all bounded, Σ-measurable functions defined on S with the sup-norm. Let X be a Banach space and let

$$ca(\Sigma, X : \mu)$$

be the space of all countably additive set functions $\nu : \Sigma \to X$ which are μ continuous

$$(i.e., \quad \lim_{\mu(A) \to 0} \nu(A) = 0).$$

We define a complete norm on $ca(\Sigma, X : \mu)$ by setting

$$\|\nu\| = \sup\{\|\nu(A)\| : A \in \Sigma\}$$

(there is an equivalent norm using the semi-variation of ν; see Chapter 1). Then

$$(ca(\Sigma, X : \mu), B(\Sigma) : X)$$

is an abstract triple with respect to the pairing $\langle \nu, f \rangle \to \int_S f d\nu$ (see Chapter 1.27 for integration of scalar functions with respect to vector valued measures). Define projections P_j on $(ca(\Sigma, X : \mu), B(\Sigma) : X)$ by

$$P_j \nu(\cdot) = \nu(A_j \cap \cdot).$$

We also have

$$ca(\Sigma, X : \mu)^\beta = B(\Sigma)$$

by the countable additivity of the integral (Theorem 2.12). We claim that

$$(ca(\Sigma, X : \mu), B(\Sigma) : X)$$

is monotone and, therefore, has WGHP. Let $\nu \in ca(\Sigma, X : \mu)$ and $I = \{n_j\} \subset \mathbb{N}$. If $f \in B(\Sigma)$, by the countable additivity of the integral,

$$\sum_{j \in I} \langle P_j \nu, f \rangle = \sum_{j=1}^{\infty} \int_{A_{n_j}} f d\nu = \int_{\cup_{j=1}^{\infty} A_{n_j}} f d\nu.$$

Thus, the series

$$\sum_{j \in I} P_j \nu$$

converges in $w(ca(\Sigma, X : \mu), B(\Sigma))$ to

$$P_I \nu = \nu((\cup_{j=1}^{\infty} A_{n_j}) \cap \cdot)$$

and $ca(\Sigma, X : \mu)$ is monotone. We next claim that $(ca(\Sigma, X : \mu), B(\Sigma) : X)$ has 0-GHP. Suppose $\|\nu_k\| \to 0$ and $\{I_k\}$ is an increasing sequence of intervals. Pick n_k such that $\|\nu_{n_k}\| \leq 1/2^k$. For every $A \in \Sigma$,

$$\sum_{k=1}^{\infty} \|\nu_{n_k}(A)\| \leq \sum_{k=1}^{\infty} \|\nu_{n_k}\| \leq \sum_{k=1}^{\infty} 1/2^k < \infty$$

so the series $\sum_{k=1}^{\infty} \nu_{n_k}(A)$ is absolutely convergent. Let

$$(*) \quad \nu(A) = \sum_{k=1}^{\infty} \nu_{n_k}(A).$$

By the Vitali–Hahn–Saks Theorem (Chapter 2, Theorem 2.44: [DU] I.4.10), ν is countably additive and μ continuous so $\nu \in ca(\Sigma, X : \mu)$. Moreover, if $f \in B(\Sigma)$,

$$\left\| \sum_{k=N}^{\infty} \int_S f d\nu_{n_k} \right\| \leq \sum_{k=N}^{\infty} \left\| \int_S f d\nu_{n_k} \right\| \leq \sum_{k=N}^{\infty} \|f\|_{\infty} 2 \|\nu_{n_k}\| \leq 2 \|f\|_{\infty} \sum_{k=N}^{\infty} 1/2^k$$

(see Chapter 1, 1.26 and 1.27 for the inequality above) so the series

$$\sum_{k=1}^{\infty} \nu_{n_k}$$

converges to ν in $w(ca(\Sigma, X : \mu), B(\Sigma))$. This establishes the claim.

From Example 40 and Theorem 36 we have a result similar in spirit to the Nikodym Convergence Theorem (Chapter 2) except that we have a stronger hypothesis and a stronger conclusion.

Corollary 8.39. *Suppose $\nu_k \to 0$ in $ca(\Sigma, X : \mu)$ and $\{f_k\} \subset B(\Sigma)$ is such that*

$$\lim_k \int_S f_k d\nu$$

exists for every $\nu \in ca(\Sigma, X : \mu)$. Then the family of vector measures

$$H = \left\{ \int_. f_k d\nu_l : k, l \in \mathbb{N} \right\}$$

is uniformly countably additive and uniformly μ continuous.

Proof. The first conclusion follows from Theorem 36. The last conclusion follows from Theorem 3.14.1 of [Sw3].   □

Similarly, if $ca(\Sigma, X)$ is the space of countably additive, X valued measures with the norm as defined above,

$$(ca(\Sigma, X), B(\Sigma) : X)$$

forms an abstract triple under the map $\langle \nu, f \rangle \to \int_S f d\nu$ and with projections as defined above. The proof above shows the triple $(ca(\Sigma, X), B(\Sigma) : X)$ is monotone and has 0-GHP (in the proof of the fact that $\nu \in ca(\Sigma, X)$ one employs the Nikodym Convergence Theorem (Chapter 2, 2.36; [DS] IV.10.6) in place of the Vitali–Hahn–Saks Theorem).

Example 8.40. The triple $(ca(\Sigma, X), B(\Sigma) : X)$ is monotone and, therefore, has WGHP and has 0-GHP.

Thus, a result analogous to Corollary 41 holds in this case.

Corollary 8.41. *Suppose* $\nu_k \to 0$ *in* $ca(\Sigma, X)$ *and* $\{f_k\} \subset B(\Sigma)$ *is such that*

$$\lim_k \int_S f_k d\nu$$

exists for every $\nu \in ca(\Sigma, X)$. *Then the family of vector measures*

$$H = \left\{ \int f_k d\nu_l : k, l \in \mathbb{N} \right\}$$

is uniformly countably additive.

Similar results hold for the triple

$$(rca(\mathcal{B}, X), C(S) : X)$$

under the integration map $(\nu, f) \to \int_S f d\nu$ when S is a compact Hausdorff space and $rca(\mathcal{B}, X)$ is the space of regular, countably additive set functions defined on the Borel sets, $\mathcal{B}$, of S.

The results in Theorems 20 and 33 give conditions for the series to converge uniformly over subsets of either E or E^β. We can obtain a similar result where the series converge uniformly over subsets of both E and E^β as in Theorem 36 by imposing another gliding hump condition.

Definition 8.42. The space E (or the triple $(E, F : G)$ with projections $\{P_j\}$) has the signed strong gliding hump property (signed-SGHP) if whenever $\{x^k\}$ is bounded in E and $\{I_k\}$ is an increasing sequence of intervals, there exist a subsequence $\{p_k\}$ and a sequence of signs $\{s_k\}$ such that the series

$$\sum_{k=1}^{\infty} s_k P_{I_{p_k}} x^{p_k}$$

is $w(E, E^\beta)$ convergent in E. If all the signs can be chosen equal to 1, then E has the strong gliding hump property (SGHP).

Note that the SGHP depends on the abstract triple but also on the topology of the space E. See Appendix B and [Sw4] for the scalar and vector space definitions and examples. A different definition is given in [ZCL] where the sequence $\{x^k\}$ is required to be $w(E, E^\beta)$ bounded.

Theorem 8.43. *Assume E has signed-SGHP. If $y \in E^\beta$ and $B \subset E$ is bounded, then the series*

$$\sum_{j=1}^{\infty} \langle P_j x, y \rangle$$

converge uniformly for $x \in B$.

Proof. If the conclusion fails, then as in previous arguments there exist a symmetric neighborhood U in $G, m_{k+1} > n_k > m_k, x^k \in B$ such that

$$(*) \qquad \sum_{j=m_k}^{n_k} \langle P_j x^k, y \rangle \notin U.$$

Put $I_k = [m_k, n_k]$. By the signed-SGHP, there exist an increasing sequence $\{p_k\}$ and signs $\{s_k\}$ such that

$$x = \sum_{k=1}^{\infty} s_k P_{I_{p_k}} x^{p_k} \in E$$

with the series being $w(E, E^\beta)$ convergent. But then the series

$$\sum_{j=1}^{\infty} \langle P_j x, y \rangle = \sum_{k=1}^{\infty} s_k \sum_{j=m_k}^{n_k} \langle P_j x^k, y \rangle$$

fails the Cauchy criterion by $(*)$. $\qquad \square$

A similar result is obtained in Theorem 8 of [ZCL] under different hypothesis. See also Theorem 2.16 of [Sw4].

Using Theorem 45 we can obtain a more general result where the series converge uniformly over subsets of both E and E^β.

Theorem 8.44. *Assume that E has signed-SGHP. If $\{y^k\} \subset E^\beta$ is such that*

$$\lim_k \langle y^k, x \rangle$$

exists for every $x \in E$ and $B \subset E$ is bounded, then the series

$$\sum_{j=1}^{\infty} \langle P_j x, y^k \rangle$$

converge uniformly for $k \in \mathbb{N}, x \in B$.

Proof. If the conclusion fails, then as in previous arguments there exists a symmetric neighborhood U in G such that for every k there exist $p_k > k, m_{k+1} > n_k > m_k, x^k \in B$ such that

$$(\#) \quad \sum_{j=m_k}^{n_k} \langle P_j x^k, y^{p_k} \rangle \notin U.$$

For $k = 1$ this condition gives

$$\sum_{j=m_1}^{n_1} \langle P_j x^1, y^{p_1} \rangle \notin U.$$

By Theorem 45 there exists $m' > n_1$ such that

$$\sum_{j=p}^{q} \langle P_j x, y^k \rangle \in U$$

for $1 \le k \le p_1, x \in B, q > p > m'$. By $(\#)$ for $k = m'$ we have

$$\sum_{j=m_2}^{n_2} \langle P_j x^2, y^{p_2} \rangle \notin U.$$

Thus $p_2 > p_1$. Continuing this construction produces increasing sequences $\{p_k\}, \{m_k\}, \{n_k\}, m_{k+1} > n_k > m_k, \{x^k\} \subset B$ such that

$$(\#\#) \quad \sum_{j \in I_k} \langle P_j x^k, y^{p_k} \rangle \notin U,$$

where $I_k = [m_k, n_k]$. Now define a matrix

$$M = [m_{ij}] = \left[\sum_{l \in I_j} \langle P_l x^j, y^{p_i} \rangle \right].$$

As in the proof of Theorem 36, M is a signed $\mathcal{K}$ matrix so by the signed version of the Antosik–Mikusinski Matrix Theorem (Appendix E) the diagonal of M converges to 0 contradicting $(\#\#)$. $\qquad \square$

This result can be compared to Theorem 2.35 of [Sw4]. The conclusion may fail to hold without assumptions on E.

Example 8.45. Consider the dual pair l^2, l^2. The sequence $\{e^k\}$ converges to 0 weakly in l^2 and is also l^2 bounded. But the series $\sum_{j=1}^{\infty} P_j e^k \cdot e^l$ do not converge uniformly.

A uniform boundedness result as in Theorem 16 can be obtained from Theorem 46.

Theorem 8.46. *Assume E has signed-SGHP and $P_j E$ is barrelled for every j. If $A \subset E$ is bounded and $B \subset E^{\beta}$ is pointwise bounded on E, then*

$$\left\{ x \cdot y = \sum_{j=1}^{\infty} \langle P_j x, y \rangle : y \in B, x \in A \right\} = A \cdot B$$

is bounded.

Proof. For suppose $x_j \in A, y_j \in B$ and $t_j \to 0$. Let U be a neighborhood of 0 and V a neighborhood of 0 such that $V + V \subset U$. Since $t_k y_k \to 0$ in $w(F, E)$, by Theorem 46 there exists N such that

$$\sum_{j=N}^{\infty} t_k \langle P_j x_k, y_k \rangle \in V$$

for all k. Now $\{P_j x_k : k\}$ is bounded and $\{y_k\}$ is pointwise bounded so $\{\langle P_j x_k, y_k \rangle : k\}$ is bounded by the barrelledness assumption. Therefore, there exists K such that

$$\sum_{j=1}^{N-1} t_k \langle P_j x_k, y_k \rangle \in V$$

for $k \geq K$. Then if $k \geq K$,

$$\sum_{j=1}^{\infty} t_k \langle P_j x_k, y_k \rangle = \sum_{j=1}^{N-1} t_k \langle P_j x_k, y_k \rangle + \sum_{j=N}^{\infty} t_k \langle P_j x_k, y_k \rangle \in V + V \subset U$$

and the result follows. $\qquad\square$

Note that the 0-GHP and SGHP assumptions are independent of one another (consider c_0 and l^{∞}) so the result above and the result in Theorem 16 are independent.

Theorems 46 and 48 can be applied to sequence spaces as in Example 18. Let X be a Banach space and Y be a normed space. Consider the triple

$(l^\infty(X), l^1(L(X,Y)) : Y)$ under the map $(\{x_j\}, \{T_j\}) \to \sum_{j=1}^\infty T_j x_j$. It is easily seen that $l^\infty(X)$ has SGHP in this triple so Theorems 46 and 48 are applicable. In particular, if $A \subset l^\infty(X)$ is bounded and $B \subset l^1(L(X,Y))$ is pointwise bounded on $l^\infty(X)$, then

$$A \cdot B = \left\{ \sum_{j=1}^\infty T_j x_j : x \in A, T \in B \right\}$$

is bounded. Moreover, if $T^j \in l^1(L(X,Y))$ is such that $\lim_j T^j \cdot x$ exists for every $x \in l^\infty(X)$ and $A \subset l^\infty(X)$ is bounded, the series $\sum_{j=1}^\infty T_j^k x_j$ converge uniformly for $k \in \mathbb{N}, x \in A$.

We give an application of Theorem 46 to weak topologies on L^1. First we show $L^\infty(\mu)$ has SGHP in the triple $(L^\infty(\mu), L^1(\mu, X) : X)$ when $L^\infty(\mu)$ has the essential-sup norm $\|\cdot\|_\infty$.

Example 8.47. $L^\infty(\mu)$ has SGHP in the triple

$$(L^\infty(\mu), L^1(\mu, X) : X)$$

when $L^\infty(\mu)$ has the essential-sup norm $\|\cdot\|_\infty$. First, note that $L^\infty(\mu)^\beta = L^1(\mu, X)$ by the countable additivity of the integral. Let $\{g_k\}$ be bounded in $L^\infty(\mu)$ and $\{I_k\}$ be an increasing sequence of intervals. For convenience, assume $\|g_k\|_\infty \leq 1$ for every k. The series

$$\sum_k P_{I_k} g_k = \sum_k \chi_{\cup_{j \in I_k} A_j} g_k$$

converges pointwise to a function g which is essentially bounded and measurable and so belongs to $L^\infty(\mu)$. We claim the series converges to g with respect to $w(L^\infty(\mu), L^1(\mu, X))$; this will establish the result. Let $f \in L^1(\mu, X)$. Then for every n, we have

$$\left\| \sum_{k=1}^n \chi_{\cup_{j \in I_k} A_j} g_k(\cdot) f(\cdot) \right\| \leq \|f(\cdot)\|$$

so by the Dominated Convergence Theorem for the Bochner integral (Chapter 1, Theorem 1.18)

$$\sum_{k=1}^\infty \int_{\cup_{j \in I_k} A_j} g_k f d\mu = \int_S g f d\mu$$

so $g_k \to g$ in $w(L^\infty(\mu), L^1(\mu, X))$.

We use Theorem 46 and Example 49 to establish a result for weak convergence in $L^1(\mu, X)$.

Theorem 8.48. *Let $\{f_k\} \subset L^1(\mu, X)$ be such that*

$$\lim \int_S g f_k d\mu$$

exists for every $g \in L^\infty(\mu)$, i.e., $\{f_k\}$ is "weak" Cauchy. If $B \subset L^\infty(\mu)$ is bounded, then the family of vector measures

$$H = \left\{ \int g f_k d\mu : k \in \mathbb{N}, g \in B \right\}$$

is uniformly countably additive.

This result should be compared to Theorem IV.8.9 of [DS] which implies that if $\{f_k\} \subset L^1(\mu)$ is a weak Cauchy sequence, then $\{\int f_k d\mu : k \in \mathbb{N}\}$ is uniformly countably additivity. In Theorem 50 the uniform countable additivity is additionally over bounded subsets of $L^\infty(\mu)$. The conclusion of Theorem 50 can also be rephrased to read that the elements of the set H are uniformly μ continuous (see [DS], IV.8, [Sw3]3.14.1).

From Theorem 48 we also have a uniform boundedness result for this triple.

We can also obtain results like those in Example 49 and Theorem 46 for vector and operator valued functions. Let X, Y be Banach spaces and consider the pair

$$L^\infty(\mu, X), L^1(\mu, L(X, Y)).$$

If $f \in L^1(\mu, L(X, Y))$ and $g \in L^\infty(\mu, X)$, we first observe that the function $t \to f(t)(g(t))$ is strongly measurable. Suppose first that g is a simple function, $g = \sum_{j=1}^n \chi_{B_j} x_j$, with $\{B_j\}, B_j \in \Sigma$, a partition of S. Then

$$f(\cdot)(g(\cdot)) = \sum_{j=1}^n \chi_{B_j}(\cdot) f(\cdot)(x_j)$$

so $f(\cdot)(g(\cdot))$ is a measurable function. If $g \in L^\infty(\mu, X)$, there exists a sequence $\{g_k\}$ of simple functions which converges pointwise almost everywhere to g ([DU] II.1). Then

$$f(\cdot)(g_k(\cdot)) \to f(\cdot)(g(\cdot))$$

almost everywhere so $f(\cdot)(g(\cdot))$ is measurable. Moreover,

$$\|f(t)(g(t))\| \le \|f(t)\| \, \|g(t)\| \le \|g\|_\infty \|f(t)\|$$

implies $f(\cdot)(g(\cdot))$ is Bochner integrable with

$$\left\| \int_S f(\cdot)(g(\cdot))d\mu \right\| \le \|g\|_\infty \|f\|_1 .$$

Thus,

$$(L^\infty(\mu, X), L^1(\mu, L(X, Y)) : Y)$$

is an abstract triple under the mapping

$$(g, f) \to \int_S f(\cdot)(g(\cdot))d\mu$$

and if $\{A_j\} \subset \Sigma$ is pairwise disjoint, $P_j g = \chi_{A_j} g$ defines projections on $L^\infty(\mu, X)$. The proof of Example 49 shows that the triple

$$(L^\infty(\mu, X), L^1(\mu, L(X, Y)) : Y)$$

has SGHP so a result as in Theorem 50 holds in this case.

Theorem 8.49. *Let $\{f_k\} \subset L^1(\mu, L(X, Y))$ be such that*

$$\lim \int_S f_k \circ g \, d\mu$$

exists for every $g \in L^\infty(\mu, X)$. If $B \subset L^\infty(\mu)$ is bounded, then the family of vector measures

$$H = \left\{ \int_{\cdot} f_k \circ g \, d\mu : k \in \mathbb{N}, g \in B \right\}$$

is uniformly countably additive.

From Theorem 48 we also have a uniform boundedness result for this triple.

It should also be noted that dually, we have the triple

$$(L^1(\mu, X), L^\infty(\mu, L(X, Y)) : Y)$$

under the same type of mapping and projections and as in Example 10 and Example 29 the triple has 0-GHP and is monotone so a result like that stated following Theorem 51 holds for this triple. Note that this triple in general does not have SGHP.

Corollary 8.50. *Suppose $\{g_k\} \subset L^\infty(\mu, X)$ is such that*

$$\lim \int_S f \circ g_k \, d\mu$$

exists for every $f \in L^1(\mu, L(X, Y))$. Then if $\{f_k\} \subset L^1(\mu, L(X, Y))$ converges to 0 in $L^1(\mu, L(X, Y))$, the family of vector measures

$$H = \left\{ \int_{\cdot} f_j \circ g_k \, d\mu : j, k \in \mathbb{N} \right\}$$

is uniformly countably additive.

Similarly, if $1 < p < \infty$ and $\frac{1}{p} + \frac{1}{q} = 1$, then the triple

$$(L^p(\mu, X), L^q(L(X, Y)) : Y)$$

has 0-GHP and is monotone so a result as stated in Corollary 52 also holds for this triple. Note that in general this triple does not have SGHP.

Theorem 46 can also be used to establish a Schur type result for $l^1(X)$ when X is a Banach space. Recall the Schur Theorem asserts that a sequence in l^1 which is weakly convergent converges in norm and a normed space with this property is called a *Schur space*. Consider the dual pair $l^\infty(X'), l^1(X)$ under the pairing

$$\langle x, y \rangle = \sum_{j=1}^{\infty} (x_j, y_j),$$

where $x = \{x_j\} \in l^\infty(X'), y = \{y_j\} \in l^1(X)$ and $(\cdot, \cdot)$ is the duality between X' and X. The triple $(l^\infty(X'), l^1(X))$ with the projections as in Example 2 has SGHP so Theorem 46 applies.

Theorem 8.51. *Suppose $y^k \to 0$ with respect to $\sigma(l^1(X), l^\infty(X'))$. If X is a Schur space, then $\left\| y^k \right\|_1 \to 0$ so $l^1(X)$ is a "Schur type space".*

Proof. Let $\epsilon > 0$ and N be such that

$$\left| \sum_{j=N}^{\infty} \langle P_j x, y^k \rangle \right| = \left| \sum_{j=N}^{\infty} (x_j, y_j^k) \right| < \epsilon$$

for all k and $\|x\|_\infty \le 1$ (Theorem 46). For each j, $\lim_k y_j^k = 0$ with respect to $\sigma(X, X')$ [consider $\langle P_j(e^j \otimes x'), y^k \rangle = (x', y_j^k)$ for $x' \in X'$] so $\lim_k \left\| y_j^k \right\| = 0$ for each j. Therefore, there exists M such that

$$\left| \sum_{j=1}^{N-1} (x_j, y_j^k) \right| < \epsilon$$

for $\|x\|_\infty \le 1$ and $k \ge M$. Hence, if $k \ge M$, then

$$\left| \sum_{j=1}^{\infty} (x_j, y_j^k) \right| \le \left| \sum_{j=1}^{N-1} (x_j, y_j^k) \right| + \left| \sum_{j=N}^{\infty} (x_j, y_j^k) \right| < 2\epsilon$$

when $\|x\|_\infty \le 1$. Fix $k \ge M$ and pick $x_j' \in X'$ such that $\left\| x_j' \right\| = 1$ and $\left| (x_j', y_j^k) \right| = \left\| y_j^k \right\|$. Then

$$z = \sum_{j=1}^{\infty} e^j \otimes x_j' \in l^\infty(X'),$$

$\|z\|_\infty \le 1$, so

$$\left| \sum_{j=1}^{\infty} \langle P_j z, y^k \rangle \right| = \left| \sum_{j=1}^{\infty} (x'_j, y^k_j) \right| = \sum_{j=1}^{\infty} \|y^k_j\| = \|y^k\|_1 < 2\epsilon$$

for $k \ge M$. $\square$

A result for vector measures like that in Theorem 50 can also be obtained from Theorem 46.

Example 8.52. Let $B(\Sigma)$ be the space of all bounded, Σ measurable functions defined on S with the sup-norm, $\|\cdot\|_\infty$ and let $ca(\Sigma, X)$ be the space of all countably additive set functions $\nu : \Sigma \to X$ with the complete norm

$$\|\nu\| = \sup\{\|\nu(A)\| : A \in \Sigma\}$$

(there is an equivalent norm using the semi-variation of ν; see Chapter 1,Theorem 1.26 or [DS]IV.10.4). Then

$$(B(\Sigma), ca(\Sigma, X) : X)$$

forms an abstract triple under the pairing $\langle f, \nu \rangle \to \int_S f d\nu$ (see Chapter 1.27). Let $\{A_j\} \subset \Sigma$ be pairwise disjoint and define projections $P_j : B(\Sigma) \to B(\Sigma)$ by $P_j f = \chi_{A_j} f$ so $(B(\Sigma), ca(\Sigma, X) : X)$ is a triple with projections when $B(\Sigma)$ has the sup-norm. Note $B(\Sigma)^\beta = ca(\Sigma, X)$ by the countable additivity of the elements of $ca(\Sigma, X)$ (Theorem 2.12). We show that $(B(\Sigma), ca(\Sigma, X) : X)$ has SGHP. Suppose $\{f_k\}$ is a bounded subset of $B(\Sigma)$ and $\{I_k\}$ is an increasing sequence of intervals. For convenience set $B_k = \cup_{j \in I_k} A_j$ so $P_{I_k} f = \chi_{B_k} f$. The series

$$\sum_{k=1}^{\infty} P_{I_k} f_k = \sum_{k=1}^{\infty} \chi_{B_k} f_k$$

converges pointwise to a function f which is bounded and measurable and so belongs to $B(\Sigma)$. We claim the series $\sum_{k=1}^{\infty} P_{I_k} f_k$ is $w(B(\Sigma), ca(\Sigma, X))$ convergent to f. For this let $\nu \in ca(\Sigma, X)$. Then

$$\sum_{k=1}^{\infty} P_{I_k} f_k \cdot \nu = \sum_{k=1}^{\infty} \int_{B_k} f_k d\nu = \int_S f d\nu$$

by the Bounded Convergence Theorem (see Chapter 1, Theorem 1.34).

From Theorem 46 and Example 54, we can obtain an improvement in the conclusion of the Nikodym Convergence Theorem (Chapter 2, Theorem 2.36, [DS] IV.10.6).

Theorem 8.53. *Let* $\{\nu_k\} \subset ca(\Sigma, X)$ *be such that*

$$\lim_k \nu_k(A) = \nu(A)$$

exists for every $A \in \Sigma$. *If* $B \subset B(\Sigma)$ *is bounded, then the family of vector measures*

$$H = \left\{ \int fd\nu_k : k \in \mathbb{N}, f \in B \right\}$$

is uniformly countably additive.

Proof. We claim that $\lim_k \int_S fd\nu_k$ exists for every $f \in B(\Sigma)$. Let $f \in B(\Sigma)$, $\epsilon > 0$ and pick a simple function g such that $\|f - g\|_\infty < \epsilon$ and choose n such that $k \geq n$ implies

$$\left\| \int_S gd\nu_k - \int_S gd\nu \right\| < \epsilon.$$

By the Nikodym Convergence Theorem 2.36, ν is countably additive and by the Nikodym Boundedness Theorem,

$$\sup\{\|\nu_k\| : k \in \mathbb{N}\} = M < \infty$$

(Chapter 2, Theorem 2.45). Using the inequalities in Chapter 1, Proposition 26, if $k \geq n$, we have

$$\left\| \int_S fd\nu_k - \int_S fd\nu \right\| \leq \left\| \int_S (f-g)d\nu_k \right\| + \left\| \int_S (f-g)d\nu \right\| + \left\| \int_S gd\nu_k - \int_S gd\nu \right\|$$

$$\leq 2\|f-g\|_\infty (M + \|\nu\|) + \left\| \int_S gd\nu_k - \int_S gd\nu \right\|$$

$$< \epsilon(2(M + \|\nu\|) + 1)$$

justifying the claim. The result now follows from Theorem 46 and Example 54. $\square$

In the classical Nikodym Convergence Theorem (Chapter 2, Theorem 2.36; [DS] IV.10.6) the uniform countable additivity is for the measures $\{\nu_k\}$ and in the result above the uniform additivity is for the indefinite integrals over bounded subsets of $B(\Sigma)$. For an application of uniform countable additivity in this setting, see [DS] IV.13.22.

Theorem 55 also gives an improvement to the Vitali–Hahn–Saks Theorem (Chapter 2, Theorem 2.44; [DS] IV.7.2). If each of the measures ν_k

is μ-continuous with respect to a positive measure μ, then each indefinite integral in H is μ continuous so the family of indefinite integrals in H is uniformly μ continuous by Theorem 55 and Theorem 3.14.1 of [Sw3] (if $\nu \in ca(\Sigma, X)$ is μ continuous and g is simple, then clearly $\int g d\nu$ is μ continuous, and if $f \in B(\Sigma)$, then $\int f d\nu$ is the pointwise limit of a sequence $\int g_k d\nu$ of indefinite integrals of simple functions g_k so $\int f d\nu$ is μ continuous by the Vitali–Hahn–Saks Theorem (Chapter 2, Theorem 2.44; [DS] IV.7.2)).

We can obtain a similar result for the space $L^1(m)$ of Chapter 1.

Example 8.54. Consider the triple

$$(B(\Sigma), L^1(m) : X)$$

under the integration map $(g, f) \to \int_S g f dm$ with the projections $P_j g = \chi_{A_j} g$. We claim this triple has SGHP. Let $\{I_k\}$ be an increasing sequence of intervals and $\{g_k\} \subset B(\Sigma)$ be bounded. Put $B_k = \cup_{j=1}^k A_j$. Let h be the pointwise limit of the series

$$\sum_{k=1}^{\infty} P_{I_k} g_k = \sum_{k=1}^{\infty} \chi_{B_k} g_k$$

so $h \in B(\Sigma)$. We claim the series $\sum_{k=1}^{\infty} \chi_{B_k} g_k$ converges to h in $w(B(\Sigma), L^1(m))$. Let $f \in L^1(m)$. Then

$$\left| \sum_{k=1}^{n} \chi_{B_k} g_k f \right| \le \|g_k\|_{\infty} |f| \le \sup_k \|g_k\|_{\infty} |f|$$

so by the Dominated Convergence Theorem for $L^1(m)$ (Chapter 1, Theorem 1.34),

$$\sum_{k=1}^{\infty} \int_{B_k} g_k f dm = \int_S h f dm$$

and the claim is satisfied.

From Theorem 46 and Example 56, we have

Theorem 8.55. *Let* $\{f_k\} \subset L^1(m)$ *be such that*

$$\lim_k \int_S f_k g dm$$

exists for every $g \in B(\Sigma)$. *If* $B \subset B(\Sigma)$ *is bounded, then the family of vector measures*

$$H = \left\{ \int f_k g dm : k \in \mathbb{N}, g \in B \right\}$$

is uniformly countably additive.

We indicate two more applications where the integration theories have not been discussed in detail in the book. We will indicate references where the details may be seen.

We can also obtain a result similar to Theorem 57 for operator valued measures and vector valued functions. Let X, Y be Banach spaces, $B(\Sigma, X)$ the space of all bounded, X valued, Σ measurable functions with the sup-norm, $\lambda : \Sigma \to [0, \infty)$ a finite measure, and

$$ca(\Sigma, L(X, Y) : \lambda)$$

the space of all $\nu : \Sigma \to L(X, Y)$ which are countably additive and λ continuous. The space $ca(\Sigma, L(X, Y) : \lambda)$ is given the (operator) semi-variation norm $\|\nu\|$. If $A \in \Sigma$, the semi-variation of ν at A is defined to be

$$semi - var(\nu)(A) = \sup\left\{\left\|\sum_{j=1}^{n} \nu(A_j)x_j\right\|\right\},$$

where the supremum is taken over all partitions $\{A_j : j = 1, ..., n\}$ of A and all $\|x_j\| \leq 1$. Then

$$\|\nu\| = semi - var(\nu)(S)$$

and we have

$$\left\|\int_S f d\nu\right\| \leq \|f\|_\infty \|\nu\|$$

for $f \in B(\Sigma)$ (see [Bar] for the integration of vector valued functions with respect to operator valued measures and their properties,). Then

$$(B(\Sigma, X), ca(\Sigma, L(X, Y) : \lambda) : Y)$$

forms an abstract triple under the continuous bilinear map $(f, \nu) \to \int_S f d\nu$ and projections P_j can be defined as above. Under these assumptions, the integral is countably additive, bounded measurable functions are integrable and the Bounded Convergence Theorem holds for the integral ([Bar]). Hence, the proof of Example 56 can be repeated to show that the triple $(B(\Sigma, X), ca(\Sigma, L(X, Y) : \lambda) : Y)$ has SGHP and Theorem 46 gives a result like Theorem 57.

Theorem 8.56. *Let* $\{\nu_k\} \subset ca(\Sigma, L(X, Y) : \lambda)$ *be such that*

$$\lim_k \int_S f d\nu_k$$

exists for every $f \in B(\Sigma, X)$. If $B \subset B(\Sigma, X)$ is bounded, then the family of Y valued measures

$$H = \left\{ \int_{\cdot} f d\nu_k : k \in \mathbb{N}, f \in B \right\}$$

is uniformly countably additive.

Finally, we consider spaces of integrable functions with respect to a measure with values in a locally convex space. We discussed the integral of bounded, measurable functions with respect to countably additive, Banach space valued measures in Chapter 1. The integral when the measures have values in LCTVS is more technical. We will give references when the properties are used. Assume G is a sequentially complete Hausdorff locally convex space and $\nu : \Sigma \to G$ is a countably additive vector measure. A Σ measurable function $f : S \to \mathbb{R}$ is *scalarly integrable* with respect to ν if f is $x'\nu$ integrable for every $x' \in G'$ and f is ν *integrable* if f is scalarly ν integrable and for every $A \in \Sigma$ there exists $x_A \in G$ such that

$$\int_A f dx'\nu = x'(x_A);$$

we write

$$\int_A f d\nu = x_A.$$

(For the integral and for properties of the integral, we refer to [KK]; see also, [Pa].) Let

$$L^1(\nu)$$

be the space of all ν integrable functions. We will describe the topology of $L^1(\nu)$. Let $\mathcal{P}$ be a family of semi-norms which generate the topology of G and if $p \in \mathcal{P}$, let

$$U_p = \{x \in G : p(x) \leq 1\}$$

and U_p^0 be the polar of U_p. Define a semi-norm $\widehat{p}$ on $L^1(\nu)$ by

$$\widehat{p}(f) = \left\{ \int_S |f| \, d \, |x'\nu| : x' \in U_p^0 \right\};$$

the topology of $L^1(\nu)$ is defined to be the topology generated by the semi-norms $\{\widehat{p} : p \in \mathcal{P}\}$. Since G is sequentially complete, the product of bounded measurable functions and ν integrable functions are ν integrable ([KK] Theorem II.3.1) and we can define an abstract triple

$$(B(\Sigma), L^1(\nu) : G)$$

under the integration map $(f, g) \to \int_S fg d\nu$ when $B(\Sigma)$ has the sup-norm and this bilinear map is continuous since

$$p\left(\int_S fg d\nu\right) \le \|f\|_\infty \, \widehat{p}(g).$$

The Dominated Convergence Theorem holds for the integral ([KK] Theorem II.4.2) so the proof in Example 56 shows that the triple $(B(\Sigma), L^1(\nu) : G)$ with the projections defined as before has SGHP and a weak convergence result as in Theorem 57 holds for the space $L^1(\nu)$.

Corollary 8.57. *Suppose* $\{f_k\} \subset L^1(\nu)$ *is such that*

$$\lim_k \int_S f_k g d\nu$$

exists for every $g \in B(\Sigma)$ *and* $B \subset B(\Sigma)$ *is bounded. Then the family of vector measures*

$$H = \left\{\int f_k g d\nu : k \in \mathbb{N}, g \in B\right\}$$

is uniformly countably additive.

Dually, consider the triple

$$(L^1(\nu), B(\Sigma) : G)$$

under the integration map. Since the indefinite integral is countably additive, the proof in Example 32 shows that the triple is monotone and, therefore, has WGHP. We also consider the 0-GHP for the triple. In order to do this we first establish an abstract result for triples and then apply the result to $(L^1(\nu), B(\Sigma) : G)$.

Theorem 8.58. *Let* $(E, F : G)$ *be an abstract triple with projections* $\{P_j\}$ *and* E *a complete, metrizable locally convex space whose topology is generated by the semi-norms* $p_1 \le p_2 \le \dots$. *Assume that*

$$p_l(P_I x) \le p_l(x)$$

for every $l \in \mathbb{N}$ *and finite* $I \subset \mathbb{N}$. *Then* E *has 0-GHP.*

Proof. Let $x_k \to 0$ in E and $\{I_k\}$ be an increasing sequence of intervals. Pick an increasing sequence of integers $\{n_k\}$ such that

$$p_{n_k}(x_j) < 1/2^k$$

for $j \geq n_k$. Consider the series

$$\sum_{k=1}^{\infty} P_{I_{n_k}} x_{n_k}.$$

We claim the series is absolutely convergent in E. Fix $l \in \mathbb{N}$. Then by hypothesis

$$\sum_{\{k:n_k \geq l\}} p_l(P_{I_{n_k}} x_{n_k}) \leq \sum_{\{k:n_k \geq l\}} p_{n_k}(P_{I_{n_k}} x_{n_k})$$

$$\leq \sum_{\{k:n_k \geq l\}} p_{n_k}(x_{n_k}) \leq \sum_{\{k:n_k \geq l\}} 1/2^k$$

so the series is absolutely convergent and, therefore, convergent to some x in E since E is complete. Hence, the series $\sum_{k=1}^{\infty} P_{I_{n_k}} x_{n_k}$ is $w(E,F)$ convergent to x and E has 0-GHP. $\qquad\square$

If G is sequentially complete, metrizable, then $L^1(\nu)$ is complete (see [KK] IV4.1 and IV.7.1 for this result) and the projections satisfy the condition that

$$\widehat{p}(P_I f) \leq \widehat{p}(f)$$

for $f \in L^1(\nu)$ and I finite. Thus, the theorem above applies and the triple

$$(L^1(\nu), B(\Sigma) : G)$$

has 0-GHP. A weak convergence result somewhat similar to Corollary 58 follows from Theorem 36 since the triple has both WGHP and 0-GHP.

Corollary 8.59. *Suppose* $\{g_k\} \subset B(\Sigma)$ *is such that*

$$\lim_k \int_S f g_k d\nu$$

exists for every $f \in L^1(\nu)$. *If* $f_k \to 0$ *in* $L^1(\nu)$, *then the family of vector measures*

$$H = \left\{ \int_{\cdot} f_l g_k d\nu : k, l \in \mathbb{N} \right\}$$

is uniformly countably additive.

Chapter 9

Weak Compactness in Triples

Let E, F be (real) vector spaces and G a topological vector space and assume that there is a bilinear map $b : E \times F \to G$. We consider the abstract triple

$$(E, F : G).$$

We study sequential compactness and sequential completeness for the topology $w(E, F)$ when E is a space of vector valued, bounded, finitely additive set functions or the space of Bochner or Pettis integrable functions, F is a space of bounded measurable functions, G is a Banach space and the bilinear map is defined via an integral.

In what follows X will denote a Banach space and Σ a σ-algebra of subsets of a set S. We will utilize integration results for bounded, measurable functions with respect to bounded, finitely additive, scalar set functions and bounded, finitely additive X-valued set functions. These results are covered in Chapter 1 and will be referred to as needed.

Recall $ba(\Sigma, X)$ is the space of all bounded, finitely additive set functions from Σ into X and we equip $ba(\Sigma, X)$ with the semi-variation norm (Definition 1.25) or the equivalent norm,

$$\|m\| (S) = \|m\| ,$$

where

$$\|m\| (A) = \sup\{\|m(B)\| : B \subset A, B \in \Sigma\}$$

(see Theorem 1.26 for the norms). Let $B(\Sigma)$ be the space of all bounded, Σ measurable functions with the sup-norm $\|\cdot\|_{\infty}$.

We establish a lemma which will be used in the sequel.

Lemma 9.1. *Suppose* $m_k \in ba(\Sigma, X)$ *and*

$$m(A) = \lim m_k(A)$$

exists for every $A \in \Sigma$. *Then*

(i) $m \in ba(\Sigma, X)$ and

(ii) for every $f \in B(\Sigma)$ and $A \in \Sigma$, $\lim \int_A f dm_k = \int_A f dm$.

Proof. (i): That m is finitely additive is clear and m is bounded by the Nikodym Boundedness Theorem (2.52).

(ii): Let $\epsilon > 0$. Pick g simple such that $\|f - g\|_\infty \le \epsilon$ and let

$$M = \sup\{\|m_k(A)\| : k \in \mathbb{N}, A \in \Sigma\};$$

note $M < \infty$ by the Nikodym Boundedness Theorem. Pick n such that

$$\left\| \int_A g dm_k - \int_A g dm \right\| \le \epsilon$$

for $k \ge n$ (hypothesis). For $k \ge n$,

$$\left\| \int_A f dm_k - \int_A f dm \right\|$$

$$\le \left\| \int_A (f - g) dm_k \right\| + \left\| \int_A (f - g) dm \right\|$$

$$+ \left\| \int_A g dm_k - \int_A g dm \right\|$$

$$\le \|f - g\|_\infty 2 \|m_k\| (A) + \|f - g\|_\infty 2 \|m\| (A) + \epsilon$$

$$\le \epsilon 4M + \epsilon$$

by Proposition 1.27.   □

Let $S(\Sigma)$ be the subspace of $B(\Sigma)$ consisting of the simple functions. Consider the abstract triple

$$(ba(\Sigma, X), S(\Sigma) : X)$$

under the integration map $(m, f) \to \int_S f dm$. The hypothesis of Lemma 1 asserts that the sequence $\{m_k\}$ is

$$w(ba(\Sigma, X), S(\Sigma))$$

Cauchy. Now consider the abstract triple $(ba(\Sigma, X), B(\Sigma) : X)$ under the integration map. The conclusion of Lemma 1 is that there exists $m \in ba(\Sigma, X)$ such that the sequence $\{m_k\}$ converges to m in the stronger topology $w(ba(\Sigma, X), B(\Sigma))$ for the triple

$$(ba(\Sigma, X), B(\Sigma) : X).$$

Thus, Lemma 1 can be viewed as a "Schur-type" result; i.e., in l^1 a sequence which is Cauchy in the weak topology actually is convergent in the stronger norm topology (Theorem 2.56; [Sw4] 7.1).

We consider compactness results for spaces of vector valued set functions. Recall a subset K of a topological vector space E is relatively sequentially compact if every sequence $\{x_k\} \subset K$ has a subsequence $\{x_{n_k}\}$ which converges to an element in E; K is conditionally sequentially compact if every sequence $\{x_k\} \subset K$ has a subsequence $\{x_{n_k}\}$ which is Cauchy.

We first consider a boundedness result.

Proposition 9.2. *Suppose $K \subset ba(\Sigma, X)$ is $w(ba(\Sigma, X), S(\Sigma))$ conditionally sequentially compact. Then K is norm bounded in $ba(\Sigma, X)$.*

Proof. $\{m(A) : m \in K, A \in \Sigma\}$ is bounded iff $\{m_k(A_k)\}$ is bounded for every $\{m_k\} \subset K$ and every pairwise disjoint sequence $\{A_k\} \subset \Sigma$ (Appendix C). By $w(ba(\Sigma, X), S(\Sigma))$ conditionally sequentially compactness we may assume that $\lim m_k(A)$ exists for every $A \in \Sigma$. Then $\{m_k(A)\}_k$ is bounded for every $A \in \Sigma$ and $\{m_k(A_k)\}_k$ is bounded by the Nikodym Boundedness Theorem. K is norm bounded by Theorem 1.26. $\qquad\square$

Recall a finitely additive set function $m : \Sigma \to X$ is strongly additive (strongly bounded) if the series $\sum_{k=1}^{\infty} m(A_k)$ converges for every pairwise disjoint sequence $\{A_k\} \subset \Sigma$ (see Proposition 2.38). A strongly additive set function is bounded (Corollary 2.48; Appendix C).

Let

$$sba(\Sigma, X)$$

be the space of all strongly additive elements of $ba(\Sigma, X)$. A subset $M \subset sba(\Sigma, X)$ is uniformly strongly additive if the series $\sum_{k=1}^{\infty} m(A_k)$ converge uniformly for $m \in M$ and every pairwise disjoint sequence $\{A_k\} \subset \Sigma$ (Proposition 2.39).

Consider the abstract triple

$$(sba(\Sigma, X), S(\Sigma) : X)$$

under the integration map $(m, f) \to \int_S f dm$.

Proposition 9.3. *If $K \subset sba(\Sigma, X)$ is $w(sba(\Sigma, X), S(\Sigma))$ conditionally sequentially compact, then*

(I) K *is uniformly strongly additive.*

Proof. If the conclusion fails, there exist $m_k \in K, \{A_k\} \subset \Sigma$ pairwise disjoint, an increasing sequence of intervals, $\{I_k\}$, in $\mathbb{N}$ and $\epsilon > 0$ such that

$$(\&) \qquad \left\| \sum_{j \in I_k} m_k(A_j) \right\| > \epsilon.$$

By $w(sba(\Sigma, X), S(\Sigma))$ conditionally sequentially compactness we may assume

$$\lim m_k(A) = m(A)$$

exists for every $A \in \Sigma$. By the Nikodym Convergence Theorem (2.41), $\{m_k\}$ is uniformly strongly additive contradicting (&). $\qquad\square$

Proposition 9.4. *Suppose $K \subset ba(\Sigma, X)$ is $w(ba(\Sigma, X), B(\Sigma))$ conditionally (relatively) sequentially compact. Then for every $f \in B(\Sigma)$,*

$$(II) \ K_f = \left\{ \int_S f dm : m \in K \right\} \text{ is } \|\cdot\| \text{ conditionally}$$
$$(relatively) \text{ sequentially compact.}$$

In particular, for every $A \in \Sigma$,

$$(III) \ K_A = \{m(A) : m \in K\} \text{ is } \|\cdot\| \text{ conditionally}$$
$$(relatively) \text{ sequentially compact.}$$

Proof. The linear map $m \to \int_S f dm$ from $ba(\Sigma, X) \to X$ is

$$w(ba(\Sigma, X), B(\Sigma)) - \|\cdot\|$$

sequentially continuous so the result follows. $\qquad\square$

Conditions (I), (II) and (III) give necessary conditions for $w(sba(\Sigma, X), B(\Sigma))$ relative sequential compactness. We now consider sufficient conditions.

Theorem 9.5. *Let $K \subset sba(\Sigma, X)$. Assume that Σ is generated by a countable algebra $\mathcal{A}$. Then conditions (I) and (III) imply that K is $w(sba(\Sigma, X), B(\Sigma))$ relatively sequentially compact.*

Proof. Let $\{m_k\} \subset sba(\Sigma, X)$. By (III) and the diagonalization method ([Ke] 7.D, p.238), there exists a subsequence $\{q_k\}$ of $\{m_k\}$ such that

$$\|\cdot\| - \lim q_k(A) = m(A)$$

exists for every $A \in \mathcal{A}$. We claim that

$$\|\cdot\| - \lim q_k(A) = m(A)$$

exists for every $A \in \Sigma$. For this let

$$\Sigma_1 = \left\{ A \in \sum : \|\cdot\| - \lim_k q_k(A) = m(A) \text{ exists} \right\}.$$

Then $\mathcal{A} \subset \Sigma_1$ and we claim that Σ_1 is a monotone class. Suppose $B_j \in \Sigma$ and $B_j \uparrow B$. By definition of Σ_1,

$$\|\cdot\| - \lim_k q_k(B_j) = m(B_j)$$

exists for every j. Now $B_j = B_1 \cup (\cup_{i=1}^{j-1}(B_{i+1} \setminus B_i))$ so by (I),

$$\|\cdot\| - \lim_j q_k(B_j) = z_k$$

exists uniformly for $k \in \mathbb{N}$ (2.38, 2.39; [DU] I.1.17, I.1.18). By the Iterated Limit Theorem (Appendix A),

$$\lim_j \lim_k q_k(B_j) = \lim_k \lim_j q_k(B_j) = \lim_k z_k = \lim_j m(B_j),$$

where all limits are with respect to the norm. Hence, $B \in \Sigma_1$. A similar computation holds for decreasing sequences from Σ_1. Hence, Σ_1 is a monotone class and the Monotone Class Theorem ([Hal], [Sw3] 2.1.16) implies that $\Sigma_1 = \Sigma$ justifying the claim. By the Nikodym Convergence Theorem $m \in sba(\Sigma, X)$. Now by Lemma 1

$$\lim \int_S f dq_k = \int_S f dm$$

holds for every $f \in B(\Sigma)$ so $q_k \to m$ in $w(sba(\Sigma, X), B(\Sigma))$. $\qquad\square$

Corollary 9.6. *Let $K \subset sba(\Sigma, X)$. Assume that Σ is generated by a countable algebra $\mathcal{A}$. Then K is $w(sba(\Sigma, X), B(\Sigma))$ relatively sequentially compact iff (I) and (III) hold.*

In the remarks below assume that Σ is generated by a countable algebra $\mathcal{A}$.

The same proof shows that if $K \subset ba(\Sigma, X)$ satisfies (I) and (III), then K is $w(ba(\Sigma, X), B(\Sigma))$ relatively sequentially compact in the triple

$$(ba(\Sigma, X), B(\Sigma) : X).$$

For if

$$\|\cdot\| - \lim q_k(A) = m(A)$$

exists for every $A \in \Sigma$, then $m \in ba(\Sigma, X)$ by applying the Nikodym Boundedness Theorem in place of the Nikodym Convergence Theorem.

Similarly, if $ca(\Sigma, X)$ denotes the space of all countably additive members of $sba(\Sigma, X)$ and $K \subset ca(\Sigma, X)$, then K is $w(ca(\Sigma, X), B(\Sigma))$ relatively sequentially compact in the triple

$$(ca(\Sigma, X), B(\Sigma) : X)$$

iff (I) and (III) hold. For the limit of countably additive set function is countably additive by the Nikodym Convergence Theorem (Theorem 2.36).

Let $\lambda : \Sigma \to [0, \infty)$ be a measure and let

$$ca(\Sigma, X : \lambda)$$

be the elements of $ca(\Sigma, X)$ which are continuous with respect to λ

$$\left(i.e., \lim_{\lambda(A) \to 0} m(A) = 0 \right).$$

Then K is $w(ca(\Sigma, X : \lambda), B(\Sigma))$ is relatively sequentially compact in the triple

$$(ca(\Sigma, X : \lambda), B(\Sigma) : X)$$

iff (I) and (III) hold. For if

$$\|\cdot\| - \lim q_k(A) = m(A)$$

exists for every $A \in \Sigma$, then $m \in ca(\Sigma, X : \lambda)$ by the Vitali–Hahn–Saks Theorem (Theorem 2.44).

- It is worthwhile observing that the methods above imply that if $\{m_k\}$ is a $w(Y, S(\Sigma))$ Cauchy sequence when

  $$Y = ba(\Sigma, X), sba(\Sigma, X), ca(\Sigma, X), ca(\Sigma, X : \lambda),$$

  then there exists $m \in Y$ such that $m_k \to m$ with respect to $w(Y, B(\Sigma))$. In particular, $w(Y, S(\Sigma))$ is sequentially complete.

We next consider abstract triples of vector valued integrable functions. Let $L^1(\lambda, X)$ be the space of Bochner λ integrable X valued functions (see Chapter 1) with the complete norm

$$\|f\|_1 = \int_S \|f(s)\| \, d\lambda(s)$$

and consider the triples

$$(L^1(\lambda, X), S(\Sigma) : X) \text{ and } (L^1(\lambda, X), L^\infty(\lambda) : X)$$

under the integration map $(f, g) \to \int_S gf d\lambda$. We establish the analogues of Propositions 3 and 4 for this triple.

Proposition 9.7. *Let $K \subset L^1(\lambda, X)$ and suppose K is $w(L^1(\lambda, X), S(\Sigma))$ conditionally sequentially compact. Then*

$$\left\{ \int f d\lambda : f \in K \right\}$$

is uniformly countably additive.

Proof. If the conclusion fails, there exist $\{f_k\} \subset K$, $\{A_k\} \subset \Sigma$ pairwise disjoint, an increasing sequence of intervals $\{I_k\}$ and $\epsilon > 0$ such that

$$(*) \qquad \left\| \int_{\cup_{j \in I_k} A_j} f_k d\lambda \right\| > \epsilon.$$

We may assume that $\lim_k \int_A f_k d\lambda$ exists for every $A \in \Sigma$. The Nikodym Convergence Theorem (2.36) asserts that the measures $\{\int f_k d\lambda : f \in K\}$ are uniformly countably additive. This contradicts $(*)$. $\qquad\square$

Proposition 9.8. *Let $K \subset L^1(\lambda, X)$ and suppose K is $w(L^1(\lambda, X), S(\Sigma))$ conditionally sequentially compact. Then for every $A \in \Sigma$,*

$$K_A = \left\{ \int_A f d\lambda : f \in K \right\}$$

is $\|\cdot\|$ conditionally sequentially compact. If K is $w(L^1(\lambda, X), L^\infty(\lambda))$ conditionally sequentially compact,then

$$K_h = \left\{ \int_S h f d\lambda : f \in K \right\}$$

is $\|\cdot\|$ conditionally sequentially compact for every $h \in L^\infty(\lambda)$.

Proof. The linear map

$$H : L^1(\lambda, X) \to X, \quad f \to \int_S h f d\lambda,$$

is $w(L^1(\lambda, X), L^\infty(\lambda)) - \|\cdot\|$ continuous so the second conclusion follows. The first statement can be treated similarly. $\qquad\square$

The conclusions in Propositions 7 and 8 are necessary conditions for weak conditional sequential compactness. As in Theorem 5 we consider sufficient conditions.

Theorem 9.9. *Let $K \subset L^1(\lambda, X)$ be bounded. Assume that Σ is generated by a countable algebra $\mathcal{A}$. If*

$$\left\{ \int f d\lambda : f \in K \right\}$$

is uniformly countably additive and

$$K_h = \left\{ \int_S h f d\lambda : f \in K \right\}$$

is $\|\cdot\|$ conditionally sequentially compact for every $h \in L^\infty(\lambda)$, then K is

$$w(L^1(\lambda, X), L^\infty(\lambda))$$

conditionally sequentially compact. If X has the Radon-Nikodym property with respect to λ (see [DU]), then K is $w(L^1(\lambda, X), L^\infty(\lambda))$ relatively sequentially compact.

Proof. Let $\{f_k\} \subset K$ and set $M = \sup\{\|f_k\|_1 : k\}$. As in the proof of Theorem 5 we claim that there exists a subsequence $\{g_k\}$ of $\{f_k\}$ such that $\lim \int_A g_k d\lambda$ exists for every $A \in \Sigma$. By conditional sequential compactness of $K_A = \{\int_A f d\lambda : f \in K\}$ and the diagonalization method ([Ke] 7.D, p.238), there exists a subsequence $\{g_k\}$ of $\{f_k\}$ such that

$$\|\cdot\| - \lim \int_A g_k d\lambda = m(A)$$

exists for every $A \in \mathcal{A}$. We claim that

$$\|\cdot\| - \lim \int_A g_k d\lambda = m(A)$$

exists for every $A \in \Sigma$. For this let

$$\Sigma_1 = \left\{ A \in \sum : \|\cdot\| - \lim_k \int_A g_k d\lambda = m(A) \text{ exists} \right\}.$$

Then $\mathcal{A} \subset \Sigma_1$ and we claim that Σ_1 is a monotone class. Suppose $B_j \in \Sigma$ and $B_j \uparrow B$. By definition of Σ_1,

$$\|\cdot\| - \lim_k \int_{B_j} g_k d\lambda = m(B_j)$$

exists for every j. Now by the uniform countable additivity,

$$\|\cdot\| - \lim_j \int_{B_j} g_k d\lambda = \int_B g_k d\lambda$$

exists uniformly for $k \in \mathbb{N}$. By the Iterated Limit Theorem (Appendix A),

$$\lim_j \lim_k \int_{B_j} g_k d\lambda = \lim_k \lim_j \int_{B_j} g_k d\lambda = \lim_k \int_B g_k d\lambda = \lim_j m(B_j),$$

where all limits are with respect to the norm. Hence, $B \in \Sigma_1$. A similar computation holds for decreasing sequences from Σ_1. Hence, Σ_1 is a monotone class and the Monotone Class Theorem ([Hal], [Sw3] 2.1.16) implies that $\Sigma_1 = \Sigma$ justifying the claim

Now we claim that $\lim \int_S h g_k d\lambda$ exists for every $h \in L^\infty(\lambda)$. The linear maps

$$G_k : L^\infty(\lambda) \to X, \quad G_k(h) = \int_S h g_k d\lambda,$$

are continuous and uniformly bounded since

$$\|G_k(h)\| \le \|g_k\|_1 \|h\|_\infty \le M \|h\|_\infty.$$

Now $\lim \int_S h g_k d\lambda$ exists for every simple function h and the simple functions are dense in $L^\infty(\lambda)$ so $\lim \int_S h g_k d\lambda$ exists for every $h \in L^\infty(\lambda)$ by the uniform boundedness (equicontinuity) of the $\{G_k\}$ ([DS] II.1.18). Hence, K is $w(L^1(\lambda, X), L^\infty(\lambda))$ conditionally sequentially compact.

Assume the Radon–Nikodym property. Set

$$m(A) = \lim \int_A g_k d\lambda \ for \ A \in \Sigma.$$

Then m is countably additive by the Nikodym Convergence Theorem (2.36) and is λ continuous by the Vitali–Hahn–Saks Theorem (2.44) and we claim that m has bounded variation. Let $\{A_j : j = 1, ..., n\} \subset \Sigma$ be a partition of S. Then

$$\sum_{j=1}^n \|m(A_j)\| = \lim_k \sum_{j=1}^n \left\| \int_{A_j} g_k d\lambda \right\| \le \limsup_k \sum_{j=1}^n \int_{A_j} \|g_k(s)\| \, d\lambda(s) \le M$$

so the claim is justified. By the Radon–Nikodym property there exists $g \in L^1(\lambda, X)$ such that

$$m(A) = \int_A g d\lambda$$

for $A \in \Sigma$. By what was established above $g_k \to g$ in $w(L^1(\lambda, X), L^\infty(\lambda))$ and K is $w(L^1(\lambda, X), L^\infty(\lambda))$ relatively sequentially compact. $\qquad\square$

For the scalar version of these results for $L^1(\lambda)$, see [DS] IV.8.9. The proof of IV.8.9 of [DS] shows that in the scalar case the assumption that Σ is generated by a countable algebra can be omitted.

The proof of Theorem 9 also gives analogues of Theorems IV.8.6 and IV.8.7 of [DS]. Namely, we have:

Let $\{f_k\} \subset L^1(\lambda, X)$ be bounded. Then

- $\{f_k\}$ is $w(L^1(\lambda, X), L^\infty(\lambda))$ Cauchy iff $\lim_k \int_A f_k d\lambda = m(A)$ exists for every $A \in \Sigma$,
- $\{f_k\}$ is $w(L^1(\lambda, X), L^\infty(\lambda))$ convergent to $f \in L^1(\lambda, X)$ iff

$$\lim_k \int_A f_k d\lambda = \int_A f d\lambda$$

for every $A \in \Sigma$,

- if X has the Radon–Nikodym Property and $\{f_k\}$ is $w(L^1(\lambda, X), L^\infty(\lambda))$ Cauchy, then there exists $f \in L^1(\lambda, X)$ such that $f_k \to f$ with respect to $w(L^1(\lambda, X), L^\infty(\lambda))$.

Next, we consider abstract triples involving the Pettis integral. A function $f : S \to X$ is *Pettis integrable* with respect to λ if the function $x'f$ is λ

integrable for every $x' \in X'$ and for every $A \in \Sigma$ there exists $x_A \in X$ such that

$$\int_A x' f d\lambda = x'(x_A);$$

x_A is called the *Pettis integral* of f and is denoted by

$$\int_A f d\lambda.$$

For the properties of the Pettis integral, see Chapter 1. In particular, the indefinite integral of a Pettis integrable function is countably additive (2.11). Let

$$P(\lambda, X)$$

be the space of Pettis integrable functions; this space has two equivalent norms,

$$\|f\|_P = \sup \left\{ \left| \int_S |x' f| \, d\lambda \right| : \|x'\| \le 1 \right\}$$

and

$$\|f\|'_P = \sup \left\{ \left\| \int_A f d\lambda \right\| : A \in \Sigma \right\}$$

which in general are not complete (see Chapter 1). Consider the abstract triples

$$(P(\lambda, X), S(\Sigma) : X), \quad (P(\lambda, X), B(\Sigma) : X)$$

under the integral map $(f, g) \to \int_S g f d\lambda$ (1.14). The analogue of Proposition 7 for this triple is established as in the case of the Bochner integral above; the proof uses the norm countable additivity of the Pettis integral so the Nikodym Convergence Theorem can be applied.

Proposition 9.10. *Let $K \subset P(\lambda, X)$ and suppose K is $w(P(\lambda, X), S(\Sigma))$ conditionally sequentially compact. Then*

$$\left\{ \int f d\lambda : f \in K \right\}$$

is uniformly countably additive.

Proposition 9.11. *Let $K \subset P(\lambda, X)$ and suppose K is $w(P(\lambda, X), S(\Sigma))$ conditionally sequentially compact. Then for every $A \in \Sigma$,*

$$K_A = \left\{ \int_A f d\lambda : f \in K \right\}$$

is $\|\cdot\|$ conditionally sequentially compact. If K is $w(P(\lambda, X), B(\Sigma))$ conditionally sequentially compact, then

$$K_h = \left\{ \int_S h f d\lambda : f \in K \right\}$$

is $\|\cdot\|$ conditionally sequentially compact for every $h \in B(\Sigma)$.

Proof. The linear map

$$H : P(\lambda, X) \to X, \quad f \to \int_S hf d\lambda,$$

is $w(P(\lambda, X), B(\Sigma)) - \|\cdot\|$ continuous so the second statement follows and similarly for the first statement. $\square$

The conclusions in Propositions 10 and 11 are necessary conditions for weak conditional sequential compactness. As in Theorem 9 we consider sufficient conditions.

Theorem 9.12. *Let $K \subset P(\lambda, X)$. Assume that Σ is generated by a countable algebra $\mathcal{A}$. If*

$$\left\{ \int f d\lambda : f \in K \right\}$$

is uniformly countably additive and

$$K_h = \left\{ \int_S hf d\lambda : f \in K \right\}$$

is $\|\cdot\|$ conditionally sequentially compact for every $h \in B(\Sigma)$, then K is $w(P(\lambda, X), B(\Sigma))$ conditionally sequentially compact.

Proof. Let $\{f_k\} \subset K$. As in the proof of Theorem 9 we may assume that there exists a subsequence $\{g_k\}$ of $\{f_k\}$ such that $\lim \int_A g_k d\lambda$ exists for every $A \in \Sigma$ (this uses the conditional sequential compactness of $K_A = \{\int_A f d\lambda : f \in K\}$ and the uniform countable additivity).

Now we claim that

$$\lim \int_S hg_k d\lambda$$

exists for every $h \in B(\Sigma)$. Since

$$K_h = \left\{ \int_S hf d\lambda : f \in K \right\}$$

is $\|\cdot\|$ conditionally sequentially compact, every subsequence of $\{g_k\}$ has a subsequence $\{g_{n_k}\}$ such that

$$\lim \int_S hg_{n_k} d\lambda$$

exists. Since X is complete, $\lim \int_S hg_k d\lambda$ exists. Thus, K is $w(P(\lambda, X), B(\Sigma))$ conditionally sequentially compact. $\square$

In the proof of Theorem 12 we have that

$$\lim \int_A g_k d\lambda = m(A)$$

exists for every $A \in \Sigma$, but a lack of Radon–Nikodym Theorems for the Pettis integral means that we do not have a Pettis integrable function g such that $\int g d\lambda = m(\cdot)$ (see [Mu]). This is the reason for the condition sequential compactness statement instead of a relative sequential compactness statement in the theorem.

- The methods above can be used to show that a sequence $\{f_k\}$ in $P(\lambda, X)$ is $w(P(\lambda, X), B(\Sigma))$ convergent to 0 iff $\lim_k \int_A f_k d\lambda = 0$ for every $A \in \Sigma$. Indeed, if this condition is satisfied and $h \in B(\Sigma)$, let $\epsilon > 0$ and pick a simple function g such that $\|h - g\|_\infty < \epsilon$. There exists n such that $k \geq n$ implies

$$\left\| \int_S g f_k d\lambda \right\| < \epsilon.$$

Then if $k \geq n$,

$$\left\| \int_S h f_k d\lambda \right\| \leq \left\| \int_S (h - g) f_k d\lambda \right\| + \left\| \int_S g f_k d\lambda \right\| \leq \|h - g\|_\infty \|f_k\| + \epsilon$$

and

$$\sup_k \|f_k\| < \infty$$

by the Nikodym Boundedness Theorem (see Chapter1 for the equivalent norms on $P(\lambda, X)$). The other implication is clear.

We can obtain similar results for the space $L^1(m)$ of scalar functions which are integrable with respect to a countably additive set function $m : \Sigma \to X$, where X is a Banach space; see Chapter 1. The space $L^1(m)$ has the norm

$$\|f\| = \sup \left\{ \int_S |f| \, d \, |x'm| : \|x'\| \leq 1 \right\}$$

(or the equivalent norm

$$\|f\|' = \sup \left\{ \left\| \int_A f dm \right\| : A \in \Sigma \right\};$$

see 1.30). Consider the triples

$$(L^1(m), B(\Sigma) : X), \ (L^1(m), S(\Sigma) : X)$$

under the integration map $(f, g) \to \int_S f g \, dm$ (1.31). As in Proposition 10 since the indefinite integrals $\int f \, dm$ are countably additive (2.12), we have

Proposition 9.13. *Let $K \subset L^1(m)$ be $w(L^1(m), S(\Sigma))$ conditionally sequentially compact. Then*

$$\left\{ \int_{\cdot} f dm : f \in K \right\}$$

is uniformly countably additive.

We also have the analogue of Proposition 11.

Proposition 9.14. *Let $K \subset L^1(m)$ and suppose K is $w(L^1(m), S(\Sigma))$ conditionally sequentially compact. Then for every $A \in \Sigma$,*

$$K_A = \left\{ \int_A f d\lambda : f \in K \right\}$$

is $\|\cdot\|$ conditionally sequentially compact. If K is $w(L^1(m), B(\Sigma))$ conditionally sequentially compact, then

$$K_h = \left\{ \int_S h f d\lambda : f \in K \right\}$$

is $\|\cdot\|$ conditionally sequentially compact for every $h \in B(\Sigma)$.

Proof. As before the linear map

$$H : L^1(m) \to X, \quad f \to \int_S h f dm,$$

is

$$w(L^1(m), B(\Sigma)) - \|\cdot\|$$

continuous. □

As was the case for the Pettis integral, we have

Theorem 9.15. *Let $K \subset L^1(m)$. Assume that Σ is generated by a countable algebra $\mathcal{A}$. If*

$$\left\{ \int_{\cdot} f d\lambda : f \in K \right\}$$

is uniformly countably additive and

$$K_h = \left\{ \int_S h f d\lambda : f \in K \right\}$$

is $\|\cdot\|$ conditionally sequentially compact for every $h \in B(\Sigma)$, then K is $w(L^1(m), B(\Sigma))$ conditionally sequentially compact.

- As was the case for the Pettis integral, a sequence $\{f_k\} \subset L^1(m)$ is $w(L^1(m), B(\Sigma))$ convergent to 0 iff $\lim_k \int_A f_k dm = 0$ for every $A \in \Sigma$. Indeed, if this condition holds, let $h \in B(\Sigma)$. Let $\epsilon > 0$ and pick a simple function g such that $\|g - h\|_\infty < \epsilon$. There exists n such that

$$\left\| \int_S g f_k dm \right\| < \epsilon$$

for $k \geq n$. If $k \geq n$,

$$\left\| \int_S h f_k dm \right\| \leq \left\| \int_S (h - g) f_k dm \right\| + \left\| \int_S g f_k dm \right\|$$

$$\leq \sup \left\{ \left| \int_S (h - g) f_k dx' m \right| : \|x'\| \leq 1 \right\} + \epsilon$$

$$\leq \sup \left\{ \int_S |(h - g) f_k| \, d\,|x'm| : \|x'\| \leq 1 \right\} + \epsilon$$

$$\leq \|g - h\|_\infty \sup \left\{ \int_S |f_k| \, d\,|x'm| : \|x'\| \leq 1 \right\} + \epsilon$$

$$= \|g - h\|_\infty \|f_k\| + \epsilon$$

and

$$\sup_k \|f_k\| < \infty$$

by the Nikodym Boundedness Theorem (see Theorem 1.30 for equivalent norms on $L^1(m)$).

We consider triples involving vector valued sequence spaces and their β duals. Let E be a vector space of X valued sequences which contains the space $c_{00}(X)$ of sequences which are eventually 0. Let Y be a Banach space and $L(X, Y)$ the space of continuous linear operators from X into Y with the strong operator topology τ. The β dual of E with respect to Y is defined to be

$$E^{\beta Y} = \left\{ \{T_j\} \subset L(X, Y) : \sum_{j=1}^\infty T_j x_j \text{ converges for every } x = \{x_j\} \in E \right\}.$$

We write

$$T \cdot x = \sum_{j=1}^\infty T_j x_j$$

when $T = \{T_j\} \in E^{\beta Y}$ and $x = \{x_j\} \in E$. Consider the triple

$$(E^{\beta Y}, E : Y)$$

under the bilinear map $(T, x) \to T \cdot x$.

We consider the analogues of Propositions 3 and 4 for this triple.

Proposition 9.16. *Suppose* $K \subset E^{\beta Y}$ *is* $w(E^{\beta Y}, E)$ *relatively sequentially compact. Then for every* j,

(a) $\{T_j : T = \{T_k\} \in K\}$ *is* τ *relatively sequentially compact.*

Proof. The linear map

$$G_j : E^{\beta Y} \to L(X, Y), \quad T \to T_j,$$

is $w(E^{\beta Y}, E) - \tau$ continuous so the result follows. $\qquad\square$

The space E has the signed weak gliding hump property (signed-WGHP) if for every $x = \{x_j\} \in E$ and every increasing sequence of intervals $\{I_j\}$ there exist a subsequence $\{n_j\}$ and a sequence of signs $\{s_j\}$ such that the coordinate sum of the series

$$\sum_{j=1}^{\infty} s_j \chi_{I_{n_j}} x_j \in E,$$

where χ_I is the characteristic function of I and $\chi_I x$ is the coordinate product of χ_I and x. See Chapter 8 and Appendix B. From Theorem 8.33, we have

Proposition 9.17. *Assume* E *has signed-WGHP and* $K \subset E^{\beta Y}$ *is* $w(E^{\beta Y}, E)$ *conditionally sequentially compact. Then for every* $x \in E$ *the series*

$$(b) \sum_{j=1}^{\infty} T_j x_j \text{ converge uniformly in } Y \text{ for } T \in K.$$

Conditions (a) and (b) are necessary conditions for relative sequential compactness. We now show that they are sufficient as in Theorem 5.

Theorem 9.18. *Let* $K \subset E^{\beta Y}$ *satisfy conditions* (a) *and* (b). *Then* K *is* $w(E^{\beta Y}, E)$ *relatively sequentially compact.*

Proof. Let $\{T^k\} \in K$. By (a) and the diagonalization method ([Ke] p.238), there exists a subsequence $\{T^{n_k}\}$ such that

$$\tau - \lim_k T_j^{n_k} = T_j \in L(X, Y)$$

for every j (Banach–Steinhaus Theorem). Let U be a closed neighborhood of 0 in Y and $x \in E$. By (b), there exists N such that

$$\sum_{j=n}^{p} T_j^{n_k} x_j \in U$$

for $p > n \geq N$, $k \in \mathbb{N}$. Then

$$\sum_{j=n}^{p} T_j x_j \in U$$

for $p > n \geq N$ so the series $\sum_{j=1}^{\infty} T_j x_j$ converges or $T = \{T_j\} \in E^{\beta Y}$. Now, if $x \in E$,

$$(\&) \ (T^{n_k} - T) \cdot x = \sum_{j=1}^{M} (T_j^{n_k} - T_j) x_j + \sum_{j=M+1}^{\infty} (T_j^{n_k} - T_j) x_j.$$

There exists M such that the last term on the right hand side of (&) will belong to U for every k by (b). With this M fixed the first term on the right hand side of (&) will belong to U for large k. Hence, $T^{n_k} \to T$ with respect to τ and K is $w(E^{\beta Y}, E)$ relatively sequentially compact. $\qquad \square$

- Note that the proof above also shows that a sequence $\{T^k\}$ in $E^{\beta Y}$ is $w(E^{\beta Y}, E)$ convergent to $T \in E^{\beta Y}$ iff $\lim_k T_j^k x = T_j x$ for every $x \in X$, $j \in \mathbb{N}$ and for every $x = \{x_j\} \in E$, the series $\sum_{j=1}^{\infty} T_j^k x_j$ converge uniformly for $k \in \mathbb{N}$. If $\{T^k\}$ is $w(E^{\beta Y}, E)$ Cauchy, then $\lim_k T_j^k x = T_j x$ exists for every j and $x \in X$. Then T_j is linear and continuous by the Banach–Steinhaus Theorem and $T = \{T_j\} \in E^{\beta Y}$ by the proof of the theorem above. Then $T^k \to T$ with respect to $w(E^{\beta Y}, E)$. Thus, $w(E^{\beta Y}, E)$ is sequentially complete when E has signed-WGHP and X, Y are Banach spaces (see [St1],[St2], [Sw4]11.18).

We consider the case of multiplier convergent series. Let G be a LCTVS and λ a K-space with $\Lambda \subset \lambda$ having the signed-SGHP (Appendix B). Let

$$\Lambda(G) = \Lambda^{\beta G}$$

be the space of all G valued, Λ multiplier convergent series with the locally convex topology, $\tau_{\Lambda(G)}$, generated by the semi-norms

$$\widehat{p_B}(\{x_j\}) = \sup \left\{ p\left(\sum_{j=1}^{\infty} t_j x_j \right) : \{t_j\} \in B \right\}$$

when p runs through the continuous semi-norms p generating the topology of G and B runs through the bounded subsets of Λ. It follows from the corollary below that each $\widehat{p_B}(\{x_j\}) < \infty$.

Consider the triple

$$(\Lambda(G), \Lambda : G)$$

under the map $(x, t) \to \sum_{j=1}^{\infty} t_j x_j = x \cdot t$. We consider weak compactness for this triple.

We assume that Λ has signed-SGHP.

Proposition 9.19. *If* $K \subset \Lambda(G)$ *is* $w(\Lambda(G), \Lambda)$ *sequentially conditionally compact, then the series*

$$\sum_{j=1}^{\infty} t_j x_j$$

converge uniformly for t *belonging to bounded subsets of* Λ *and* $x = \{x_j\} \in K$.

Proof. If the conclusion fails to hold, there exist a bounded set $B \subset \Lambda$ and a neighborhood, U, of 0 in G such that for every k there exist $t^k \in B, x^k \in K$ and an interval I_k with $\min I_k > k$ and

$$\sum_{j \in I_k} t_j^k x_j^k \notin U.$$

For $k = 1$ we have $t^1 \in B, x^1 \in K$ and I_1 with $\min I_1 > 1$ and

$$\sum_{j \in I_1} t_j^1 x_j^1 \notin U.$$

Put $N_1 = \max I_1$. Then there exist $t^2 \in B, x^2 \in K$ and I_2 with $\min I_2 > N_1$ and

$$\sum_{j \in I_2} t_j^2 x_j^2 \notin U.$$

Continuing this construction produces $\{t^k\} \subset B, \{x^k\} \subset K$ and an increasing sequence of intervals $\{I_k\}$ with

$$(\&) \quad \sum_{j \in I_k} t_j^k x_j^k \notin U.$$

There exists a subsequence $\{x^{n_k}\}$ such that

$$\lim_k x^{n_k} \cdot t$$

exists for every $t \in \Lambda$. The Hahn–Schur Theorem (4.35) implies that the series

$$\sum_{j=1}^{\infty} t_j^{n_k} x_j^{n_k}$$

converge uniformly for $k \in \mathbb{N}$. This contradicts (&).   □

Corollary 9.20. *If $x \in \Lambda(G)$ and $B \subset \Lambda$ is bounded, then*

$$\left\{ \sum_{j=1}^{\infty} t_j x_j : t \in B \right\}$$

is bounded.

Proof. Let U be a balanced neighborhood of 0 on G. Pick a balanced neighborhood, V, such that $V + V \subset U$. By the proposition above there exists N such that

$$\sum_{j=N}^{\infty} t_j x_j \in V$$

for $t \in B$. Since each $\{t_j x_j : t \in B\}$ is bounded, there exists $s > 1$ with

$$\sum_{j=1}^{N-1} t_j x_j \in sV$$

for $t \in B$. Then

$$\sum_{j=1}^{\infty} t_j x_j = \sum_{j=1}^{N-1} t_j x_j + \sum_{j=N}^{\infty} t_j x_j \in sV + V \subset sU$$

for $t \in B$.   □

Thus, each $\widehat{p_B}(\{x_j\}) < \infty$.

Proposition 9.21. *If $K \subset \Lambda(G)$ is $w(\Lambda(G), \Lambda)$ sequentially conditionally (relatively) compact, then for each j the set $\{x_j : x \in K\}$ is sequentially conditionally (relatively) compact.*

Proof. The linear map

$$F_j : \Lambda(G) \to G, \ x = \{x_k\} \to x_j$$

is $w(\Lambda(G), \Lambda) - G$ sequentially continuous.   □

The propositions above give necessary conditions for weak compactness. We now consider sufficient conditions.

Theorem 9.22. *Let G be sequentially complete and $K \subset \Lambda(G)$. If the series $\sum_{j=1}^{\infty} t_j x_j$ converge uniformly for t belonging to bounded subsets of Λ, $x \in K$ and for each j the set $\{x_j : x \in K\}$ is sequentially relatively compact, then K is $w(\Lambda(G), \Lambda)$ sequentially relatively compact.*

Proof. Let $\{x^k\} \subset K$. For each j the sequence $\{x_j^k : k\}$ has a subsequence which converges. The diagonalization method ([Ke] p.238) gives a subsequence $\{n_k\}$ such that

$$\lim_k x_j^{n_k} = x_j$$

exists for each j. The converse of the Hahn–Schur Theorem (4.39) implies that $\{x_j\} \in \Lambda(G)$ and

$$\lim_k \sum_{j=1}^{\infty} t_j x_j^{n_k} = \sum_{j=1}^{\infty} t_j x_j$$

for each $t \in \Lambda$. $\qquad \square$

Remark 9.23. If Λ has signed-SGHP, then by the Hahn–Schur Theorem (4.35) a sequence in $\Lambda(G)$ is $w(\Lambda(G), \Lambda)$ convergent iff the sequence is $\tau_{\Lambda(G)}$ convergent so a subset of $\Lambda(G)$ is $w(\Lambda(G), \Lambda)$ sequentially relatively compact iff the set is $\tau_{\Lambda(G)}$ sequentially relatively compact.

The results and remark above apply to the spaces of subseries and bounded multiplier convergent series since both $\Lambda = \{\chi_\sigma : \sigma \subset \mathbb{N}\} \subset m_0$ and $\Lambda = l^\infty$ have SGHP.

If $\lambda = \Lambda = l^\infty$ and

$$bmc(G) = (l^\infty)^{\beta G}$$

is the space of G valued, bounded multiplier convergent series, we have the following corollaries for the triple

$$(bmc(G), l^\infty : G)$$

under the map $(x, t) \to x \cdot t$.

Corollary 9.24. *Suppose $K \subset bmc(G)$ is $w(bmc(G), l^\infty)$ sequentially conditionally (relatively) compact. Then the series $\sum_{j=1}^{\infty} t_j x_j$ converge uniformly for $x \in K$, $\|\{t_j\}\|_\infty \leq 1$ and for each j the set $\{x_j : x \in K\}$ is sequentially conditionally (relatively) compact.*

Corollary 9.25. *Assume G is sequentially complete. If $K \subset bmc(G)$ is such that the series $\sum_{j=1}^{\infty} t_j x_j$ converge uniformly for $x \in K$, $\|\{t_j\}\|_{\infty} \leq 1$ and for each j the set $\{x_j : x \in K\}$ is sequentially relatively compact, then K is $w(bmc(G), l^{\infty})$ sequentially relatively compact.*

Similarly, if $\lambda = m_0$, $\Lambda = \{\chi_\sigma : \sigma \subset \mathbb{N}\}$ and

$$ss(G) = m_0^{\beta G}$$

is the space of G valued, subseries convergent series, we consider the triple

$$(ss(G), \{\chi_\sigma : \sigma \subset \mathbb{N}\} : G)$$

under the map $(x, t) \to x \cdot t$ and we have the corollaries.

Corollary 9.26. *Suppose $K \subset ss(G)$ is $w(ss(G), \{\chi_\sigma : \sigma \subset \mathbb{N}\})$ sequentially conditionally (relatively) compact. Then the series $\sum_{j \in \sigma} t_j x_j$ converge uniformly for $x \in K$, $\sigma \subset \mathbb{N}$ and for each j the set $\{x_j : x \in K\}$ is sequentially conditionally (relatively) compact.*

Corollary 9.27. *Assume G is sequentially complete. If $K \subset ss(G)$ is such that the series $\sum_{j \in \sigma} t_j x_j$ converge uniformly for $x \in K$, $\sigma \subset \mathbb{N}$ and for each j the set $\{x_j : x \in K\}$ is sequentially relatively compact, then K is*

$$w(ss(G), \{\chi_\sigma : \sigma \subset \mathbb{N}\})$$

sequentially relatively compact.

By the remark above a subset of $bmc(G)$ ($ss(G)$) is

$$w(bmc(G), l^{\infty}) \quad (w(ss(G), \{\chi_\sigma : \sigma \subset \mathbb{N}\}))$$

sequentially relatively compact iff the subset is $\tau_{bmc(G)}$ ($\tau_{ss(G)}$) sequentially relatively compact.

Appendix A

Topology

In this appendix we will record some of the results pertaining to topological groups and vector spaces (TVS) which will be used throughout the text. For convenience we assume that all vector spaces are real.

An *Abelian topological group* is an Abelian group which has a topology under which the algebraic operations of addition and inversion are continuous. A *quasi-norm* on an Abelian group G is a function $|\cdot| : G \to [0, \infty)$ satisfying $|0| = 0$, $|x + y| \leq |x| + |y|$ for $x, y \in G$, $|x| = |-x|$ for $x \in G$. A quasi-norm induces a semi-metric d on G defined by $d(x, y) = |x - y|$ which is *translation invariant* in the sense that $d(z + x, z + y) = d(x, y)$ for $x, y, z \in G$. The semi-metric d is a metric iff the quasi-norm $|\cdot|$ is *total*, i.e., $|x| = 0$ iff $x = 0$. An Abelian topological group whose topology is induced by a quasi-norm is called a *quasi-normed group*.

It is an interesting and useful fact due to Burzyk and P. Mikusinski that the topology of any Abelian topological group is always induced by a family of quasi-norms ([BM]).

A *topological vector space* (TVS) is a vector space X supplied with a topology τ such that the operations of addition and scalar multiplication are continuous with respect to τ. A subset U of a TVS X is *symmetric (balanced)* if $x \in U$ implies $-x \in U$ ($x \in U$ implies $tx \in U$ for $|t| \leq 1$). Any TVS has a neighborhood base at 0 which consists of symmetric (balanced, closed) sets. See [Sch], [Sw2] or [Wi2] for discussions of TVS.

A (semi-) *metric linear space* is a TVS whose topology is induced by a translation invariant (semi-) metric. We establish a useful property of metric linear spaces which will be used later. A null sequence $\{x_j\}$ in a TVS E is *Mackey convergent* if there exist scalars $t_j \to \infty$ such that $t_j x_j \to 0$. A TVS E is a *braked* space if every null sequence is Mackey convergent. Every metric linear space is a braked space.

Theorem A.1. *If E is a metric linear space, then E is a braked space.*

Proof. Let $\{U_k : k \in \mathbb{N}\}$ be a base at 0 of balanced sets with $U_{k+1} \subset U_k$ for all k and let $x_k \to 0$ in E. There exist n_1 such that $x_j \in U_1$ for $j \geq n_1$. There exists $n_2 > n_1$ such that $x_j \in (1/2)U_2$ for $j \geq n_2$. Continuing, there exists an increasing sequence $\{n_k\}$ such that

$$x_j \in (1/k)U_k$$

for $j \geq n_k$. Define $t_j \uparrow \infty$ by $t_j = k$ if $n_k \leq j < n_{k+1}$. Then $t_j x_j \to 0$ since if given k and $j \geq n_k$, $t_j x_j \in U_k$. $\qquad\qquad\square$

A TVS X is *locally convex (LCTVS)* if X has a neighborhood base at 0 consisting of convex sets. Any LCTVS also has a base at 0 consisting of closed, absolutely convex sets. The topology τ of any LCTVS is generated by a family of semi-norms $\{p_a : a \in A\}$. See [Sch], [Sw2] or [Wi2] for the basic properties of LCTVS.

We now give a description of polar topologies which will play an important role when we discuss Orlicz–Pettis Theorems. A pair of vector spaces X, X' are said to be in *duality* if there is a bilinear map $\langle \cdot, \cdot \rangle : X' \times X \to \mathbb{R}$ such that

(i) $\{\langle \cdot, x \rangle : x \in X, x \neq 0\}$ separates the points of X and

(ii) $\{\langle x', \cdot \rangle : x' \in X', x' \neq 0\}$ separates the points of X'.

If X, X' are in duality, the weak topology of X (X'), $\sigma(X, X')$ $(\sigma(X'X))$, is the locally convex vector topology generated by the semi-norms $p(x) = |< x', x >|$, $x' \in X'$ $(p(x') = |< x', x >|)$, $x \in X$. A subset $A \subset X$ is $\sigma(X, X')$ *bounded* iff

$$\sup\{|< x', x >| : x \in A\} < \infty$$

for every $x' \in X'$.

Let $\mathcal{A}$ be a family of $\sigma(X', X)$ bounded subsets of X'. For $A \in \mathcal{A}$, set

$$p_A(x) = \sup\{|\langle x, x' \rangle| : x' \in A\}.$$

The semi-norms $\{p_A : A \in \mathcal{A}\}$ generate a locally convex topology, $\tau_{\mathcal{A}}$, on X called the *polar topology* of uniform convergence on $\mathcal{A}$ (for the reason the topology is called a polar topology, see [Sw2] 17). Thus, a net $\{x^\delta\}$ converges to 0 in $\tau_{\mathcal{A}}$ iff $\langle x', x^\delta \rangle \to 0$ uniformly for $x' \in A$ for every $A \in \mathcal{A}$. A subset $B \subset X$ is $\tau_{\mathcal{A}}$ bounded iff

$$\sup\{|\langle x', x \rangle| : x' \in A, x \in B\} < \infty$$

for every $A \in \mathcal{A}$.

We will use the following polar topologies in the text.

(1) The *weak topology* $\sigma(X, X')$ is generated by the family $\mathcal{A}$ of all finite subsets of X'.

(2) The *strong topology* of X, denoted by $\beta(X, X')$, is generated by the family of all $\sigma(X', X)$ bounded subsets of X'.

(3) The *Mackey topology*, denoted by $\tau(X, X')$, is generated by the family of all absolutely convex, $\sigma(X', X)$ compact subsets of X'.

(4) The polar topology generated by the family of all $\sigma(X', X)$ compact subsets of X' is denoted by $\lambda(X, X')$.

(5) A subset $A \subset X'$ is said to be *conditionally* $\sigma(X', X)$ *sequentially compact* if every sequence $\{x'_j\} \subset A$ has a subsequence $\{x'_{n_j}\}$ which is $\sigma(X', X)$ Cauchy, i.e., $\lim \langle x'_{n_j}, x \rangle$ exists for every $x \in X$. The polar topology generated by the family of conditionally $\sigma(X', X)$ sequentially compact sets is denoted by $\gamma(X, X')$.

(6) $\beta^*(X, X')$ is the topology of uniform convergence on the $\beta(X', X)$ bounded subsets of X'.

The topology $\lambda(X, X')$ was introduced by G. Bennett and Kalton ([BK]) and is obviously stronger than the Mackey topology $\tau(X, X')$; it can be strictly stronger ([Köl] 21.4).

Let $w(X, X')$ be a polar topology defined for all dual pairs X, X'. We have the following useful notion introduced by Wilansky ([Wi2]).

Definition A.2. Let Y, Y' be a dual pair. The topology $w(\cdot, \cdot)$ is a Hellinger–Toeplitz topology if whenever

$$T : (X, \sigma(X, X')) \to (Y, \sigma(Y, Y'))$$

is linear and continuous, then

$$T : (X, w(X, X')) \to (Y, w(Y, Y'))$$

is continuous.

Wilansky has given a very useful criterion for Hellinger–Toeplitz topologies ([Wi2] 11.2.2).

If $T : (X, \sigma(X, X')) \to (Y, \sigma(Y, Y'))$ is linear and continuous, then the *adjoint (transpose)* operator of T is the linear operator $T' : Y' \to X'$ defined by

$$\langle T'y', x \rangle = \langle y', Tx \rangle$$

for $x \in X, y' \in Y'$. The adjoint T' is $\sigma(Y', Y) - \sigma(X', X)$ continuous.

Let $\mathcal{A}(X', X)$ be a family of $\sigma(X', X)$ bounded subsets which is defined for all dual pairs X, X'. Let $w(X, X')$ be the polar topology generated by the elements of $\mathcal{A}(X', X)$. We have

Theorem A.3. *The topology $w(X, X')$ is a Hellinger–Toeplitz topology if whenever $T : (X, \sigma(X, X')) \to (Y, \sigma(Y, Y'))$ is linear and continuous, then*

$$T'\mathcal{A}(Y', Y) \subset \mathcal{A}(X', X).$$

Proof. Let $\{x^\delta\}$ be a net in X which converges to 0 in $w(X, X')$. Let $A \in \mathcal{A}(Y', Y)$. Then $\{T'y' : y' \in A\} \in \mathcal{A}(X', X)$ so

$$\langle y', Tx^\delta \rangle = \langle T'y', x^\delta \rangle \to 0$$

uniformly for $y' \in A$. That is, $Tx^\delta \to 0$ in $w(Y, Y')$. $\qquad\square$

Theorem 3 clearly implies that the polar topologies given in (1)–(5) are all Hellinger–Toeplitz topologies.

We will encounter spaces with a Schauder basis at various places in the text. A sequence $\{b_j\}$ is a *Schauder basis* for a TVS E if each $x \in E$ has a unique series expansion

$$x = \sum_{j=1}^{\infty} t_j b_j.$$

The linear functionals f_j on E defined by $f_j(x) = t_j$ are called the *coordinate functionals* associated with the basis $\{b_j\}$; if E is a complete metric linear space, the coordinate functionals are continuous ([Sw2] 10.1.13) but not in general.

A result which will be used is the Iterated Limit Theorem.

Let D_1, D_2 be directed sets and partially order $D_1 \times D_2$ by $(d_1, d_2) \leq (d_1', d_2')$ iff $d_1 \leq d_1', d_2 \leq d_2'$ so $D_1 \times D_2$ is a directed set. Let (X, d) be a metric space.

Theorem A.4. (Iterated Limit) *Let $f : D_1 \times D_2 \to X$ be a net. Assume*

(i) *for each $d_2 \in D_2$, $\lim\limits_{D_1} f(d_1, d_2) = g(d_2)$ exists, and*

(ii) $\lim\limits_{D_2} f(d_1, d_2) = h(d_1)$ *exists uniformly on D_2.*

Then the 3 limits

$$\lim_{D_2} g(d_2), \quad \lim_{D_1} h(d_1), \quad \lim_{D_1 \times D_2} f(d_1, d_2)$$

all exist and are equal.

See [DS] I.7.6.

Let (X, d) be a metric space and $x_{ij} \in X$ for $i, j \in \mathbb{N}$. The *double limit*,

$$\lim_{i,j} x_{ij},$$

exists and equals x if for every $\epsilon > 0$, there exists N such that $i \geq N, j \geq N$ implies $d(x_{ij}, x) < \epsilon$. From the result above we have the following well known criteria for double limits.

Theorem A.5. (Iterated Limit) *Assume*

(i) *for every j, $\lim\limits_{i} x_{ij} = x_j$ exists and*

(ii) $\lim\limits_{j} x_{ij} = y_i$ *converge uniformly for $i \in \mathbb{N}$.*

Then the 3 limits

$$\lim_{i} \lim_{j} x_{ij} = \lim_{i} y_i, \quad \lim_{j} \lim_{i} x_{ij} = \lim_{j} x_j, \quad \lim_{i,j} x_{ij}$$

exist and are equal.

Appendix B

Sequence Spaces

In this appendix we will list the sequence spaces and their properties which will be used in the text. If λ is a vector space of (real) sequences containing c_{00}, the space of all sequences which are eventually 0, the β-dual of λ is defined to be

$$\lambda^\beta = \left\{ s = \{s_j\} : \sum_{j=1}^{\infty} s_j t_j = \{s_j\} \cdot \{t_j\} = s \cdot t \text{ converges for every } t = \{t_j\} \in \lambda \right\}.$$

Since $\lambda \supset c_{00}$, the pair λ, λ^β are in duality with respect to the pairing

$$s \cdot t = \{s_j\} \cdot \{t_j\}$$

for $s \in \lambda^\beta, t \in \lambda$.

We now list some of the scalar valued sequence spaces which will be encountered in the text.

- $c_{00} = \{\{t_j\} : t_j = 0 \text{ eventually}\}$
- $c_0 = \{\{t_j\} : \lim t_j = 0\}$
- $c_c = \{\{t_j\} : t_j \text{ is eventually constant}\}$
- $c = \{\{t_j\} : \lim t_j \text{ exists}\}$
- $m_0 = \{\{t_j\} : \text{the range of } \{t_j\} \text{ is finite}\} = \text{span}\{\chi_\sigma : \sigma \subset \mathbb{N}\}$
- $l^\infty = \{\{t_j\} : \sup_j\{|t_j|\} = \|\{t_j\}\|_\infty < \infty\}$

All of the sequence spaces above are usually equipped with the sup-norm, $\|\cdot\|_\infty$, defined above.

For $0 < p < 1$,

- $l^p = \{\{t_j\} : \sum_{j=1}^{\infty} |t_j|^p = |\{t_j\}|_p < \infty\}$

The space l^p $(0 < p < 1)$ is usually equipped with the quasi-norm $|\cdot|_p$ which generates the metric $d_p(s,t) = |s - t|_p$ under which it is complete.

For $1 \leq p < \infty$,

- $l^p = \{\{t_j\} : (\sum_{j=1}^{\infty} |t_j|^p)^{1/p} = \|\{t_j\}\|_p < \infty\}$

The space l^p is usually equipped with the norm $\|\cdot\|_p$ under which it is a Banach space.

- $bs = \left\{ \{t_j\} : \sup_n \left\{ \left| \sum_{j=1}^n t_j \right| = \|t\|_{bs} < \infty \right\} \right\}$

The space bs is called the space of *bounded series* and is usually equipped with the norm $\|\cdot\|_{bs}$ under which it is a Banach space.

- $cs = \{\{t_j\} : \sum_{j=1}^{\infty} t_j \text{ converges}\}$

The space cs is a subspace of bs and is called the space of *convergent series*; cs is a closed subspace of bs under the norm $\|\cdot\|_{bs}$.

A list of sequence spaces and their β-duals can be found in [KG].

We now list some of the properties of sequence spaces which will be encountered in the sequel. Throughout the remainder of this appendix λ will denote a sequence space containing c_{00}. Suppose that λ is equipped with a Hausdorff vector topology.

Definition B.1. The space λ is a K-space if the coordinate functionals

$$t = \{t_j\} \to t_j$$

are continuous from λ into $\mathbb{R}$ for every j. If the K-space λ is a Banach (Frechet) space, λ is called a BK-space (FK-space).

All of the spaces listed above are K-spaces under their natural topologies.

Let e^j be the sequence with a 1 in the jth coordinate and 0 in the other coordinates.

Definition B.2. The K-space λ is an AK-space if the $\{e^j\}$ form a Schauder basis for λ, i.e., if $t = \{t_j\} \in \lambda$, then

$$t = \lim_n \sum_{j=1}^n t_j e^j,$$

where the convergence is in λ.

The spaces c_{00}, c_0, l^p $(0 < p < \infty), cs$ are AK-spaces. The spaces m_0 and l^∞ are not AK-spaces.

Throughout this text numerous gliding hump properties are employed. We now list these gliding hump properties and give examples of sequence spaces which satisfy the various gliding hump properties. If $\sigma \subset \mathbb{N}$, χ_σ will denote the characteristic function of σ and if $t = \{t_j\}$ is any sequence (scalar or vector), $\chi_\sigma t$ will denote the coordinatewise product of χ_σ and t.

A sequence space λ is *monotone* if $\chi_\sigma t \in \lambda$ for every $\sigma \subset \mathbb{N}$ and $t \in \lambda$. A sequence space λ is *normal (solid)* if $t \in \lambda$ and $|s_j| \leq |t_j|$ implies that $s = \{s_j\} \in \lambda$. Obviously, a normal space is monotone; the space m_0 is monotone but not normal. The spaces c_{00}, c_0, l^p $(0 < p \leq \infty)$ and s, the space of all sequences, are normal whereas c_c, c, bs, cs are not monotone.

An *interval* in $\mathbb{N}$ is a subset of the form

$$[m, n] = \{j \in \mathbb{N} : m \leq j \leq n\},$$

where $m, n \in \mathbb{N}$ with $m \leq n$. A sequence of intervals $\{I_j\}$ is *increasing* if $\max I_j < \min I_{j+1}$ for every j. A sequence of *signs* is a sequence $\{s_j\}$ with $s_j = \pm 1$ for every j.

We begin with two gliding hump properties which are algebraic and require no topology on the sequence space λ.

Definition B.3. Let $\Lambda \subset \lambda$. Then Λ has the signed weak gliding hump property (signed-WGHP) if for every $t \in \Lambda$ and every increasing sequence of intervals $\{I_j\}$, there is a subsequence $\{n_j\}$ and a sequence of signs $\{s_j\}$ such that the coordinatewise sum of the series

$$\sum_{j=1}^{\infty} s_j \chi_{I_{n_j}} t$$

belongs to Λ. If the signs s_j can all be chosen to be equal to 1 for every $t \in \Lambda$, then Λ has the weak gliding hump property (WGHP).

The weak gliding hump property was introduced by Noll ([No]) and the signed weak gliding hump property was introduced by Stuart ([St1],[St2]). We now give some examples of sequence spaces with WGHP and signed-WGHP.

Example B.4. Any monotone space has WGHP.

We now show that the space cs of convergent series is not monotone but has WGHP.

Example B.5. *cs* has WGHP but is not monotone. Let $t \in cs$ and $\{I_j\}$ be an increasing sequence of intervals. Since the series $\sum_j t_j$ converges,

$$\chi_{I_j \cap J} \cdot t = \sum_{k \in I_j \cap J} t_k \to 0$$

for any interval J. Pick a subsequence $\{n_j\}$ such that

$$\left| \chi_{I_{n_j} \cap J} \cdot t \right| < 1/2^j$$

for every interval J. Then

$$\sum_{j=1}^{\infty} \chi_{I_{n_j}} t \in cs$$

since this series satisfies the Cauchy criterion.

If $t = \{(-1)^j / j\}$, then $t \in cs$, but if $\sigma = \{1, 3, 5, ...\}$, then $\chi_\sigma t \notin cs$ so *cs* is not monotone.

Example B.6. The space c of convergent sequences does not have WGHP. For example, $t = \{1, 1, ...\} \in c$ and if $I_j = \{2j\}$, then $\sum_{j=1}^{\infty} \chi_{I_{n_j}} t \notin c$ for any subsequence $\{n_j\}$.

For further examples of spaces with WGHP, see Appendix B of [Sw4]. Let λ be a K-space.

Definition B.7. Let $\Lambda \subset \lambda$. Then Λ has the signed strong gliding hump property (signed-SGHP) if for every bounded sequence $\{t^j\}$ and every increasing sequence of intervals $\{I_j\}$, there exist a subsequence $\{n_j\}$ and a sequence of signs $\{s_j\}$ such that the coordinate sum of the series

$$\sum_{j=1}^{\infty} s_j \chi_{I_{n_j}} t^{n_j} \in \Lambda.$$

If the signs s_j can be chosen to be equal to 1 for each j, then Λ has the strong gliding hump property (SGHP)

The strong gliding hump property was introduced by Noll ([No]) and the signed strong gliding hump property was introduced in [Sw5]. Note the SGHP depends on the topology of λ.

Example B.8. The space l^∞ has SGHP and the space *bs* has signed-SGHP but not SGHP (see Appendix B of [Sw4] for the proof). The spaces l^p, $0 < p < \infty$, c_0 do not have SGHP (consider $\{e^j\}$ and $I_j = \{j\}$).

Example B.9. The space $\Lambda = \{\chi_\sigma : \sigma \subset \mathbb{N}\}$ has SGHP while $\lambda = m_0 =$ span Λ does not have SGHP.

We give another gliding hump property which depends on the topology of the space.

Definition B.10. The K-space λ has the zero gliding hump property (0-GHP) if whenever $t^j \to 0$ in λ and $\{I_j\}$ is an increasing sequence of intervals, there is a subsequence $\{n_j\}$ such that the coordinate sum of the series

$$\sum_{j=1}^{\infty} \chi_{I_{n_j}} t^{n_j} \in \lambda.$$

The 0-GHP was essentially introduced by Lee Peng Yee ([LPY]); see also [LPYS].

Example B.11. The spaces l^p, $0 < p < \infty$, c_0, c, cs and bs have 0-GHP. The space c_{00} does not have 0-GHP.

The properties 0-GHP and WGHP are independent. The space c has 0-GHP but not WGHP while the space c_{00} has WGHP but not 0-GHP.

Several spaces of vector valued sequences will also be encountered. We give a short list.

Let X be a normed space.

- $c_{00}(X)$ is the space of all X valued sequences which are eventually 0.
- $c_0(X)$ is the space of all X valued sequences which converge to 0.
- $c(X)$ is the space of all X valued sequences which are convergent.
- $l^\infty(X)$ is the space of all X valued sequences which are bounded.

These spaces are usually equipped with the sup-norm,

$$\|\{x_j\}\|_\infty = \sup\{\|x_j\| : j \in \mathbb{N}\}.$$

Let $0 < p < \infty$.

- $l^p(X)$ is the space of all X valued sequences such that

$$\|\{x_j\}\| = \sum_{j=1}^{\infty} \|x_j\|^p < \infty.$$

If $0 < p < 1$, the space $l^p(X)$ is equipped with the quasi-norm $\|\cdot\|$ while if $1 \le p < \infty$, then $l^p(X)$ is equipped with the norm

$$\|\{x_j\}\|_p = \|\{x_j\}\|^{1/p}.$$

Let E be a vector space of X valued sequences which contains $c_{00}(X)$. Then E is a *K-space* if the coordinate maps $\{x_j\} \to x_j$ from E into X are continuous for each j. For $j \in \mathbb{N}$ and $x \in X$ denote the sequence with x in the jth coordinate and 0 in the other coordinates by

$$e^j \otimes x.$$

Then E is an *AK-space* if each $\{x_j\} \in E$ has an expansion

$$\{x_j\} = \sum_{j=1}^{\infty} e^j \otimes x_j,$$

where the series converges in E.

All of the spaces listed above are K-spaces in their natural topologies. The spaces $c(X)$ and $l^\infty(X)$ are not AK-spaces while the other spaces are AK-spaces.

Proofs and further examples can be found in Appendix B of [Sw4].

Appendix C

Boundedness Criterion

In this appendix we will establish a boundedness criterion for a family of finitely additive set functions with values in a semi-convex topological vector space which is pointwise bounded to be uniformly bounded on a σ-algebra. This criterion is often used in proofs of the Nikodym Boundedness Theorem in which gliding hump methods are employed.

A subset U of a vector space E is *semi-convex* if there exists a such that $U + U \subset aU$ ([Rob1],[Rob2]). For example, if U is convex we may take $a = 2$. In l^p, $0 < p < 1$, the spheres about the origin are semi-convex with $a = 2^{\frac{1}{p}}$. A topological vector space (TVS) is *semi-convex* if it has a neighborhood base consisting of semi-convex sets ([Rob1],[Rob2]). The spaces l^p, $0 < p < 1$, are semi-convex but not locally convex.

Let E be a semi-convex TVS and Σ a σ-algebra of subsets of a set S. Let $\mathcal{M}$ be a family of finitely additive set functions defined on Σ with values in E. If $\mathcal{M}$ is pointwise bounded on Σ, we give necessary and sufficient conditions for $\mathcal{M}$ to be uniformly bounded on Σ.

Lemma C.1. *Suppose $\mathcal{M}$ is pointwise bounded on Σ. Then $\mathcal{M}$ is uniformly bounded on Σ iff $\{m_k(E_k)\}$ is bounded for every pairwise disjoint sequence $\{E_k\}$ from Σ and $\{m_k\} \subset \mathcal{M}$.*

Proof. For $A \in \Sigma$ set $\Sigma_A = \{B \in \Sigma : B \subset A\}$.

Suppose $\mathcal{M}(\Sigma_A)$ is not absorbed by the semi-convex neighborhood U with $U + U \subset aU$ and V is a symmetric, semi-convex neighborhood of 0 with $V + V \subset U$. We claim that for every k there exist a partition (A_k, B_k) of A, $n_k > k$ and $m_k \in \mathcal{M}$ such that

$$m_k(A_k) \notin n_k V, \ m_k(B_k) \notin n_k V.$$

By the pointwise bounded assumption there exists $n_k > k$ such that

$\mathcal{M}(A) \subset n_k V$. But,

$$\mathcal{M}(\Sigma_A) \not\subset n_k(V+V)$$

since $V + V \subset U$ so there exist $A_k \in \Sigma, A_k \subset A, m_k \in \mathcal{M}$ such that

$$m_k(A_k) \notin n_k(V+V).$$

Hence, $m_k(A_k) \notin n_k V$. Set $B_k = A \setminus A_k$. Then $m_n(B_k) \notin n_k V$ since otherwise

$$m_k(A_k) = m_k(A) - m_k(B_k) \in n_k V + n_k V = n_k(V+V).$$

Now assume $\mathcal{M}(\Sigma)$ is not bounded. Then there exists a semi-convex neighborhood of 0, U, with $U + U \subset aU$ which does not absorb $\mathcal{M}(\Sigma)$. Pick a symmetric neighborhood of 0, V, such that $V + V \subset U$. By the observation above, there exist a partition (A_1, B_1) of S, $n_1 > 1$ and $m_1 \in \mathcal{M}$ such that

$$m_1(A_1) \notin n_1 V, \; m_1(B_1) \notin n_1 V.$$

Now either $\mathcal{M}(\Sigma_{A_1})$ or $\mathcal{M}(\Sigma_{B_1})$ is not absorbed by U; for if both were absorbed by U there exists m such that

$$\mathcal{M}(\Sigma_S) \subset \mathcal{M}(\Sigma_{A_1}) + \mathcal{M}(\Sigma_{B_1}) \subset mU + mU = m(U+U) \subset m(aU)$$

by the semi-convexity of U so $\mathcal{M}(\Sigma_S)$ would be absorbed by U. Pick whichever of A_1 or B_1 is such that $\mathcal{M}(\Sigma_{A_1})$ or $\mathcal{M}(\Sigma_{B_1})$ which is not absorbed by U and label it F_1 and set $E_1 = S \setminus F_1$. Now treat F_1 as above to obtain a partition (E_2, F_2) of F_1, $n_2 > n_1$ and $m_2 \in \mathcal{M}$ such that

$$m_2(E_2) \notin n_2 V, \; m_2(F_2) \notin n_2 V$$

and $\mathcal{M}(\Sigma_{F_2})$ is not absorbed by U. Continuing this construction produces a pairwise disjoint sequence $\{E_k\}$ and a sequence $\{m_k\} \subset \mathcal{M}$ such that $\{m_k(E_k)\}$ is not absorbed by U. Thus, $\{m_k(E_k)\}$ is not bounded.

The other implication is obvious. $\qquad\qquad\square$

A finitely additive set function $m : \Sigma \to E$ is *strongly additive* if $m(A_k) \to 0$ for every pairwise disjoint sequence $\{A_k\}$ from Σ. It is clear that any countably additive set function defined on a σ-algebra is strongly additive. It follows from the lemma that any strongly additive (countably additive) set function with values in a semi-convex TVS is bounded.

Theorem C.2. *If E is semi-convex and $m : \Sigma \to E$ is strongly additive (countably additive), then m is bounded.*

We will study strongly additive set functions in the text.

Appendix D

Drewnowski

In this appendix we establish a remarkable result of Drewnowski which asserts that a strongly additive set function defined on a σ-algebra is in some sense not "too far" from being countably additive ([Dr]). This result is very useful in treating finitely additive set functions.

Let Σ be a σ-algebra of subsets of a set S, X be an Abelian topological group whose topology is generated by the quasi-norm $|\cdot|$ and let $\mu : \Sigma \to X$ be finitely additive and strongly additive. Recall μ is *strongly additive* if $\mu(E_j) \to 0$ whenever $\{E_j\}$ is a pairwise disjoint sequence from Σ. For $E \in \Sigma$, set

$$\mu'(E) = \sup\{|\mu(A)| : A \subset E, A \in \Sigma\};$$

μ' is called the *submeasure majorant* of μ and μ' is also strongly additive in the sense that $\mu'(E_j) \to 0$ whenever $\{E_j\}$ is a pairwise disjoint sequence from Σ.

Lemma D.1. (Drewnowski) *If $\mu : \Sigma \to X$ is finitely additive and strongly additive and $\{E_j\}$ is a pairwise disjoint sequence from Σ, then $\{E_j\}$ has a subsequence $\{E_{n_j}\}$ such that μ is countably additive on the σ-algebra generated by $\{E_{n_j}\}$.*

Proof. Partition $\mathbb{N}$ into a pairwise disjoint sequence of infinite sets $\{K_j^1\}_{j=1}^{\infty}$. By the strong additivity of μ',

$$\mu'(\cup_{j \in K_i^1} E_j) \to 0$$

as $i \to \infty$ so there exists i such that

$$\mu'(\cup_{j \in K_i^1} E_j) < 1/2.$$

Set $N_1 = K_i^1$ and $n_1 = \inf N_1$. Now partition $N_1 \setminus \{n_1\}$ into a pairwise disjoint sequence of infinite subsets $\{K_j^2\}_{j=1}^{\infty}$. As above there exists i such that

$$\mu'(\cup_{j \in K_i^2} E_j) < 1/2^2.$$

Let $N_2 = K_i^2$ and $n_2 = \inf N_2$. Note $N_2 \subset N_1$ and $n_2 > n_1$. Continuing this construction produces a subsequence $n_j \uparrow \infty$ and a sequence of infinite subsets of $\mathbb{N}$, $\{N_j\}$, such that $N_{j+1} \subset N_j$ and

$$\mu'(\cup_{i \in N_j} E_i) < 1/2^j.$$

If Σ_0 is the σ-algebra generated by $\{E_{n_j}\}$, then μ is countably additive on Σ_0. $\qquad \square$

We also have a version of Drewnowski's Lemma for a sequence of finitely additive set functions.

Corollary D.2. *Let $\mu_i : \Sigma \to X$ be finitely additive and strongly additive for each $i \in \mathbb{N}$. If $\{E_j\}$ is a pairwise disjoint sequence from Σ, then there is a subsequence $\{E_{n_j}\}$ such that each μ_i is countably additive on the σ-algebra generated by $\{E_{n_j}\}$.*

Proof. Define a quasi-norm $|\cdot|'$ on $X^{\mathbb{N}}$ by

$$|x|' = |\{x_i\}|' = \sum_{i=1}^{\infty} \frac{|x_i|}{(1 + |x_i|)2^i}.$$

Define $\mu : \Sigma \to X^{\mathbb{N}}$ by $\mu(E) = \{\mu_i(E)\}$. Then μ is finitely additive and strongly additive with respect to $|\cdot|'$ so by Lemma 1 there is a subsequence $\{E_{n_j}\}$ such that μ is countably additive on the σ-algebra Σ_0 generated by $\{E_{n_j}\}$. Thus, each μ_i is countably additive on Σ_0. $\qquad \square$

Appendix E

Antosik–Mikusinski Matrix Theorems

In this appendix we will present two versions of the Antosik–Mikusinski Matrix Theorems. These matrix theorems have proven to be very useful in treating applications in functional analysis and measure theory where gliding hump techniques are employed (see [Sw1] for more versions of the matrix theorem and applications). These theorems are used at various points in the text in gliding hump proofs.

Let X be a (Hausdorff) Abelian topological group. We begin with a simple lemma.

Lemma E.1. *Let $x_{ij} \in X$ for $i, j \in \mathbb{N}$. If $\lim_i x_{ij} = 0$ for every j and $\lim_j x_{ij} = 0$ for every i and if $\{U_k\}$ is a sequence of neighborhoods of 0 in X, then there exists an increasing sequence $\{p_i\}$ such that $x_{p_i p_j} \in U_j$ for $j > i$.*

Proof. Set $p_1 = 1$. There exists $p_2 > p_1$ such that $x_{ip_1} \in U_2$, $x_{p_1 j} \in U_2$ for $i, j \geq p_2$. Then there exists $p_3 > p_2$ such that $x_{ip_1}, x_{ip_2}, x_{p_1 j}, x_{p_2 j} \in U$ for $i, j \geq p_3$. Now just continue the construction. □

We now establish our version of the Antosik–Mikusinski Matrix Theorem.

Theorem E.2. (Antosik–Mikusinski) *Let $x_{ij} \in X$ for $i, j \in \mathbb{N}$. Suppose*

(I) $\lim_i x_{ij} = x_j$ *exists for each j and*
(II) *for each increasing sequence of positive integers $\{m_j\}$ there is a subsequence $\{n_j\}$ of $\{m_j\}$ such that $\{\sum_{j=1}^{\infty} x_{in_j}\}_i$ is Cauchy.*

Then

$$\lim_i x_{ij} = x_j$$

uniformly for $j \in \mathbb{N}$. In particular,

$$\lim_{i,j} x_{ij} = \lim_i \lim_j x_{ij} = \lim_j \lim_i x_{ij} = 0 \text{ and } \lim_i x_{ii} = 0.$$

Proof. If the conclusion fails, there is a closed, symmetric neighborhood, U_0, of 0 and increasing sequences of positive integers $\{m_k\}$ and $\{n_k\}$ such that

$$x_{m_k n_k} - x_{n_k} \notin U_0$$

for all k. Pick a closed, symmetric neighborhood U_1 of 0 such that $U_1 + U_1 \subseteq U_0$ and set $i_1 = m_1, j_1 = n_1$. Since

$$x_{i_1 j_1} - x_{j_1} = (x_{i_1 j_1} - x_{ij_1}) + (x_{ij_1} - x_{j_1}),$$

there exists i_0 such that

$$x_{i_1 j_1} - x_{ij_1} \notin U_1$$

for $i \geq i_0$. Choose k_0 such that

$$m_{k_0} > \max\{i_1, i_0\}, n_{k_0} > j_1 \text{ and set } i_2 = m_{k_0}, \ j_2 = n_{k_0}.$$

Then

$$x_{i_1 j_1} - x_{i_2 j_1} \notin U_1 \text{ and } x_{i_2 j_2} - x_{j_2} \notin U_0.$$

Proceeding in this manner produces increasing sequences $\{i_k\}, \{j_k\}$ such that

$$x_{i_k j_k} - x_{j_k} \notin U_0 \text{ and } x_{i_k j_k} - x_{i_{k+1} j_k} \notin U_1.$$

For convenience, set $z_{k,l} = x_{i_k j_l} - x_{i_{k+1} j_l}$ so

$$z_{k,k} \notin U_1.$$

Choose a sequence of closed, symmetric neighborhoods of 0, $\{U_n\}$, such that $U_n + U_n \subseteq U_{n-1}$ for $n \geq 1$. Note that

$$U_3 + U_4 + \cdots + U_m = \sum_{j=3}^m U_j \subseteq U_2$$

for each $m \geq 3$. By (I) and (II), $\lim_k z_{kl} = 0$ for each l and $\lim_l z_{kl} = 0$ for each k so by Lemma 1 there is an increasing sequence of positive integers $\{p_k\}$ such that

$$z_{p_k p_l}, z_{p_l p_k} \in U_{k+2}$$

for $k > l$. By (II), $\{p_k\}$ has a subsequence $\{q_k\}$ such that $\{\sum_{k=1}^{\infty} x_{iq_k}\}_{i=1}^{\infty}$ is Cauchy so

$$\lim_k \sum_{l=1}^{\infty} z_{q_k q_l} = 0.$$

Thus, there exists k_0 such that $\sum_{l=1}^{\infty} z_{q_{k_0} q_l} \in U_2$. Then for $m > k_0$,

$$\sum_{l=1, l \neq k_0}^{m} z_{q_{k_0} q_l} = \sum_{l=1}^{k_0-1} z_{q_{k_0} q_l} + \sum_{l=k_0+1}^{m} z_{q_{k_0} q_l} \in \sum_{l=1}^{k_0-1} U_{k_0+2}$$

$$+ \sum_{l=k_0+1}^{m} U_{l+2} \subseteq \sum_{l=3}^{m+2} U_l \subseteq U_2$$

so

$$z_{k_0} = \sum_{l=1, l \neq k_0}^{\infty} z_{q_{k_0} q_l} \in U_2.$$

Thus,

$$z_{q_{k_0} q_{k_0}} = \sum_{l=1}^{\infty} z_{q_{k_0} q_l} - z_{k_0} \in U_2 + U_2 \subseteq U_1$$

This is a contradiction and establishes the result. The last statement follows from the Iterated Limit Theorem (Appendix A). □

A matrix $[x_{ij}]$ satisfying conditions (I) and (II) of Theorem 2 is called a *K-matrix*. [The appellation "$\mathcal{K}$" here refers to the Katowice branch of the Mathematical Institute of the Polish Academy of Science where the matrix theorems and applications were developed by Antosik, Mikusinski and other members of the institute.]

At other points in the text we will also require another version of the matrix theorem which was developed by Stuart ([St1],[St2]) to treat weak sequential completeness of β-duals.

Let X be a Hausdorff TVS.

Theorem E.3. (Stuart) *Let $x_{ij} \in X$ for $i, j \in \mathbb{N}$. Suppose*

(I) $\lim_i x_{ij} = x_j$ *exists for all j and*
(II) *for each increasing sequence of positive integers $\{m_j\}$ there is a subsequence $\{n_j\}$ and a choice of signs $s_j \in \{-1, 1\}$ such that $\{\sum_{j=1}^{\infty} s_j x_{in_j}\}_{i=1}^{\infty}$ is Cauchy.*

Then

$$\lim_i x_{ij} = x_j$$

uniformly for $j \in \mathbb{N}$. In particular,

$$\lim_{i,j} x_{ij} = \lim_i \lim_j x_{ij} = \lim_j \lim_i x_{ij} = 0 \quad \text{and} \quad \lim_i x_{ii} = 0.$$

Proof. If the conclusion fails, there is a closed, symmetric neighborhood of 0, U_0, and increasing sequences of positive integers $\{m_k\}$ and $\{n_k\}$ such that

$$x_{m_k n_k} - x_{n_k} \notin U_0$$

for all k. Pick a closed, symmetric neighborhood of 0, U_1, such that $U_1 + U_1 \subset U_0$ and set $i_1 = m_1, j_1 = n_1$. Since

$$x_{i_1 j_1} - x_{j_1} = (x_{i_1 j_1} - x_{i j_1}) + (x_{i j_1} - x_{j_1}),$$

there exists i_0 such that

$$x_{i_1 j_1} - x_{i j_1} \notin U_1$$

for $i \geq i_0$. Choose k_0 such that $m_{k_0} > \max\{i_1, i_0\}, n_{k_0} > j_1$ and set $i_2 = m_{k_0}, j_2 = n_{k_0}$. Then

$$x_{i_1 j_1} - x_{i_2 j_1} \notin U_1 \text{ and } x_{i_2 j_2} - x_{j_2} \notin U_0.$$

Proceeding in this manner produces increasing sequences $\{i_k\}$ and $\{j_k\}$ such that

$$x_{i_k j_k} - x_{j_k} \notin U_0 \text{ and } x_{i_k j_k} - x_{i_{k+1} j_k} \notin U_1.$$

For convenience, set $z_{kl} = x_{i_k j_l} - x_{i_{k+1} j_l}$ so $z_{kk} \notin U_1$.

Choose a sequence of closed, symmetric neighborhoods of 0, $\{U_n\}$, such that $U_n + U_n \subset U_{n-1}$ for $n \geq 1$. Note that

$$U_3 + U_4 + \ldots + U_m = \sum_{j=3}^m U_j \subset U_2 \text{ for each } m \geq 3.$$

By (I), $\lim_k z_{kl} = 0$ for each l and by (II), $\lim_l z_{kl} = 0$ for each k so by Lemma 1 there is an increasing sequence of positive integers $\{p_k\}$ such that

$$z_{p_k p_l}, z_{p_l p_k} \in U_{k+2}$$

for $k > l$. By (II) there is a subsequence $\{q_k\}$ of $\{p_k\}$ and a choice of signs s_k such that

$$\left\{ \sum_{k=1}^\infty s_k x_{i q_k} \right\}_{i=1}^\infty$$

is Cauchy so

$$\lim_k \sum_{l=1}^{\infty} s_l z_{q_k q_l} = 0.$$

Thus, there exists k_0 such that

$$\sum_{l=1}^{\infty} s_l z_{q_{k_0} q_l} \in U_2.$$

Then for $m > k_0$,

$$\sum_{l=1, l \neq k_0}^{m} s_l z_{q_{k_0} q_l} = \sum_{l=1}^{k_0-1} s_l z_{q_{k_0} q_l} + \sum_{l=k_0+1}^{m} s_l z_{q_{k_0} q_l} \in \sum_{l=1}^{k_0-1} U_{k_0+2} + \sum_{l=k_0+1}^{m} U_l \subset U_2$$

so

$$z_{k_0} = \sum_{l=1, l \neq k_0}^{\infty} s_l z_{q_{k_0} q_l} \in U_2.$$

Thus,

$$s_{k_0} z_{q_{k_0} q_{k_0}} = \sum_{l=1}^{\infty} s_l z_{q_{k_0} q_l} - z_{k_0} \in U_2 + U_2 \subset U_1$$

since U_1 is symmetric

$$z_{q_{k_0} q_{k_0}} \in U_1$$

as well. This is a contradiction. The last statement follows from the Iterated Limit Theorem. $\qquad \square$

A matrix which satisfies conditions (I) and (II) of Theorem 3 will be called a *signed $\mathcal{K}$-matrix* and Theorem 3 will be referred to as the signed version of the Antosik–Mikusinski Matrix Theorem.

We give an example of a matrix which is a signed $\mathcal{K}$-matrix but is not a $\mathcal{K}$-matrix.

Example E.4. Let X be bs, the space of bounded series, equipped with the topology of coordinatewise convergence, $\sigma(bs, c_{00})$ [Appendix B]. Define a matrix $M = [m_{ij}]$ with entries from X by $m_{ij} = e^j$. Then no row of M has a subseries which converges in X so M is not a $\mathcal{K}$-matrix. However, given any subsequence $\{n_j\}$ there is a sequence of signs $\{s_j\}$ such that the series $\sum_{j=1}^{\infty} s_j e^{n_j}$ converges in X so M is a signed $\mathcal{K}$-matrix.

Other refinements and comments on the matrix theorems can be found in [Sw1] 2.2 and Appendix D of [Sw4]. The text [Sw1] contains numerous applications of the matrix theorems to topics in topological vector spaces, measure and integration theory and sequence spaces.

References

[Al] A. Alexiwicz, On Sequences of Operations I, Studia Math., 11,(1950), 1–30.

[A] P. Antosik, On interchange of limits, Generalized Functions, Convergence Strucctures and Their Applications, Plenum Press, N.Y., 1988, p. 367–374.

[AS1] P. Antosik and C. Swartz, Matrix Methods in Analysis, Springer Lecture Notes in Mathematics 1113, Heidelberg, 1985.

[Ba] S. Banach, Theorie des Operations Lineaires, Warsaw, 1932.

[Bar] R. Bartle, A general bilinear vector integral, Studia Math., 15(1956), 337–352.

[Bs] B. Basit, On a Theorem of Gelfand and a new proof of the Orlicz–Pettis Theorem, Rend. Inst. Matem. Univ. di Trieste, 18(1986), 159–162.

[Be] G. Bennett, Some inclusion theorems for sequence spaces, Pacific J. Math. 46(1973), 17–30.

[BK] G. Bennett and N. Kalton, FK spaces containing c_0, Duke Math. J., 39(1972), 561–582.

[BP] C. Bessaga and A. Pelczynski, On Bases and Unconditional Convergence of series in Banach Space, Studia Math., 17(1958), 151–164.

[BCS] O. Blasco, J. M. Calabuig and T. Signes, A bilinear version of Orlicz–Pettis Theorem, J. Math. Anal. Appl., 348(2008), 150–164.

[Bo] J. Boos, Classical and Modern Methods in Summability, Oxford University Press, Oxford, 2000.

[BS] C. Bosch and C. Swartz, Functional Calculi, World Sci. Publ. Singapore, 2015.

[BW] Qingying Bu and Cong Xin Wu, Unconditionally Convergent Series of Operators on Banach Spaces, J. Math. Anal. Appl., 207(1997),

291–299.

[BKL] J. Burzyk, C. Clis and Z. Lipecki, On Metrizable Abelian Groups with Completeness-type Property, Colloq. Math., 49(1984), 33–39.

[BM] J. Burzyk and P. Mikusinski, On Normability of Semigroups, Bull. Polon.Acad. Sci., 28(1980), 33–35.

[CL] A. Chen and R. Li, A Version of Orlicz–Pettis Theorem for Quasi-homogeneous Operator Space, J. Math. Anal. Appl., 373(2011), 127–133.

[CLS1] M. Cho, R. Li and C. Swartz, Subseries Convergence in Abstract Duality Pairs, Proy. J. Math., 3392014), 447–470.

[CLS2] M. Cho, R. Li and C. Swartz, The Banach–Steinhaus Theorem in Abstract Duality Pairs, Proy. J. Math., 34(2015), 391–399.

[Co] C. Constantinescu, On Nikodym Boundedness Theorem, Libertas Math., 1(1981), 51–73.

[DeS] J. DePree and C. Swartz, Introduction to Real Analysis, Wiley, N.Y., 1987.

[Die] P. Dierolf, Theorems of Orlicz–Pettis type for locally convex spaces, Man. Math., 20(1977), 73–94.

[DF] J. Diestel and F. Faires, On Vector Measures, Trans. Amer. Math. Soc., 198(1974), 253–271.

[DU] J. Diestel and J. Uhl, Vector Measures, Amer. Math. Soc. Surveys #15, Providence, 1977.

[Di] J. Dieudonne, History of Functional Analysis, North Holland, Amsterdam, 1981.

[Din] N. Dinculeanu, Weak Compactness and Uniform Convergence of Operators in Space of Bochner Integrable Functions, J. Math. Anal. Appl., 1090(1985), 372–387.

[Do] I. Dobrakov, On Integration in Banach Spaces I, Czech. Math. J., 20(1970), 511–536.

[Dr] L. Drewnowski, Equivalence of Brooks–Jewett, Vitali–Hahn–Saks and Nikodym Theorems, Bull. Acad. Polon. Sci., 20(1972), 725–731.

[DFP] L. Drewnowski, M. Florencio and P. Paul, The Space of Pettis Integrable Functions is Barrelled, Proc. Amer. Math. Soc., 114(1992), 687–694.

[DS] N. Dunford and J. Schwartz, Linear Operators I, Interscience, N.Y., 1958.

[FL] W. Filter and I. Labudu, Essays on the Orlicz–Pettis Theorem I, Real. Anal. Exch., 16(1990/91), 393–403.

[GDS] H.G. Garnir, M. DeWilde and J. Schmets, Analyse Fontionnelle I,

Birkhauser, Basel, 1968.

[GR] W. Graves, Proceedings of the Conference on Integration, Topology, and Geometry in Linear Spaces, Amer. Math. Soc., Providence, 1980.

[Ha] H. Hahn, Uber Folgen Linearen Operationen, Monatsch. fur Math. und Phys. 32(1922), 1–88.

[Hal] P. Halmos, Measure Theorey, Van Nostrand, Princeton, 1950.

[HT] E. Hellinger and O. Toeplitz, Grundlagen fur eine Theorie den unendlichen Matrizen, Math. Ann., 69(1910), 289–330.

[Ho] J. Howard, The Comparison of an Unconditionally Converging Operator, Studia Math., 33(1969), 295–298.

[Ka] N. J. Kalton, Spaces of Compact Operators, Math. Ann., 208(1974), 267–278.

[Ka2] N. Kalton, Subseries Convergence in Topological Groups and Vector Spaces, Isreal J. Math., 10(1971), 402–412.

[Ka3] N.J. Kalton, The Orlicz–Pettis Theorem, Contemporary Math., Amer. Math. Soc., Providence, 1980.

[KG] P.K. Kamthan and M. Gupta, Sequence Spaces and Series, Marcel Dekker, N.Y., 1981.

[Ke] J. Kelley, General Topology, Van Nostrand, N.Y., 1955.

[Kh] S. M. Khaleelulla, Counter Examples in Topological Vector Spaces, Springer Lecture Notes 936, Heidelberg, 1982.

[Kl] C. Clis, An Example of a Non-Complete Normed (K) Space, Bull. Acad. Polon. Sci., 26(1978), 415–420.

[KK] I. Kluvanek and G. Knowles, Vector Measures and Control Systems, North Holland, Amsterdam, 1976.

[Kö1] G. Köthe, Topological Vector Spaces I, Springer–Verlag, Berlin, 1969.

[Kö2] G. Köthe, Topological Vector Spaces II, Springer–Verlag, Berlin, 1979.

[LaW] E. Lacey and R.J. Whitley, Conditions under which all the Bounded Linear Maps are Compact, Math. Ann., 158(1965), 1–5.

[LPY] Lee Peng Yee, Sequence Spaces and the Gliding Hump Property, Southeast Asia Bull. Math., Special Issue (1993), 65–72.

[LPYS] Lee Peng Yee and C. Swartz, Continuity of Superposition Operators on Sequence Spaces, New Zealand J. Math., 24(1995), 41–52.

[LC] R. Li and M. Cho, A general Kalton-like Theorem, J. Harbin Inst. Tech., 25(1992), 100–104.

[LB] R. Li and Q. Bu, Locally Convex Spaces Containing no Copy of c_0, J. Math. Anal. Appl., 172(1993), 205–211.

[LS1] R. Li and C. Swartz, Spaces for Which the Uniform Boundedness Principle Holds, Studia Sci. Math. Hung, 27(1992), 379–384.

[LS2] R. Li and C. Swartz, A Nonlinear Schur Theorem, Acta Sci. Math., 58(1993), 497–508.

[LS3] R. Li and C. Swartz, An Abstract Orlicz–Pettis Theorem and Applications, Proy. J. Math., 27(2008), 155–169.

[LSC] R. Li, C. Swartz and M. Cho, Basic Properties of K-spaces, Systems Sci. and Math.Sci., 5(1992), 234–238.

[LW] R. Li and J.Wang, Invariants in Abstract Mapping Pairs, J. Austral. Math. Soc., 76(2004), 369–381.

[LT] J. Lindenstrauss and L. Tzafriri, Classical Banach Spaces I, Springer-Verlag, Berlin, 1977.

[MA] S.D. Madrigal and J.M.B. Arrese, Local Completeness and Series, Simon Stevin, 65(1991), 331–335.

[MO] S. Mazur and W. Orlicz, Uber Folgen linearen Operationen, Studia Math., 4(1933), 152–157.

[Mc] C. W. McArthur, On a theorem of Orlicz and Pettis, Pacific J. Math. 22(1967), 297–303.

[MR] C. McArthur and J. Rutherford, Some Applications of an Inequality in Locally Convex Spaces, Trans. Amer. Math. Soc., 137(1969), 115–123.

[Mo] A. Mohsen, Weak*-Norm Sequentially Continuous Operators, Math. Slovaca, 50(2000), 357–363.

[Mu] K. Musial, Topics in the Theory of Pettis Integration, Rend. Instituto Mat. Univ. Trieste, Vol. XXIII, 1991.

[No] D. Noll, Sequential Completeness and Spaces with the Gliding Humps Property, Manuscripta Math., 66(1990), 237–252.

[NS] D. Noll and W. Stadler, Abstract sliding hump techniques and characterizations of barrelled spaces, Studia Math., 94(1989), 103–120.

[Or] W. Orlicz, Beiträge zur Theorie der Orthogonalent Wichlungen II, Studia Math., 1(1929), 241–255.

[Pa] T. V. Panchapagesan, The Bartle–Dunford–Schwartz Integral, Birkhauser, Basel, 2008.

[P] A Pelczynski, On Stricty Singular and Strictly Cosingular Operators, Bull. Acad. Polon. Sci., 13(1965), 31–36.

[Pe] B. J. Pettis, On Integration in Vector Spaces, Trans. Amer. Math. Soc., 44(1938), 277–304.

[Pi] A. Pietsch, Nukleare Lokalconvexe Raume, Akademie Verlag, Berlin, 1965.

[Rob1] A. Robertson, Unconditional Convergence and the Vitali–Hahn–Saks Theorem, Bull. Soc. Math, France, Supp;. Mem. 31–32 (1972), 335–341.

[Rob2] A. Robertson, On Unconditional Convergence in Topological Vector Spaces, Proc. Royal Soc. Edinburgh, 68(1969), 145–157.

[Rol] S. Rolewicz, Metric Linear Spaces, Polish Sci. Publ., Warsaw, 1972.

[Sm] W. Schachermeyer, On some classical measure-theoretic theorems for non-sigma-complete Boolean algebras, Dissert. Math., Warsaw, 1982.

[Sch] H. H. Schaefer, Topological Vector Spaces, MacMillan, N.Y., 1966.

[Sr] J. Schur, Über lineare Tranformation in der Theorie die unendlichen Reihen, J. Reine Angew. Math., 151(1920), 79–111.

[Sti] W. J. Stiles, On Subseries Convergence in F-spaces, Israel J. Math., 8(1970), 53–56.

[St1] C. Stuart, Weak Sequential Completeness in Sequence Spaces, Ph.D. Dissertation, New Mexico State University, 1993.

[St2] C. Stuart, Weak Sequential Completeness of β-duals, Rocky Mt. Math. J., 26(1996), 1559–1568.

[St3] C. Stuart, Interchanging the limit in a double series, Southeast Asia Bull. Math., 18(1994), 81–84.

[SS] C. Stuart and C. Swartz, A Projection Property and Weak Sequential Completeness of α-duals, Collect. Math.l, 43(1992), 177–185.

[Sw1] C. Swartz, Infinite Matrices and the Gliding Hump, World Sci. Publ., Singapore, 1996.

[Sw2] C. Swartz, An Introduction to Functional Analysis, Marcel Dekker, N.Y., 1992.

[Sw3] C. Swartz, Measure Integration and Function Spaces, World Sci., Pub., Singapore, 1994.

[Sw4] C. Swartz, Multiplier Convergent Series, World Sci. Publ., Singapore, 2009.

[Sw5] C. Swartz, Subseries Convergence in Spaces with Schander Basis, Proc. Amer. Math. Soc., 129(1995), 455–457.

[Sw6] C. Swartz, An Abstract Orlicz–Pettis Theorem, Bull. Acad.Polon. Sci.,32(1984), 433–437.

[Sw7] C. Swartz, A Bilinear Orlicz–Pettis Theorem, J. Math. Anal. Appl., 365(2010), 332–337.

[Sw8] C. Swartz, The Evolution of the Uniform Boundedness Principle, Math. Chron, 19(1990), 1–18.

[Sw9] C. Swartz, Addendum to "The Evolution of the Uniform Boundedness Principle", Math. Chron., 20 (1991), 157–159.

[Th] G.E.F. Thomas, L'integration par rapport a une mesure de Radon vectorielle, Ann. Inst. Fourier, 20(1970), 55–191.

[Tw] I. Tweddle, Unconditional Convergence and Vector-valued Measures, J. London Math. Soc., 2(1970), 603–610.

[Wi1] A. Wilansky, Functional Analysis, Blaisdale, N.Y., 1964.

[Wi2] A. Wilansky, Modern Methods in Topological Vector Spaces, McGraw–Hill, N.Y., 1978.

[We] H. Weber, Compactness in Spaces of group valued Contents, the Vitali–Hahn–Saks Theorem and Nikodym Boundedness Theorem, Rocky Mount. J. Math., 16(1985), 253–275.

[Wu] J. Wu, The compact sets in the infinite matrix topological algebras, Acta Math. Sinica, to appear.

[WCC] J. Wu, C. Cui and M. Cho, The Abstract Uniform Boundedness Principle, Southeast Asia Bull. Math., 24(2000), 655–660.

[WLC] J. Wu, J. Luo and C. Cui, The Abstract Gliding Hump Properties and Applications, Taiwan J. Math., 10(2006), 639–649.

[Y] K. Yosida, Functional Analysis, Springer–Verlag, N.Y., 1966.

[ZCL] F. Zheng, C. Cui and R. Li, Abstract Gliding Hump Properties in the Vector Valued Dual Pairs, Acta Anal. Funct. Appl., 12(2010), 322–327.

Index